Microbiology: Concepts and Applications

Microbiology: Concepts and Applications

Editor: David Cooke

Callisto Reference

www.callistoreference.com

Callisto Reference,
118-35 Queens Blvd., Suite 400,
Forest Hills, NY 11375, USA

Visit us on the World Wide Web at:
www.callistoreference.com

ISBN: 978-1-64116-007-0 (Hardback)

The publisher's policy is to use permanent paper from mills that operate a sustainable forestry policy. Furthermore, the publisher ensures that the text paper and cover boards used have met acceptable environmental accreditation standards.

Printed in China.

Cataloging-in-Publication Data

Microbiology : concepts and applications / edited by David Cooke.
p. cm.
Includes bibliographical references and index.
ISBN 978-1-64116-007-0
1. Microbiology. 2. Microorganisms. I. Cooke, David.
QR41.2 .M53 2018
579--dc23

Table of Contents

Preface VII

Chapter 1 **Understanding Microbiology** 1
- Microorganism 1
- Microbiology 12
- Importance of Microbiology 20
- Industrial Microbiology and Microbial Ecology 23
- Scope of Microbiology 25

Chapter 2 **Microbes and its Types** 26
- Members of Microbial World 26
- Importance of Microbes 128

Chapter 3 **Prokaryotic Cell Structure** 132
- Prokaryote 132
- Bacterial Cell Structure 138
- Plasma (Cytoplasmic) Membrane 157

Chapter 4 **Microscopy: An Overview** 174
- Microscopy 174
- Optical Microscope 187
- Electron Microscope 203
- Preparation of Specimen 219
- Gram Staining 220

Chapter 5 **Microbial Nutrition and Microbial Growth** 228
- Microbial Nutrient Requirements 228
- Uptake of Nutrients 231
- Culture Media 236
- Growth of Bacterial Cultures 239
- Growth Curve (Biology) 243
- Environmental Factors on Microbial Growth 250

Permissions

Index

Preface

Microbiology studies living organisms like viruses, protozoa and bacteria. The subfields of microbiology include bacteriology, parasitology, immunology, mycology, protozoology, virology and nematology. Microbes are very beneficial as they are responsible for processes such as antibiotic production and industrial fermentation. They can also be used for biodegradation of industrial and domestic waste. This book is an essential guide for both academicians and those who wish to pursue this discipline further. Coherent flow of topics, student-friendly language and extensive use of examples make this book an invaluable source of knowledge.

Given below is the chapter wise description of the book:

Chapter 1- Microbiology is a branch of biology that studies microorganisms. These microbes are important for their relevance in various fields such as environment, medicine, biotechnology and research. They are a part of the carbon cycle and help in maintaining ecological balance. Moreover, they are used to produce antibiotics, are used in genetic engineering and help in synthesis of many organic compounds. This is an introductory chapter which will introduce briefly all the significant aspects of microbiology.

Chapter 2- With the use of the electronic microscope, microbes could be studied in detail. The microorganisms could then be divided into five major classes: Archaea, Bacteria, Algae, Protozoa, and Fungi. These microbes have had a great impact on human life. Their scope ranges from their use in food production to the manufacture of medicines and extends to their need in industries and preserving the environment. The major components of the microbial world are discussed in this chapter.

Chapter 3- Unicellular microorganisms lacking a membrane-bound nucleus are known as prokaryotes, in which bacteria are considered the most prominent. These microbes lack membrane-bound organelles and usually divide by a process known as binary fusion. Prokaryotes are the smallest microorganisms and are found in spherical, rod-like and spiral shapes. The chapter on prokaryotic cell structure offers an insightful focus, keeping in mind the complex subject matter.

Chapter 4- Since microorganisms cannot be seen with the naked eye, microscopy is used to study them using microscopes. Optical, electron, and scanning probe microscopy are the branches involved in the study. Optical or light microscope uses light and lenses to enlarge the samples. As light microscopes can achieve 200nm resolution, electron microscopes with a resolution higher than 50 pm resolution are used to study the structure of microbes. This chapter is a compilation of the various branches of microscopy that form an integral part of the broader subject matter.

Chapter 5- Microbes require nutrition to conduct their functions. Microbial nutritional requirements are divided into two parts, i.e. macronutrients and micronutrients. Microbes require macronutrients in huge amounts, which include elements such as carbon, oxygen, nitrogen, and components of carbohydrates, such as lipids and proteins. The requirement for macronutrients is comparatively less. These include cobalt, molybdenum, nickel and copper. The aspects elucidated in this chapter are of vital importance, and provide a better understanding of microbiology.

At the end, I would like to thank all those who dedicated their time and efforts for the successful completion of this book. I also wish to convey my gratitude towards my friends and family who supported me at every step.

Editor

1

Understanding Microbiology

Microbiology is a branch of biology that studies microorganisms. These microbes are important for their relevance in various fields such as environment, medicine, biotechnology and research. They are a part of the carbon cycle and help in maintaining ecological balance. Moreover, they are used to produce antibiotics, are used in genetic engineering and help in synthesis of many organic compounds. This is an introductory chapter which will introduce briefly all the significant aspects of microbiology.

Microorganism

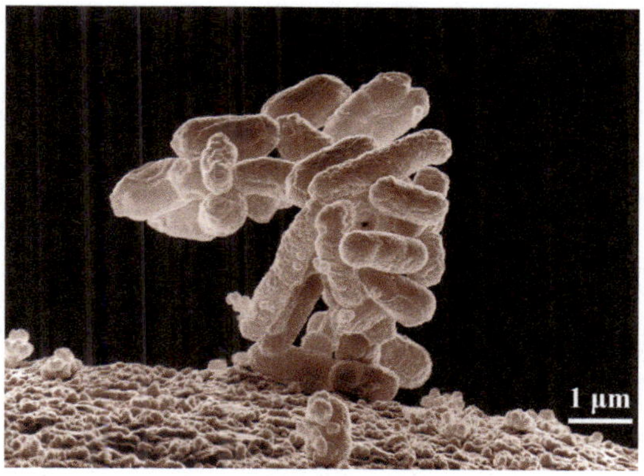

A cluster of *Escherichia coli* bacteria magnified 10,000 times

A microorganism or microbe is a microscopic organism, which may be single-celled or multicellular. The study of microorganisms is called microbiology, a subject that began with the discovery of microorganisms in the 1670s by Antonie van Leeuwenhoek. Robert Hooke coined the term "cell" after viewing plant cells under his microscope. Later, in the 19th century, Louis Pasteur found that microorganisms caused food spoilage, debunking the theory of spontaneous generation. In 1876 Robert Koch discovered that microorganisms cause diseases.

Microorganisms are very diverse and include all bacteria, archaea and most protozoa. This group also contains some fungi, algae, and some micro-animals such as rotifers. Many macroscopic animals and plants have microscopic juvenile stages. Some microbiologists classify viruses and viroids as microorganisms, but others consider these as nonliving. In July 2016, scientists identified a set of 355 genes from the last universal common ancestor of all life, including microorganisms, living on Earth.

Microorganisms live in every part of the biosphere, including soil, hot springs, inside rocks at least 19 km (12 mi) deep underground, the deepest parts of the ocean, and at least 64 km (40 mi) high in the atmosphere. Microorganisms, under certain test conditions, have been observed to thrive in the vacuum of outer space. Microorganisms likely far outweigh all other living things combined. The mass of prokaryote microorganisms including the bacteria and archaea may be as much as 0.8 trillion tons of carbon, out of the total biomass of between 1 and 4 trillion tons. Microorganisms appear to thrive in the Mariana Trench, the deepest spot in the Earth's oceans. Other researchers reported related studies that microorganisms thrive inside rocks up to 580 m (1,900 ft; 0.36 mi) below the sea floor under 2,590 m (8,500 ft; 1.61 mi) of ocean off the coast of the northwestern United States, as well as 2,400 m (7,900 ft; 1.5 mi) beneath the seabed off Japan. In August 2014, scientists confirmed the existence of microorganisms living 800 m (2,600 ft; 0.50 mi) below the ice of Antarctica. According to one researcher, "You can find microbes everywhere — they're extremely adaptable to conditions, and survive wherever they are."

Microorganisms are crucial to nutrient recycling in ecosystems as they act as decomposers. As some microorganisms can fix nitrogen, they are a vital part of the nitrogen cycle, and recent studies indicate that airborne microorganisms may play a role in precipitation and weather. Microorganisms are also exploited in biotechnology, both in traditional food and beverage preparation, and in modern technologies based on genetic engineering. A small proportion of microorganisms are pathogenic, causing disease and even death in plants and animals.

Evolution

Single-celled microorganisms were the first forms of life to develop on Earth, approximately 3–4 billion years ago. Further evolution was slow, and for about 3 billion years in the Precambrian eon, all organisms were microscopic. So, for most of the history of life on Earth, the only forms of life were microorganisms. Bacteria, algae and fungi have been identified in amber that is 220 million years old, which shows that the morphology of microorganisms has changed little since the Triassic period. The newly discovered biological role played by nickel, however — especially that brought about by volcanic eruptions from the Siberian Traps — may have accelerated the evolution of methanogens towards the end of the Permian–Triassic extinction event.

Microorganisms tend to have a relatively fast rate of evolution. Most microorganisms can reproduce rapidly, and bacteria are also able to freely exchange genes through conjugation, transformation and transduction, even between widely divergent species. This horizontal gene transfer, coupled with a high mutation rate and other means of transformation, allows microorganisms to swiftly evolve (via natural selection) to survive in new environments and respond to environmental stresses. This rapid evolution is important in medicine, as it has led to the development of multidrug resistant pathogenic bacteria, *superbugs*, that are resistant to antibiotics.

Pre-microbiology

The possible existence of microorganisms was discussed for many centuries before their discovery in the 17th century. The existence of unseen microbial life was postulated by Jainism. In the 6th century BC, Mahavira asserted the existence of unseen microbiological creatures living in earth, water, air and fire. The Jain scriptures also describe nigodas, which are sub-microscopic creatures living in large clusters and having a very short life, which are said to pervade every part of the universe, even the tissues of plants and animals. The earliest known idea to indicate the possibility of diseases spreading by yet unseen organisms was that of the Roman scholar Marcus Terentius Varro in a first century BC book titled *On Agriculture* in which he warns against locating a homestead near swamps:

> ... and because there are bred certain minute creatures that cannot be seen by the eyes, which float in the air and enter the body through the mouth and nose and they cause serious diseases.

In *The Canon of Medicine* (1020), Abū Alī ibn Sīnā (Avicenna) suggested that tuberculosis and other diseases might be contagious.

In 1546, Girolamo Fracastoro proposed that epidemic diseases were caused by transferable seed-like entities that could transmit infection by direct or indirect contact, or even without contact over long distances.

These early claims about the existence of microorganisms were speculative, and while grounded on indirect observations, they had no systematized empirical basis. Microorganisms were neither proven, observed, nor accurately described until the 17th century with the invention of the microscope.

History of Discovery

Antonie van Leeuwenhoek is acknowledged as the first to study microorganisms, using simple microscopes of his own design.

Lazzaro Spallanzani showed that boiling a broth stopped it from decaying.

Louis Pasteur showed that Spallanzani's findings held even if air could enter through a filter that kept particles out.

Robert Koch showed that microorganisms caused disease.

Antonie Van Leeuwenhoek (1632–1723) is considered to be the father of microbiology as he was the first to undisputedly discover, observe, describe, study and conduct scientific experiments with microoorganisms, using simple single-lensed microscopes of his own design. Robert Hooke, a contemporary of Leeuwenhoek, also used microscopes to observe microbial life; his 1665 book *Micrographia* describes these observations and coined the term *cell.*

Before Leeuwenhoek's discovery of microorganisms in 1675, it had been a mystery why grapes could be turned into wine, milk into cheese, or why food would spoil. Leeuwenhoek did not make the connection between these processes and microorganisms, but using a microscope, he did establish that there were signs of life that were not visible to the naked eye. Leeuwenhoek's discovery, along with subsequent observations by Spallanzani and Pasteur, ended the long-held belief that life spontaneously appeared from non-living substances during the process of spoilage.

Lazzaro Spallanzani (1729–1799) found that boiling broth would sterilise it, killing any microorganisms in it. He also found that new microorganisms could only settle in a broth if the broth was exposed to air.

Louis Pasteur (1822–1895) expanded upon Spallanzani's findings by exposing boiled broths to the air, in vessels that contained a filter to prevent all particles from passing through to the growth medium, and also in vessels with no filter at all, with air being admitted via a curved tube that would not allow dust particles to come in contact with the broth. By boiling the broth beforehand, Pasteur ensured that no microorganisms survived within the broths at the beginning of his experiment. Nothing grew in the broths in the course of Pasteur's experiment. This meant that the living organisms that grew in such broths came from outside, as spores on dust, rather than spontaneously generated within the broth. Thus, Pasteur dealt the death blow to the theory of spontaneous generation and supported germ theory.

In 1876, Robert Koch (1843–1910) established that microorganisms can cause disease. He found that the blood of cattle which were infected with anthrax always had large numbers of *Bacillus anthracis*. Koch found that he could transmit anthrax from one animal to another by taking a small sample of blood from the infected animal and injecting it into a healthy one, and this caused the healthy animal to become sick. He also found that he could grow the bacteria in a nutrient broth, then inject it into a healthy animal, and cause illness. Based on these experiments, he devised criteria for establishing a causal link between a microorganism and a disease and these are now known as Koch's postulates. Although these postulates cannot be applied in all cases, they do retain historical importance to the development of scientific thought and are still being used today.

On 8 November 2013, scientists reported the discovery of what may be the earliest signs of life on Earth—the oldest complete fossils of a microbial mat (associated with sandstone in Western Australia) estimated to be 3.48 billion years old.

Classification and Structure

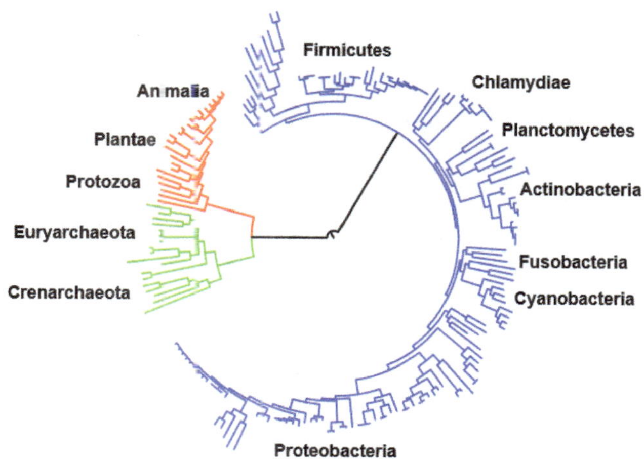

Evolutionary tree showing the common ancestry of all three domains of life. Bacteria are colored blue, eukaryotes red, and archaea green. Relative positions of some phyla are shown around the tree.

Microorganisms can be found almost anywhere on Earth. Bacteria and archaea are almost always microscopic, while a number of eukaryotes are also microscopic, including most protists, some fungi, as well as some micro-animals and plants. Viruses are generally regarded as not living and therefore not considered as microorganisms, although the field of microbiology includes virology, the study of viruses.

Prokaryotes

Prokaryotes are organisms that lack a cell nucleus and other membrane-bound organelles. They are almost always unicellular, although some species such as myxobacteria can aggregate into complex structures as part of their life cycle.

Consisting of two domains, bacteria and archaea, the prokaryotes are the most diverse and abundant group of organisms on Earth and inhabit practically all environments where the temperature is below +140 °C. They are found in water, soil, air, as the microbiome of an organism, hot springs and even deep beneath the Earth's crust in rocks. Practically all surfaces that have not been specially sterilized are covered by prokaryotes. The number of prokaryotes is estimated to be around five million trillion trillion, or 5×10^{30}, accounting for at least half the biomass on Earth.

The biodiversity of the prokaryotes is unknown, but may be very large. A May 2016 estimate, based on laws of scaling from known numbers of species against the size of organism, gives an estimate of perhaps 1 trillion species on the planet, of which most would be microorganisms. Currently, only one-thousandth of one percent of that total have been described.

Bacteria

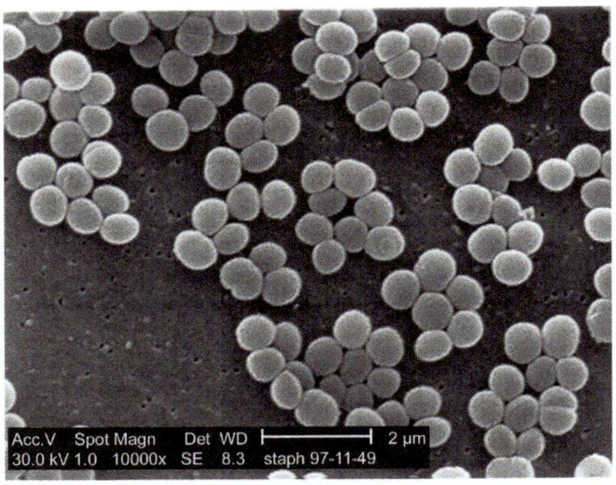

Staphylococcus aureus bacteria magnified about 10,000x

Almost all bacteria are invisible to the naked eye, with a few extremely rare exceptions, such as *Thiomargarita namibiensis*. They lack a nucleus and other membrane-bound organelles, and can function and reproduce as individual cells, but often aggregate in multicellular colonies. Their genome is usually a single loop of DNA, although they can also harbor small pieces of DNA called plasmids. These plasmids can be transferred between cells through bacterial conjugation. Bacteria are surrounded by a cell wall, which provides strength and rigidity to their cells. They reproduce by binary fission or sometimes by budding, but do not undergo meiotic sexual reproduction. However, many bacterial species can transfer DNA between individual cells by a process referred to as natural transformation. Some species form extraordinarily resilient spores, but for bacteria this is a mechanism for survival, not reproduction. Under optimal conditions bacteria can grow extremely rapidly and can double as quickly as every 20 minutes.

Archaea

Archaea are also single-celled organisms that lack nuclei. In the past, the differences between bacteria and archaea were not recognised and archaea were classified with bacteria as part of the kingdom Monera. However, in 1990 the microbiologist Carl Woese proposed the three-domain system that divided living things into bacteria, archaea and eukaryotes. Archaea differ from bacteria in both their genetics and biochemistry. For example, while bacterial cell membranes are made from phosphoglycerides with ester bonds, archaean membranes are made of ether lipids.

Archaea were originally described in extreme environments, such as hot springs, but have since been found in all types of habitats. Only now are scientists beginning to realize how common archaea are in the environment, with Crenarchaeota being the most common form of life in the ocean, dominating ecosystems below 150 m in depth. These organisms are also common in soil and play a vital role in ammonia oxidation.

Eukaryotes

Most living things that are visible to the naked eye in their adult form are eukaryotes, including humans. However, a large number of eukaryotes are also microorganisms. Unlike bacteria and archaea, eukaryotes contain organelles such as the cell nucleus, the Golgi apparatus and mitochondria in their cells. The nucleus is an organelle that houses the DNA that makes up a cell's genome. DNA (Deoxyribonuclaic acid) itself is arranged in complex chromosomes. Mitochondria are organelles vital in metabolism as they are the site of the citric acid cycle and oxidative phosphorylation. They evolved from symbiotic bacteria and retain a remnant genome. Like bacteria, plant cells have cell walls, and contain organelles such as chloroplasts in addition to the organelles in other eukaryotes. Chloroplasts produce energy from light by photosynthesis, and were also originally symbiotic bacteria.

Unicellular eukaryotes consist of a single cell throughout their life cycle. This qualification is significant since most multicellular eukaryotes consist of a single cell called a zygote only at the beginning of their life cycles. Microbial eukaryotes can be either haploid or diploid, and some organisms have multiple cell nuclei.

Unicellular eukaryotes usually reproduce asexually by mitosis under favorable conditions. However, under stressful conditions such as nutrient limitations and other conditions associated with DNA damage, they tend to reproduce sexually by meiosis and syngamy.

Protists

Of eukaryotic groups, the protists are most commonly unicellular and microscopic. This is a highly diverse group of organisms that are not easy to classify. Several algae species are multicellular protists, and slime molds have unique life cycles that involve switching between unicellular, colonial, and multicellular forms. The number of species of protists is unknown since only a small proportion has been identified. Studies from 2001-2004 have shown that a high degree of protist diversity exists in oceans, deep sea-vents, river sediment and an acidic river which suggests that a large number of eukaryotic microbial communities have yet to be discovered.

A microscopic mite *Lorryia formosa*

Animals

Some micro-animals are multicellular but at least one animal group, Myxozoa, is unicellular in its adult form. Microscopic arthropods include dust mites and spider mites. Microscopic crustaceans include copepods, some cladocera and water bears. Many nematodes are also too small to be seen with the naked eye. A common group of microscopic animals are the rotifers, which are filter feeders that are usually found in fresh water. Some micro-animals reproduce both sexually and asexually and may reach new habitats by producing eggs which can survive harsh environments that would kill the adult animal. However, some simple animals, such as rotifers, tardigrades and nematodes, can dry out completely and remain dormant for long periods of time.

Fungi

The fungi have several unicellular species, such as baker's yeast (*Saccharomyces cerevisiae*) and fission yeast (*Schizosaccharomyces pombe*). Some fungi, such as the pathogenic yeast *Candida albicans*, can undergo phenotypic switching and grow as single cells in some environments, and filamentous hyphae in others. Fungi reproduce both asexually, by budding or binary fission, as well by producing spores, which are called conidia when produced asexually, or basidiospores when produced sexually.

Plants

The green algae are a large group of photosynthetic eukaryotes that include many microscopic organisms. Although some green algae are classified as protists, others such as charophyta are classified with embryophyte plants, which are the most familiar group of land plants. Algae can grow as single cells, or in long chains of cells. The green algae include unicellular and colonial flagellates, usually but not always with two flagella per cell, as well as various colonial, coccoid, and filamentous forms. In the Charales, which are the algae most closely related to higher plants, cells differentiate into several distinct tissues within the organism. There are about 6000 species of green algae.

Ecology

Microorganisms are found in almost every habitat present in nature, including hostile environments such as the poles, deserts, geysers, rocks, and the deep sea. Some types of microorganisms have adapted to the extreme conditions and sustained colonies; these organisms are known as extremophiles. Extremophiles have been isolated from rocks as much as 7 kilometres below the Earth's surface, and it has been suggested that the amount of living organisms below the Earth's surface is comparable with the amount of life on or above the surface. Extremophiles have been known to survive for a prolonged time in a vacuum, and can be highly resistant to radiation, which may even allow them to survive in space. Many types of microorganisms have intimate symbiotic relationships with other larger organisms; some of which are mutually beneficial (mutualism), while others can be damaging to the host organism (parasitism). If microorganisms can cause disease in a host they are known as pathogens and then they are sometimes referred to as microbes. Microorganisms play critical roles in Earth's biogeochemical cycles as they are responsible for decomposition and nitrogen fixation.

Extremophiles

Extremophiles are microorganisms that have adapted so that they can survive and even thrive in conditions that are normally fatal to most life-forms. For example, some species have been found in the following extreme environments:

- Temperature: as high as 130 °C (266 °F), as low as −17 °C (1 °F)
- Acidity/alkalinity: less than pH 0, up to pH 11.5
- Salinity: up to saturation
- Pressure: up to 1,000-2,000 atm, down to 0 atm (e.g. vacuum of space)
- Radiation: up to 5kGy

Extremophiles are significant in different ways. They extend terrestrial life into much of the Earth's hydrosphere, crust and atmosphere, their specific evolutionary adaptation mechanisms to their extreme environment can be exploited in bio-technology, and their very existence under such extreme conditions increases the potential for extraterrestrial life.

In Soil

The nitrogen cycle in soils depends on the fixation of atmospheric nitrogen. This is achieved by a number of diazotrophs. One way this can occur is in the nodules in the roots of legumes that contain symbiotic bacteria of the genera *Rhizobium*, *Mesorhizobium*, *Sinorhizobium*, *Bradyrhizobium*, and *Azorhizobium*.

Symbiosis

A lichen is a symbiosis of a fungus with microbial algae. The algal partner is photosynthetic, enabling the fungus to live in habitats such as bare rocks where other sources of nutrition are not available.

Other fungi, including some edible mushrooms such as the cep, form mycorrhizal symbioses with trees. The fungi increase the supply of nutrients to the tree in return for a supply of energy.

Applications

Microorganisms are vital to humans and the environment, as they participate in the carbon and nitrogen cycles, as well as fulfilling other vital roles in virtually all ecosystems, such as recycling other organisms' dead remains and waste products through decomposition. Microorganisms also have an important place in most higher-order multicellular organisms as symbionts. Many blame the failure of Biosphere 2 on an improper balance of microorganisms.

Food Production

Microorganisms are used to make yoghurt, cheese, curd, kefir, ayran, xynogala, and other types of food. They are used to leaven bread, and to convert sugars to alcohol in wine and beer. Microorganisms are used in brewing, wine making, baking, pickling and other food-making processes.

They are also used to control the fermentation process in the production of cultured dairy products such as yogurt and cheese. The cultures also provide flavor and aroma, and inhibit undesirable organisms.

Water Treatment

The majority of all oxidative sewage treatment processes rely on a large range of microorganisms to oxidise organic constituents which are not amenable to sedimentation or flotation. Anaerobic microorganisms are also used to reduce sludge solids producing methane gas (amongst other gases) and a sterile mineralised residue. In potable water treatment, one method, the slow sand filter, employs a complex gelatinous layer composed of a wide range of microorganisms to remove both dissolved and particulate material from raw water.

Energy

Microorganisms are used in fermentation to produce ethanol, and in biogas reactors to produce methane. Scientists are researching the use of algae to produce liquid fuels, and bacteria to convert various forms of agricultural and urban waste into usable fuels.

Chemicals, Enzymes

Microorganisms are used for many commercial and industrial production of chemicals, enzymes and other bioactive molecules.

Organic acids produced by microbial fermentation include

- Acetic acid produced by the bacterium *Acetobacter aceti* and other acetic acid bacteria (AAB)
- Butyric acid (butanoic acid) produced by the bacterium *Clostridium butyricum*
- Lactic acid: *Lactobacillus* and other lactic acid bacteria (LAB)
- Citric acid: produced by the mould fungus *Aspergillus niger*

Microorganisms are used for preparation of bioactive molecules and enzymes, including:

- Streptokinase produced by the bacterium *Streptococcus* and modified by genetic engineering is used as a clot buster for removing clots from the blood vessels of patients who have undergone myocardial infarctions leading to heart attack.
- Cyclosporin A, a bioactive molecule used as an immunosuppressive agent in organ transplantation
- Statins produced by the yeast *Monascus purpureus* are commercialised as blood cholesterol lowering agents which act by competitively inhibiting the enzyme responsible for synthesis of cholesterol.

Science

Microorganisms are essential tools in biotechnology, biochemistry, genetics, and molecular biology. The yeasts (*Saccharomyces cerevisiae*) and fission yeast (*Schizosaccharomyces pombe*) are important model organisms in science, since they are simple eukaryotes that can be grown rapidly in large numbers and are easily manipulated. They are particularly valuable in genetics, genomics and proteomics. Microorganisms can be harnessed for uses such as creating steroids and treating skin diseases. Scientists are also considering using microorganisms for living fuel cells, and as a solution for pollution.

Warfare

In the Middle Ages, as an early example of biological warfare, diseased corpses were thrown into castles during sieges using catapults or other siege engines. Individuals near the corpses were exposed to the pathogen and were likely to spread that pathogen to others.

Soil

Microbes can make nutrients and minerals in the soil available to plants, produce hormones that spur growth, stimulate the plant immune system and trigger or dampen stress responses. In general a more diverse soil microbiome results in fewer plant diseases and higher yield.

Human Health

Human Bacterial Flora

Microorganisms can form an endosymbiotic relationship with other, larger organisms. For example, microbial symbiosis plays a crucial role in the immune system. The bacteria that live within the human digestive system contribute to gut immunity, synthesize vitamins such as folic acid and biotin, and ferment complex indigestible carbohydrates.

Disease

Microorganisms are the causative agents (pathogens) in many infectious diseases. The organisms involved include pathogenic bacteria, causing diseases such as plague, tuberculosis and anthrax; protozoa, causing diseases such as malaria, sleeping sickness, dysentery and toxoplasmosis; and also fungi causing diseases such as ringworm, candidiasis or histoplasmosis. However, other diseases such as influenza, yellow fever or AIDS are caused by pathogenic viruses, which are not usually classified as living organisms and are not, therefore, microorganisms by the strict definition. No clear examples of archaean pathogens are known, although a relationship has been proposed between the presence of some archaean methanogens and human periodontal disease.

Hygiene

Hygiene is a set of practices to avoid infection or food spoiling by eliminating microorganisms from the surroundings. As microorganisms, in particular bacteria, are found virtually everywhere, harmful microorganisms may be reduced to acceptable levels rather than actually eliminated. In food preparation, microorganisms are reduced by preservation methods such as cooking, cleanliness of utensils, short storage periods, or by low temperatures. If complete sterility is needed, as with surgical equipment, an autoclave is used to kill microorganisms with heat and pressure.

Microbiology

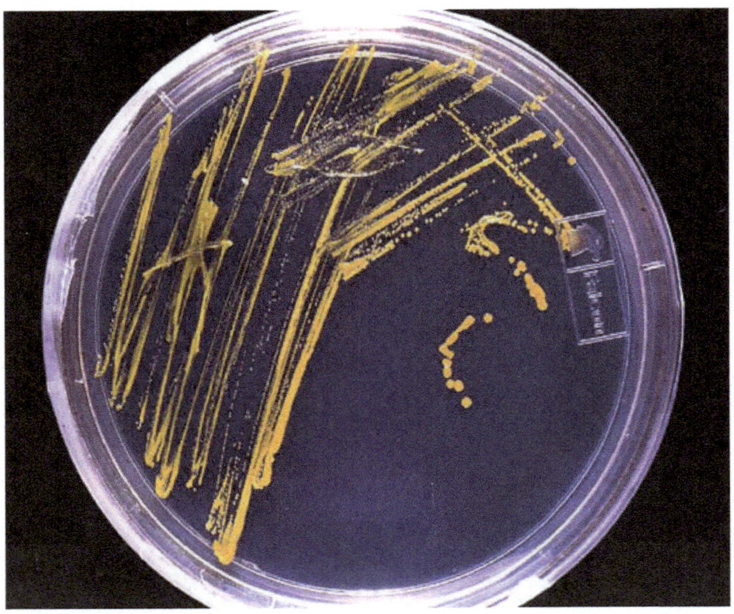

An agar plate streaked with microorganisms

Microbiology is the study of microscopic organisms, those being unicellular (single cell), multicellular (cell colony), or acellular (lacking cells). Microbiology encompasses numerous sub-disciplines including virology, mycology, parasitology, and bacteriology.

Eukaryotic micro-organisms possess membrane-bound cell organelles and include fungi and protists, whereas prokaryotic organisms—all of which are microorganisms—are conventionally classified as lacking membrane-bound organelles and include eubacteria and archaebacteria. Microbiologists traditionally relied on culture, staining, and microscopy. However, less than 1% of the microorganisms present in common environments can be cultured in isolation using current means. Microbiologists often rely on extraction or detection of nucleic acid, either DNA or RNA sequences.

Viruses have been variably classified as organisms, as they have been considered either as very simple microorganisms or very complex molecules. Prions, never considered microorganisms, have been investigated by virologists, however, as the clinical effects traced to them were originally presumed due to chronic viral infections, and virologists took search—discovering "infectious proteins".

As an application of microbiology, medical microbiology is often introduced with medical principles of immunology as *microbiology and immunology*. Otherwise, microbiology, virology, and immunology as basic sciences have greatly exceeded the medical variants, applied sciences.

Branches

The branches of microbiology can be classified into pure and applied sciences. Microbiology can be also classified based on taxonomy, in the cases of bacteriology, mycology, protozoology, and phycology. There is considerable overlap between the specific branches of microbiology with each other and with other disciplines, and certain aspects of these branches can extend beyond the traditional scope of microbiology.

Pure Microbiology

- Bacteriology: The study of bacteria
- Mycology: The study of fungi
- Protozoology: The study of protozoa
- Phycology/algology: The study of algae
- Parasitology: The study of parasites
- Immunology: The study of the immune system
- Virology: The study of viruses
- Nematology: The study of nematodes
- Microbial cytology: The study of microscopic and submicroscopic details of microorganisms

- Microbial physiology: The study of how the microbial cell functions biochemically. Includes the study of microbial growth, microbial metabolism and microbial cell structure
- Microbial ecology: The relationship between microorganisms and their environment
- Microbial genetics: The study of how genes are organized and regulated in microbes in relation to their cellular functions Closely related to the field of molecular biology
- Cellular microbiology: A discipline bridging microbiology and cell biology
- Evolutionary microbiology: the study of the evolution of microbes. This field can be subdivided into:
    - Microbial taxonomy: The naming and classification of microorganisms
    - Microbial systematic: The study of the diversity and genetic relationship of microorganisms
- Generation microbiology: The study of those microorganisms that have the same characters as their parents
- Systems microbiology: A discipline bridging systems biology and microbiology
- Molecular microbiology: The study of the molecular principles of the physiological processes in microorganisms

Other

- Nano microbiology: The study of those organisms on nano level.
- Exo microbiology (or Astro microbiology): The study of microorganisms in outer space
- Biological agent: The study of those microorganisms which are being used in weapon industries.
- Predictive microbiology: The quantification of relations between controlling factors in foods and responses of pathogenic and spoilage microorganisms using mathematical modelling

Applied Microbiology

- Medical microbiology: The study of the pathogenic microbes and the role of microbes in human illness. Includes the study of microbial pathogenesis and epidemiology and is related to the study of disease pathology and immunology. This area of microbiology also covers the study of human microbiota, cancer, and the tumor microenvironment.
- Pharmaceutical microbiology: The study of microorganisms that are related to the production of antibiotics, enzymes, vitamins,vaccines, and other pharmaceutical products and that cause pharmaceutical contamination and spoil.

Food microbiology laboratory at the Faculty of Food Technology, Latvia University of Agriculture.

- Industrial microbiology: The exploitation of microbes for use in industrial processes. Examples include industrial fermentation and wastewater treatment. Closely linked to the biotechnology industry. This field also includes brewing, an important application of microbiology.
- Microbial biotechnology: The manipulation of microorganisms at the genetic and molecular level to generate useful products.
- Food microbiology: The study of microorganisms causing food spoilage and foodborne illness. Using microorganisms to produce foods, for example by fermentation.
- Agricultural microbiology: The study of agriculturally relevant microorganisms. This field can be further classified into the following:
    - Plant microbiology and Plant pathology: The study of the interactions between microorganisms and plants and plant pathogens.
    - Soil microbiology: The study of those microorganisms that are found in soil.
- Veterinary microbiology: The study of the role of microbes in veterinary medicine or animal taxonomy.
- Environmental microbiology: The study of the function and diversity of microbes in their natural environments. This involves the characterization of key bacterial habitats such as the rhizosphere and phyllosphere, soil and groundwater ecosystems, open oceans or extreme environments (extremophiles). This field includes other branches of microbiology such as:
    - Microbial ecology
    - Microbially mediated nutrient cycling
    - Geomicrobiology
    - Microbial diversity
    - Bioremediation

- Water microbiology (or aquatic microbiology): The study of those microorganisms that are found in water.
- Aeromicrobiology (or air microbiology): The study of airborne microorganisms.
    - biotechnology: related to recombinant DNA technology or genetic engineering

Benefits

While some fear microbes due to the association of some microbes with various human illnesses, many microbes are also responsible for numerous beneficial processes such as industrial fermentation (e.g. the production of alcohol, vinegar and dairy products), antibiotic production and as vehicles for cloning in more complex organisms such as plants. Scientists have also exploited their knowledge of microbes to produce biotechnologically important enzymes such as Taq polymerase, reporter genes for use in other genetic systems and novel molecular biology techniques such as the yeast two-hybrid system.

Fermenting tanks with yeast being used to brew beer

Bacteria can be used for the industrial production of amino acids. *Corynebacterium glutamicum* is one of the most important bacterial species with an annual production of more than two million tons of amino acids, mainly L-glutamate and L-lysine. Since some bacteria have the ability to synthesize antibiotics, they are used for medicinal purposes, such as Streptomyces to make aminoglycoside antibiotics.

A variety of biopolymers, such as polysaccharides, polyesters, and polyamides, are produced by microorganisms. Microorganisms are used for the biotechnological production of biopolymers with tailored properties suitable for high-value medical application such as tissue engineering and drug delivery. Microorganisms are used for the biosynthesis of xanthan, alginate, cellulose, cyanophycin, poly(gamma-glutamic acid), levan, hyaluronic acid, organic acids, oligosaccharides and polysaccharide, and polyhydroxyalkanoates.

Microorganisms are beneficial for microbial biodegradation or bioremediation of domestic, agricultural and industrial wastes and subsurface pollution in soils, sediments and marine environments. The ability of each microorganism to degrade toxic waste depends on the nature of each contaminant. Since sites typically have multiple pollutant types, the most effective approach to microbial biodegradation is to use a mixture of bacterial and fungal species and strains, each specific to the biodegradation of one or more types of contaminants.

Symbiotic microbial communities are known to confer various benefits to their human and animal hosts health including aiding digestion, production of beneficial vitamins and amino acids, and suppression of pathogenic microbes. Some benefit may be conferred by consuming fermented foods, probiotics (bacteria potentially beneficial to the digestive system) and/or prebiotics (substances consumed to promote the growth of probiotic microorganisms). The ways the microbiome influences human and animal health, as well as methods to influence the microbiome are active areas of research.

Research has suggested that microorganisms could be useful in the treatment of cancer. Various strains of non-pathogenic clostridia can infiltrate and replicate within solid tumors. Clostridial vectors can be safely administered and their potential to deliver therapeutic proteins has been demonstrated in a variety of preclinical models.

History

Ancient Times

The existence of microorganisms was hypothesized for many centuries before their actual discovery. The existence of unseen microbiological life was postulated by Jainism which is based on Mahavira's teachings as early as 6th century BCE. Paul Dundas notes that Mahavira asserted existence of unseen microbiological creatures living in earth, water, air and fire. Jain scriptures also describe nigodas which are sub-microscopic creatures living in large clusters and having a very short life and are said to pervade each and every part of the universe, even in tissues of plants and flesh of animals. The Roman Marcus Terentius Varro made references to microbes when he warned against locating a homestead in the vicinity of swamps "because there are bred certain minute creatures which cannot be seen by the eyes, which float in the air and enter the body through the mouth and nose and there by cause serious diseases."

Medieval Islamic World

Avicenna "ibn Sina"

At the golden age of Islamic civilization, some scientists had knowledge about microorganisms, such as Ibn Sina in his book *The Canon of Medicine*, Ibn Zuhr (also known as Avenzoar) who discovered scabies mites, and Al-Razi who gave the earliest known description of smallpox in his book *The Virtuous Life* (al-Hawi).

In 1546, Girolamo Fracastoro proposed that epidemic diseases were caused by transferable seed-like entities that could transmit infection by direct or indirect contact, or vehicle transmission.

However, early claims about the existence of microorganisms were speculative, and not based on microscopic observation. Actual observation and discovery of microbes had to await the invention of the microscope in the 17th century.

Modern Times

Antonie van Leeuwenhoek, is considered to be the one of the first to observe microorganisms using a microscope.

In 1676, Anton van Leeuwenhoek, who lived most of his life in Delft, Holland, observed bacteria and other microorganisms using a single-lens microscope of his own design. While Van Leeuwenhoek is often cited as the first to observe microbes, Robert Hooke made the first recorded microscopic observation, of the fruiting bodies of moulds, in 1665. It has, however, been suggested that a Jesuit priest called Athanasius Kircher was the first to observe micro-organisms.

He was among the first to design magic lanterns for projection purposes, so he must have been well acquainted with the properties of lenses. One of his books contains a chapter in Latin, which reads in translation – "Concerning the wonderful structure of things in nature, investigated by Microscope." Here, he wrote "who would believe that vinegar and milk abound with an innumerable multitude of worms." He also noted that putrid material is full of innumerable creeping animalcule. These observations antedate Robert Hooke's *Micrographia* by nearly 20 years and were published some 29 years before van Leeuwenhoek saw protozoa and 37 years before he described having seen bacteria. Joseph Lister was the first person who said infectious diseases are caused by micro-organism and was first person who used phenol as disinfectant on the open wounds of patients.

The field of bacteriology (later a subdiscipline of microbiology) was founded in the 19th century by Ferdinand Cohn, a botanist whose studies on algae and photosynthetic bacteria led him to describe several bacteria including *Bacillus* and *Beggiatoa*. Cohn was also the first to formulate a scheme for the taxonomic classification of bacteria and discover spores. Louis Pasteur and Robert Koch were contemporaries of Cohn's and are often considered to be the father of microbiology and medical microbiology, respectively. Pasteur is most famous for his series of experiments designed to disprove the then widely held theory of spontaneous generation, thereby solidifying microbiology's identity as a biological science. Pasteur also designed methods for food preservation (pasteurization) and vaccines against several diseases such as anthrax, fowl cholera and rabies. Koch is best known for his contributions to the germ theory of disease, proving that specific diseases were caused by specific pathogenic micro-organisms. He developed a series of criteria that have become known as the Koch's postulates. Koch was one of the first scientists to focus on the isolation of bacteria in pure culture resulting in his description of several novel bacteria including *Mycobacterium tuberculosis*, the causative agent of tuberculosis.

While Pasteur and Koch are often considered the founders of microbiology, their work did not accurately reflect the true diversity of the microbial world because of their exclusive focus on micro-organisms having direct medical relevance. It was not until the late 19th century and the work of Martinus Beijerinck and Sergei Winogradsky, the founders of general microbiology (an older term encompassing aspects of microbial physiology, diversity and ecology), that the true breadth of microbiology was revealed. Beijerinck made two major contributions to microbiology: the discovery of viruses and the development of enrichment culture techniques. While his work on the Tobacco Mosaic Virus established the basic principles of virology, it was his development of enrichment culturing that had the most immediate impact on microbiology by allowing for the cultivation of a wide range of microbes with wildly different physiologies. Winogradsky was the first to develop the concept of chemolithotrophy and to thereby reveal the essential role played by micro-organisms in geochemical processes. He was responsible for the first isolation and description of both nitrifying and nitrogen-fixing bacteria. French-Canadian microbiologist Felix d'Herelle co-discovered bacteriophages and was one of the earliest applied microbiologists.

Innovative laboratory glassware and experimental methods developed by Louis Pasteur and other biologists contributed to the young field of bacteriology in the late 19th century.

Today, scientists all over the world use microbiology for basic and applied research. Jonathan Eisen used Storify to compile a slideshow of over 300 actual living microbiologists.

Importance of Microbiology

The environment

- Microbes are responsible for the cycling of carbon, nitrogen phosphorus (geochemical cycles)
- Maintain ecological balance on earth
- They are found in association with plants in symbiotic relationships, maintain soil fertility and may also be used to clean up the environment of toxic compounds (bio-remediation).
- Some are devasting plant pathogens, but others act as biological control agents against these diseases.

Medicine

- Disease causing ability of some microbes such as
- Small Pox (Variola virus)
- Cholera (Vibrio cholera)
- Malaria (Plasmodium , protozoa) etc.
- They have also provided us with the means of their control in the form of antibiotics and other medically important drugs.

Food

- Microorganisms have been used to produce food, from brewing and wine making, through cheese production and bread making, to manufacture of soy sauce.
- Microbes are also responsible for food spoilage.

Biotechnology

- Commercial applications include the synthesis of acetone, organic acids, enzymes, alcohols and many drugs.
- Genetic engineering – bacteria can produce important therapeutic substances such as insulin, human growth hormone, and interferon.

Research

- Because of their simple structure they are easier to study most life processes in simple unicellular organisms than in complex multicellular ones.

- Millions of copies of the same single cell can be produced in large numbers very quickly and at low cost to give plenty of homogenous experimental material.
- Because they reproduce very quickly, they are useful for studies involving the transfer of genetic information.

Brief History of Microbiology

- Robert Hook (1665) – reported that life's smallest structural units were 'little boxes' or 'cells'. This marked the beginning of cell theory – that all living things are composed of cells.
- Van Leuwenhoek (1673) – discovered the 'invisible' world of microorganisms 'animalcules'.
- Until second half of nineteenth century many believed that some forms of life could arise spontaneously from non-living matter – spontaneous generation.
- Francesco Redi (1668) – Strong opponent of spontaneous generation. He demonstrated that maggots appear on decaying meat only when flies are able to lay eggs on the meat.
- John Needham (1745) – claimed that microorganisms could arise spontaneously from heated nutrient broth.
- Lazzaro Spallanzani (1765) – repeated Needhams experiments and suggested that Needham's results were due to microorganisms in the air entering the broth.
- Rudolf Virchow (1858) – concept of biogenesis – living cells can arise only from preexisting cells.
- Louis Pasteur (1822-1895) – Pasteur's experiments on swan shaped necks resolved the controversy of spontaneous generation. His discoveries led to the development of aseptic techniques used in the laboratory and medical procedure to prevent contamination by microorganisms that are in the air.

Golden age of microbiology

- Rapid advances in the science of microbiology were made between 1857 and 1914.

Fermentation and Pasteurization

- Pasteur found that yeast ferments sugars to alcohols and that bacterium can oxidize the alcohol to acetic acid.
- Heating processes called pasteurization is used to kill bacteria in some alcoholic beverages and milk.

The Germ theory of disease

- Agostino Bassi (1934) and Pasteur (1865) – showed a casual relationship between microorganisms and disease.

- Joseph Lister (1860s) – introduced the use of disinfectant to clean surgical dressings in order to control infection in humans
- Robert Koch (1876) – proved that microorganisms transmit disease – Koch's postulates which are used today to prove that a particular microorganism causes a particular disease.
- Introduced pure cultures
- Koch's postulates (Henle-Koch's Postulates) are:

  1. A specific organism should be found constantly in association with the disease.
  2. The organism should be isolated and grown in a pure culture in the laboratory.
  3. The pure culture when inoculated into a healthy susceptible animal should produce symptoms/lesions of the same disease
  4. From the inoculated animal, the microorganism should be isolated in pure culture.
  5. An additional criterion introduced is that specific antibodies to the causative organism shouldbe demonstrable in patient's serum.

- Angelina – American wife of Koch's assistant suggested solidifying broths with agar as an aid to obtaining pure cultures.
- Koch also developed techniques for isolating organisms. Identified the bacillus that causes tuberculosis and anthrax, developed tuberculin and studied various diseases in Africa and Asia. His studies on Tuberculosis won him Nobel prize for philosophy and medicine in 1905.

Vaccination

- Immunity is conferred by inoculation with a vaccine.
- Edward Jenner(1798) – demonstrated that inoculations with cowpox material provides humans with immunity from small pox
- Pasteur (1880) – discovered that avirulent bacteria could be used as a vaccine for chicken cholera; he coined the word vaccine
- Modern vaccines are prepared from living avirulent microorganisms or killed pathogens, from isolated components of pathogens, and by recombinant DNA techniques.

Emergence of special fields of Microbiology

Immunology

- Immunization was first used against small pox. Edward Jenner used fluid from cowpox blisters to immunize against it.
- Pasteur developed techniques to weaken organisms so they would produce immunity without producing disease.

- Elie Metchnikoff discovered that certain cells in the body would ingest microbes and named them as phagocytes.

Industrial Microbiology and Microbial Ecology

- Pasteur – fermentation technology and pasteurization. One of his most important discoveries was that some fermentative microorganisms were anaerobic and others were able to live either aerobically or anaerobically.

Microbial ecology – Two pioneers

- Sergei N. Winogradsky (1856-1953) – Soil microbiology – discovered that soil bacteria could oxidize iron, sulfur and ammonia to obtain energy and many bacteria incorporate CO_2 into organic matter. He also isolated anaerobic nitrogen fixing soil bacteria and studied the decomposition of cellulose.
- Martinus Beijerinck (1851-1931) – He isolated aerobic nitrogen fixing bacterium *Azotobacter*, a root nodule bacterium also capable of fixing nitrogen (later renamed as *Rhizobium*); and sulfate reducing bacteria. Both of them developed enrichment culture technique and use of selective media, which have been of great importance in microbiology.

Virology

- Beijerinck characterized viruses as pathogenic molecules that could take over a host cells mechanisms for their own use.
- Wendell Stanley (1935) – crystallized TMV and crystals consisted of protein and RNA.
- Viruses were first observed with an EM in 1939.
- Alfred Hershey and Martha Chase (1952) – demonstrated that the genetic material of some viruses is DNA.
- James Watson and Francis Crick (1953) -determined the structure of DNA.

Chemotherapy

- There are two types of chemotherapeutic agents: synthetic drugs and antibiotics.
- Elrlich (1910) introduced an arsenic containing chemical called Salvarsan to treat Syphilis.
- Alexander Fleming (1928) – observed that the mold *Penicillium* inhibited the growth of bacteria and named the active ingredient as penicillin. Penicillin has been used clinically as an antibiotic since the 1940s. Domagk and others developed sulfa drugs.
- Waksman and others developed Streptomycin and other antibiotics derived from soil organisms.
- Researchers are tackling the problem of drug-resistant microbes.

Genetics and Molecular Biology

- 1900 – Modern genetics began with the rediscovery of Gregor Mendel's principles of genetics.
- Frederick Griffith (1928) - discovered that previously harmless bacteria could change their nature and become capable of causing disease
- Avery, McCarty and MacLeod (1940's) – showed that this genetic change was due to DNA. After this finding came the crucial discovery of the structure of DNA by Watson and Crick
- Edward Tatum and George Beadle – studied biochemical mutants of *Neurospora* to show how genetic information controls metabolism.
- Barbara McClintock (1950) – discovered that some genes could move from one location to another on a chromosome.
- Early 1960's witnessed a further explosion of discoveries relating to the way DNA controls protein synthesis.
- Francois Jacob and Jacques Monod (1961) – discovered mRNA and later made the first major discoveries about regulation of gene function in bacteria.
- Microorganisms can now be genetically engineered to manufacture large amounts of human hormones and other urgently needed medical substances.
- Late 1960's Paul Berg showed that fragments of human or animal DNA that code for important proteins can be attached to bacterial DNA. The resulting hybrid was the first example of recombinant DNA.

Tomorrow's history

Microbiology has been in the forefront of research in medicine and biology and continues to play a role in Genetic engineering and Gene therapy.

Genetic engineering – scientists are attempting to redesign microorganisms for a variety of purposes (drugs, hormones, vaccines and a variety of biologically important compounds)

rDNA technology – enabling us to produce improved varieties of plants and animals such as pest-resistant crops and may even enable us to correct genetic defects in human beings.

Human genome project

- Microbial genetic techniques have made possible a colossal scientific undertaking HGP. Begun in 1990 and supposed to complete by 2005 was completed in May 2000.
- Humans have just over 30,000 genes instead of estimates that ranged up to 142,000 genes. 3 billion base pairs in the human genome do not all code for useful genes (75% of them code for 'junk DNA')
- Over 100 microbial genomes have been sequenced so far.

- Approx. 113 genes have come to human genome directly from bacteria.
- Venter has sequenced mouse genome and reports that humans have only 300 genes not found in the mouse.

Scope of Microbiology

The microbiology has influence on genetics, agriculture, food science, ecology, immunology and various fields.

Genetics: Mainly involves engineered microbes to make hormones, vaccine, antibiotics and many other useful products for human being.

Agriculture: The influence of microbes on agriculture; the prevention of the diseases that mainly damage the useful crops.

Food science: It involves the prevention of spoilage of food and food borne diseases and the uses of microbes to produce cheese, yoghurt, pickles and beer.

Immunology: The study of immune system which protect the body from pathogens.

Medicine: deals with the identification of plans and measures to cure diseases of human and animals which are infectious to them.

Industry: it involves use of microbes to produce antibiotics, steroids, alcohol, vitamins and amino acids etc.

2

Microbes and its Types

With the use of the electronic microscope, microbes could be studied in detail. The microorganisms could then be divided into five major classes: Archaea, Bacteria, Algae, Protozoa, and Fungi. These microbes have had a great impact on human life. Their scope ranges from their use in food production to the manufacture of medicines and extends to their need in industries and preserving the environment. The major components of the microbial world are discussed in this chapter.

Members of Microbial World

There are five major members of microorganisms, Archaea, Bacteria, Algae, Protozoa, and Fungi. The Archaea and Bacteria are prokaryotic cells. Unicellular algae and protozoa and fungi are eukaryotic cells.

Archaea

The Archaea constitute a domain and kingdom of single-celled microorganisms. These microbes (archaea; singular archaeon) are prokaryotes, meaning that they have no cell nucleus or any other membrane-bound organelles in their cells.

Archaea were initially classified as bacteria, receiving the name archaebacteria (in the Archaebacteria kingdom), but this classification is outdated. Archaeal cells have unique properties separating them from the other two domains of life, Bacteria and Eukaryota. The Archaea are further divided into multiple recognized phyla. Classification is difficult because the majority have not been isolated in the laboratory and have only been detected by analysis of their nucleic acids in samples from their environment.

Archaea and bacteria are generally similar in size and shape, although a few archaea have very strange shapes, such as the flat and square-shaped cells of *Haloquadratum walsbyi*. Despite this morphological similarity to bacteria, archaea possess genes and several metabolic pathways that are more closely related to those of eukaryotes, notably the enzymes involved in transcription and translation. Other aspects of archaeal biochemistry are unique, such as their reliance on ether lipids in their cell membranes, including archaeols. Archaea use more energy sources than eukaryotes: these range from organic compounds, such as sugars, to ammonia, metal ions or even hydrogen gas. Salt-tolerant archaea (the Haloarchaea) use sunlight as an energy source, and other species of archaea fix carbon; however, unlike plants and cyanobacteria, no known species of archaea does both. Archaea reproduce asexually by binary fission, fragmentation, or budding; unlike bacteria and eukaryotes, no known species forms spores.

Archaea were initially viewed as extremophiles living in harsh environments, such as hot springs and salt lakes, but they have since been found in a broad range of habitats, including soils, oceans, and marshlands. They are also found in the human colon, oral cavity, and skin. Archaea are particularly numerous in the oceans, and the archaea in plankton may be one of the most abundant groups of organisms on the planet. Archaea are a major part of Earth's life and may play roles in both the carbon cycle and the nitrogen cycle. No clear examples of archaeal pathogens or parasites are known, but they are often mutualists or commensals. One example is the methanogens that inhabit human and ruminant guts, where their vast numbers aid digestion. Methanogens are also used in biogas production and sewage treatment, and enzymes from extremophile archaea that can endure high temperatures and organic solvents are exploited in biotechnology.

New Domain

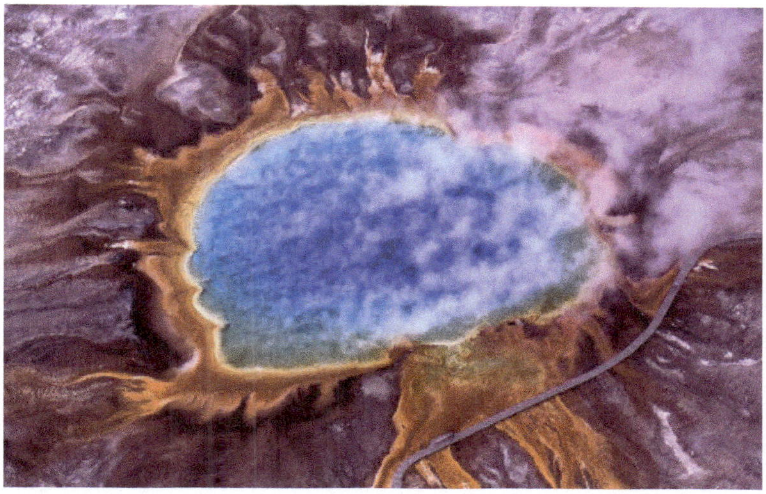

Archaea were first found in extreme environments, such as volcanic hot springs. Pictured here is Grand Prismatic Spring of Yellowstone National Park.

For much of the 20th century, prokaryotes were regarded as a single group of organisms and classified based on their biochemistry, morphology and metabolism. For example, microbiologists tried to classify microorganisms based on the structures of their cell walls, their shapes, and the substances they consume. In 1965, Emile Zuckerkandl and Linus Pauling proposed instead using the sequences of the genes in different prokaryotes to work out how they are related to each other. This approach, known as phylogenetics, is the main method used today.

Archaea were first classified as a separate group of prokaryotes in 1977 by Carl Woese and George E. Fox in phylogenetic trees based on the sequences of ribosomal RNA (rRNA) genes. These two groups were originally named the Archaebacteria and Eubacteria and treated as kingdoms or subkingdoms, which Woese and Fox termed *Urkingdoms*. Woese argued that this group of prokaryotes is a fundamentally different sort of life. To emphasize this difference, Woese later proposed a new natural system of organisms with three separate Domains: the Eukarya, the Bacteria and the Archaea, in what is now known as "The Woesian Revolution".

The word *archaea* comes from the Ancient Greek meaning "ancient things", as the first representatives of the domain Archaea were methanogens and it was assumed that their metabolism reflected Earth's primitive atmosphere and the organism's antiquity. For a long time, archaea were seen as extremophiles that only exist in extreme habitats such as hot springs and salt lakes. However, as new habitats were studied, more organisms were discovered. Extreme halophilic and hyperthermophilic microbes were also included in the Archaea. By the end of the 20th century, archaea had been identified in non-extreme environments as well. Today, they are known to be a large and diverse group of organisms that are widely distributed in nature and are common in all habitats. This new appreciation of the importance and ubiquity of archaea came from using polymerase chain reaction (PCR) to detect prokaryotes from environmental samples (such as water or soil) by multiplying their ribosomal genes. This allows the detection and identification of organisms that have not been cultured in the laboratory.

Current Classification

The ARMAN are a new group of archaea recently discovered in acid mine drainage.

The classification of archaea, and of prokaryotes in general, is a rapidly moving and contentious field. Current classification systems aim to organize archaea into groups of organisms that share structural features and common ancestors. These classifications rely heavily on the use of the sequence of ribosomal RNA genes to reveal relationships between organisms (molecular phylogenetics). Most of the culturable and well-investigated species of archaea are members of two main phyla, the Euryarchaeota and Crenarchaeota. Other groups have been tentatively created. For example, the peculiar species *Nanoarchaeum equitans*, which was discovered in 2003, has been given its own phylum, the Nanoarchaeota. A new phylum Korarchaeota has also been proposed. It contains a small group of unusual thermophilic species that shares features of both of the main phyla, but is most closely related to the Crenarchaeota. Other recently detected species of archaea are only distantly related to any of these groups, such as the Archaeal Richmond Mine acidophilic nanoorganisms (ARMAN), which were discovered in 2006 and are some of the smallest organisms known.

A superphylum - TACK - has been proposed that includes the Thaumarchaeota, Aigarchaeota, Crenarchaeota, and Korarchaeota. This superphylum may be related to the origin of eukaryotes. More recently, the superphylum Asgard has been named and proposed to be more closely related to the original eukaryote and a sister group to TACK.

Concept of Species

The classification of archaea into species is also controversial. Biology defines a species as a group of related organisms. The familiar exclusive breeding criterion (organisms that can breed with each other but not with others) is of no help here because archaea reproduce asexually.

Archaea show high levels of horizontal gene transfer between lineages. Some researchers suggest that individuals can be grouped into species-like populations given highly similar genomes and infrequent gene transfer to/from cells with less-related genomes, as in the genus *Ferroplasma*. On the other hand, studies in *Halorubrum* found significant genetic transfer to/from less-related populations, limiting the criterion's applicability. A second concern is to what extent such species designations have practical meaning.

Current knowledge on genetic diversity is fragmentary and the total number of archaeal species cannot be estimated with any accuracy. Estimates of the number of phyla range from 18 to 23, of which only 8 have representatives that have been cultured and studied directly. Many of these hypothesized groups are known from a single rRNA sequence, indicating that the diversity among these organisms remains obscure. The Bacteria also contain many uncultured microbes with similar implications for characterization.

Origin and Evolution

The age of the Earth is about 4.54 billion years. Scientific evidence suggests that life began on Earth at least 3.5 billion years ago. The earliest evidence for life on Earth is graphite found to be biogenic in 3.7 billion-year-old metasedimentary rocks discovered in Western Greenland and microbial mat fossils found in 3.48 billion-year-old sandstone discovered in Western Australia. More recently, in 2015, "remains of biotic life" were found in 4.1 billion-year-old rocks in Western Australia.

Although probable prokaryotic cell fossils date to almost 3.5 billion years ago, most prokaryotes do not have distinctive morphologies and fossil shapes cannot be used to identify them as archaea. Instead, chemical fossils of unique lipids are more informative because such compounds do not occur in other organisms. Some publications suggest that archaeal or eukaryotic lipid remains are present in shales dating from 2.7 billion years ago; such data have since been questioned. Such lipids have also been detected in even older rocks from west Greenland. The oldest such traces come from the Isua district, which include Earth's oldest known sediments, formed 3.8 billion years ago. The archaeal lineage may be the most ancient that exists on Earth.

Woese argued that the bacteria, archaea, and eukaryotes represent separate lines of descent that diverged early on from an ancestral colony of organisms. One possibility is that this occurred before the evolution of cells, when the lack of a typical cell membrane allowed unrestricted lateral gene transfer, and that the common ancestors of the three domains arose by fixation of specific subsets of genes. It is possible that the last common ancestor of the bacteria and archaea was a thermophile, which raises the possibility that lower temperatures are "extreme environments" in archaeal terms, and organisms that live in cooler environments appeared only later. Since the Archaea and Bacteria are no more related to each other than they are to eukaryotes, the term *prokaryote*'s only surviving meaning is "not a eukaryote", limiting its value.

Comparison to Other Domains

The following table compares some major characteristics of the three domains, to illustrate their similarities and differences. Many of these characteristics are also discussed below.

Property	Archaea	Bacteria	Eukarya
Cell Membrane	Ether-linked lipids, pseudopeptidoglycan	Ester-linked lipids, peptidoglycan	Ester-linked lipids, various structures
Gene Structure	Circular chromosomes, similar translation and transcription to Eukarya	Circular chromosomes, unique translation and transcription	Multiple, linear chromosomes, similar translation and transcription to Archaea
Internal Cell Structure	No membrane-bound organelles or nucleus	No membrane-bound organelles or nucleus	Membrane-bound organelles and nucleus
Metabolism	Various, with methanogenesis unique to Archaea	Various, including photosynthesis, aerobic and anaerobic respiration, fermentation, and autotrophy	Photosynthesis and cellular respiration
Reproduction	Asexual reproduction, horizontal gene transfer	Asexual reproduction, horizontal gene transfer	Sexual and asexual reproduction

Archaea were split off as a third domain because of the large differences in their ribosomal RNA structure. The particular RNA molecule sequenced, known as 16s rRNA, is present in all organisms and always has the same vital function, the production of proteins. Because this function is so central to life, organisms with mutations of its 16s rRNA are unlikely to survive, leading to great stability in the structure of this nucleotide over many generations. 16s rRNA is also large enough to retain organism-specific information, but small enough to be sequenced in a manageable amount of time. In 1977, Carl Woese, a microbiologist studying the genetic sequencing of organisms, developed a new sequencing method that involved splitting the RNA into fragments that could be sorted and compared to other fragments from other organisms. The more similar the patterns between species were, the more closely related the organisms.

Woese used his new rRNA comparison method to categorize and contrast different organisms. He sequenced a variety of different species and happened upon a group of methanogens that had vastly different patterns than any known prokaryotes or eukaryotes. These methanogens were much more similar to each other than they were to other organisms sequenced, leading Woese to propose the new domain of Archaea. One of the interesting results of his experiments was that the Archaea were more similar to eukaryotes than prokaryotes, even though they were more similar to prokaryotes in structure. This led to the conclusion that Archaea and Eukarya shared a more recent common ancestor than Eukarya and Bacteria in general. The development of the nucleus occurred after the split between Bacteria and this common ancestor. Although Archaea are prokaryotic, they are more closely related to Eukarya and thus cannot be placed within either the Bacteria or Eukarya domains.

One property unique to Archaea is the abundant use of ether-linked lipids in their cell membranes. Ether linkages are more chemically stable than the ester linkages found in Bacteria and Eukarya, which may be a contributing factor to the ability of many Archaea to survive in extreme environments that place heavy stress on cell membranes, such as extreme heat and salinity. Comparative analysis of archaeal genomes have also identified several molecular signatures in the forms of conserved signature indels and signature proteins which are uniquely present in either all Archaea or different main groups within Archaea . Another unique feature of Archaea is that no other known organisms are capable of methanogenesis (the metabolic production of methane). Methanogenic Archaea play a pivotal role in ecosystems with organisms that derive energy from oxidation of methane, many of which are Bacteria, as they are often a major source of methane in such environments and can play a role as primary producers. Methanogens also play a critical role in the carbon cycle, breaking down organic carbon into methane, which is also a major greenhouse gas.

Relationship to Bacteria

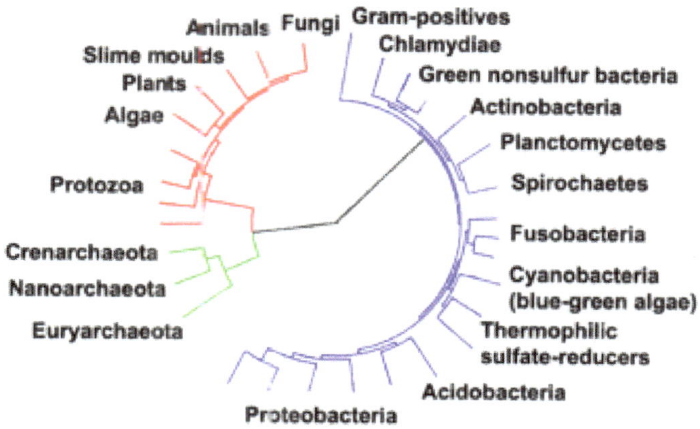

Phylogenetic tree showing the relationship between the Archaea and other domains of life. Eukaryotes are colored red, archaea green and bacteria blue. Adapted from Ciccarelli et al. (2006)

The relationship between the three domains is of central importance for understanding the origin of life. Most of the metabolic pathways, which comprise the majority of an organism's genes, are common between Archaea and Bacteria, while most genes involved in genome expression are common between Archaea and Eukarya. Within prokaryotes, archaeal cell structure is most similar to that of gram-positive bacteria, largely because both have a single lipid bilayer and usually contain a thick sacculus (exoskeleton) of varying chemical composition. In some phylogenetic trees based upon different gene/protein sequences of prokaryotic homologs, the archaeal homologs are more closely related to those of gram-positive bacteria. Archaea and gram-positive bacteria also share conserved indels in a number of important proteins, such as Hsp70 and glutamine synthetase I; however, the phylogeny of these genes was interpreted to reveal interdomain gene transfer, and might not reflect the organismal relationship(s).

It has been proposed that the archaea evolved from gram-positive bacteria in response to antibiotic selection pressure. This is suggested by the observation that archaea are resistant to a wide variety of antibiotics that are primarily produced by gram-positive bacteria, and that these antibiotics primarily act on the genes that distinguish archaea from bacteria. The proposal is that the selective pressure towards resistance generated by the gram-positive antibiotics was eventually sufficient to cause extensive changes in many of the antibiotics' target genes, and that these strains represented the common ancestors of present-day Archaea. The evolution of Archaea in response to antibiotic selection, or any other competitive selective pressure, could also explain their adaptation to extreme environments (such as high temperature or acidity) as the result of a search for unoccupied niches to escape from antibiotic-producing organisms; Cavalier-Smith has made a similar suggestion. This proposal is also supported by other work investigating protein structural relationships and studies that suggest that gram-positive bacteria may constitute the earliest branching lineages within the prokaryotes.

Relation to Eukaryotes

The evolutionary relationship between archaea and eukaryotes remains unclear. Aside from the similarities in cell structure and function that are discussed below, many genetic trees group the two.

Complicating factors include claims that the relationship between eukaryotes and the archaeal phylum Crenarchaeota is closer than the relationship between the Euryarchaeota and the phylum Crenarchaeota and the presence of archaea-like genes in certain bacteria, such as *Thermotoga maritima*, from horizontal gene transfer. The standard hypothesis states that the ancestor of the eukaryotes diverged early from the Archaea, and that eukaryotes arose through fusion of an archaean and eubacterium, which became the nucleus and cytoplasm; this explains various genetic similarities but runs into difficulties explaining cell structure. An alternative hypothesis, the eocyte hypothesis, posits that Eukaryota emerged relatively late from the Archaea.

A recently discovered lineage of archaea, *Lokiarchaeum*, named for a hydrothermal vent called Loki's Castle in the Arctic Ocean, has been found to be most closely related to eukaryotes. It has been called a transitional organism between prokaryotes and eukaryotes.

Morphology

Individual archaea range from 0.1 micrometers (μm) to over 15 μm in diameter, and occur in various shapes, commonly as spheres, rods, spirals or plates. Other morphologies in the Crenarchaeota include irregularly shaped lobed cells in *Sulfolobus*, needle-like filaments that are less than half a micrometer in diameter in *Thermofilum*, and almost perfectly rectangular rods in *Thermoproteus* and *Pyrobaculum*. Archaea in the genus *Haloquadratum* such as *Haloquadratum walsbyi* are flat, square archaea that live in hypersaline pools. These unusual shapes are probably maintained both by their cell walls and a prokaryotic cytoskeleton. Proteins related to the cytoskeleton components of other organisms exist in archaea, and filaments form within their cells, but in contrast to other organisms, these cellular structures are poorly understood. In *Thermoplasma* and *Ferroplasma* the lack of a cell wall means that the cells have irregular shapes, and can resemble amoebae.

Some species form aggregates or filaments of cells up to 200 μm long. These organisms can be prominent in biofilms. Notably, aggregates of *Thermococcus coalescens* cells fuse together in culture, forming single giant cells. Archaea in the genus *Pyrodictium* produce an elaborate multicell colony involving arrays of long, thin hollow tubes called *cannulae* that stick out from the cells' surfaces and connect them into a dense bush-like agglomeration. The function of these cannulae is not settled, but they may allow communication or nutrient exchange with neighbors. Multi-species colonies exist, such as the "string-of-pearls" community that was discovered in 2001 in a German swamp. Round whitish colonies of a novel Euryarchaeota species are spaced along thin filaments that can range up to 15 centimetres (5.9 in) long; these filaments are made of a particular bacteria species.

Structure, Composition Development, and Operation

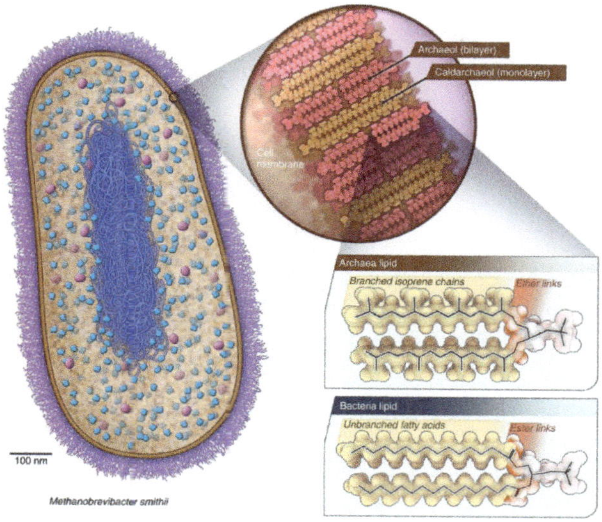

Diagrammatic view of *Methanobrevibacter smithii*, showing the cell membrane (ochre, with inset) and cell wall (purple).

Archaea and bacteria have generally similar cell structure, but cell composition and organization set the archaea apart. Like bacteria, archaea lack interior membranes and organelles. Like bacteria, the cell membranes of archaea are usually bounded by a cell wall and they swim using one or more flagella. Structurally, archaea are most similar to gram-positive bacteria. Most have a single plasma membrane and cell wall, and lack a periplasmic space; the exception to this general rule is *Ignicoccus*, which possess a particularly large periplasm that contains membrane-bound vesicles and is enclosed by an outer membrane.

Cell Wall and Flagella

Most archaea (but not *Thermoplasma* and *Ferroplasma*) possess a cell wall. In most archaea the wall is assembled from surface-layer proteins, which form an S-layer. An S-layer is a rigid array of protein molecules that cover the outside of the cell (like chain mail). This layer provides both chemical and physical protection, and can prevent macromolecules from contacting the cell membrane. Unlike bacteria, archaea lack peptidoglycan in their cell walls. Methanobacteriales do have cell walls containing pseudopeptidoglycan, which resembles eubacterial peptidoglycan in morphology, function, and physical structure, but pseudopeptidoglycan is distinct in chemical structure; it lacks D-amino acids and N-acetylmuramic acid.

Archaea flagella operate like bacterial flagella—their long stalks are driven by rotatory motors at the base. These motors are powered by the proton gradient across the membrane. However, archaeal flagella are notably different in composition and development. The two types of flagella evolved from different ancestors. The bacterial flagellum shares a common ancestor with the type III secretion system, while archaeal flagella appear to have evolved from bacterial type IV pili. In contrast to the bacterial flagellum, which is hollow and is assembled by subunits moving up the central pore to the tip of the flagella, archaeal flagella are synthesized by adding subunits at the base.

Membranes

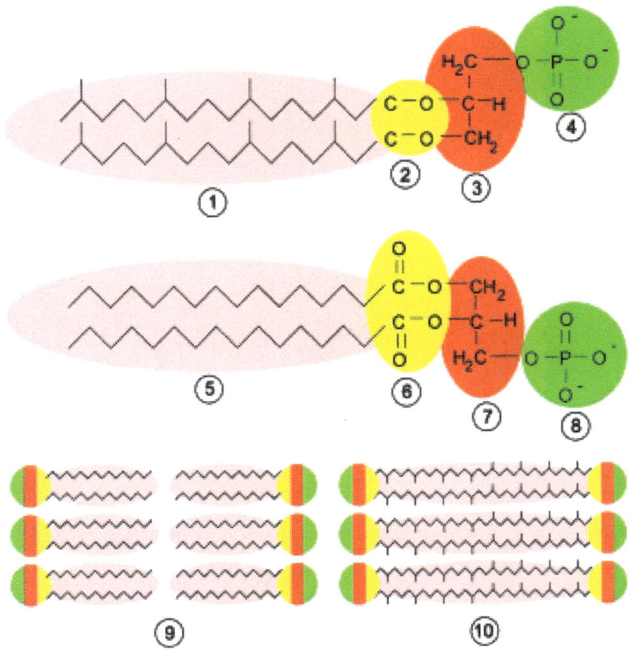

Membrane structures. Top, an archaeal phospholipid: 1, isoprene chains; 2, ether linkages; 3, L-glycerol moiety; 4, phosphate group. Middle, a bacterial or eukaryotic phospholipid: 5, fatty acid chains; 6, ester linkages; 7, D-glycerol moiety; 8, phosphate group. Bottom: 9, lipid bilayer of bacteria and eukaryotes; 10, lipid monolayer of some archaea.

Archaeal membranes are made of molecules that are distinctly different from those in all other life forms, showing that archaea are related only distantly to bacteria and eukaryotes. In all organisms, cell membranes are made of molecules known as phospholipids. These molecules possess both a polar part that dissolves in water (the phosphate "head"), and a "greasy" non-polar part that does not (the lipid tail). These dissimilar parts are connected by a glycerol moiety. In water, phospholipids cluster, with the heads facing the water and the tails facing away from it. The major structure in cell membranes is a double layer of these phospholipids, which is called a lipid bilayer.

The phospholipids of archaea are unusual in four ways:

- They have membranes composed of glycerol-ether lipids, whereas bacteria and eukaryotes have membranes composed mainly of glycerol-ester lipids. The difference is the type of bond that joins the lipids to the glycerol moiety; the two types are shown in yellow in the figure. In ester lipids this is an ester bond, whereas in ether lipids this is an ether bond. Ether bonds are chemically more resistant than ester bonds. This stability might help archaea to survive extreme temperatures and very acidic or alkaline environments. Bacteria and eukaryotes do contain some ether lipids, but in contrast to archaea these lipids are not a major part of their membranes.
- The stereochemistry of the archaeal glycerol moiety is the mirror image of that found in

other organisms. The glycerol moiety can occur in two forms that are mirror images of one another, called *enantiomers*. Just as a right hand does not fit easily into a left-handed glove, enantiomers of one type generally cannot be used or made by enzymes adapted for the other. The archaeal phospholipids are built on a backbone of *sn*-glycerol-1-phosphate, which is an enantiomer of *sn*-glycerol-3-phosphate, the phospholipid backbone found in bacteria and eucaryotes. This suggests that archaea use entirely different enzymes for synthesizing phospholipids than do bacteria and eukaryotes. Such enzymes developed very early in life's history, indicating an early split from the other two domains.

- Archaeal lipid tails differ from those of other organisms in that they are based upon long isoprenoid chains with multiple side-branches, sometimes with cyclopropane or cyclohexane rings. By contrast, the fatty acids in the membranes of other organisms have straight chains without side branches or rings. Although isoprenoids play an important role in the biochemistry of many organisms, only the archaea use them to make phospholipids. These branched chains may help prevent archaeal membranes from leaking at high temperatures.
- In some archaea, the lipid bilayer is replaced by a monolayer. In effect, the archaea fuse the tails of two phospholipid molecules into a single molecule with two polar heads (a bolaamphiphile); this fusion may make their membranes more rigid and better able to resist harsh environments. For example, the lipids in *Ferroplasma* are of this type, which is thought to aid this organism's survival in its highly acidic habitat.

Metabolism

Archaea exhibit a great variety of chemical reactions in their metabolism and use many sources of energy. These reactions are classified into nutritional groups, depending on energy and carbon sources. Some archaea obtain energy from inorganic compounds such as sulfur or ammonia (they are lithotrophs). These include nitrifiers, methanogens and anaerobic methane oxidisers. In these reactions one compound passes electrons to another (in a redox reaction), releasing energy to fuel the cell's activities. One compound acts as an electron donor and one as an electron acceptor. The energy released generates adenosine triphosphate (ATP) through chemiosmosis, in the same basic process that happens in the mitochondrion of eukaryotic cells.

Other groups of archaea use sunlight as a source of energy (they are phototrophs). However, oxygen–generating photosynthesis does not occur in any of these organisms. Many basic metabolic pathways are shared between all forms of life; for example, archaea use a modified form of glycolysis (the Entner–Doudoroff pathway) and either a complete or partial citric acid cycle. These similarities to other organisms probably reflect both early origins in the history of life and their high level of efficiency.

Nutritional types in archaeal metabolism			
Nutritional type	**Source of energy**	**Source of carbon**	**Examples**
Phototrophs	Sunlight	Organic compounds	*Halobacterium*
Lithotrophs	Inorganic compounds	Organic compounds or carbon fixation	*Ferroglobus*, *Methanobacteria* or *Pyrolobus*
Organotrophs	Organic compounds	Organic compounds or carbon fixation	*Pyrococcus*, *Sulfolobus* or *Methanosarcinales*

Some Euryarchaeota are methanogens living in anaerobic environments, such as swamps. This form of metabolism evolved early, and it is even possible that the first free-living organism was a methanogen. A common reaction involves the use of carbon dioxide as an electron acceptor to oxidize hydrogen. Methanogenesis involves a range of coenzymes that are unique to these archaea, such as coenzyme M and methanofuran. Other organic compounds such as alcohols, acetic acid or formic acid are used as alternative electron acceptors by methanogens. These reactions are common in gut-dwelling archaea. Acetic acid is also broken down into methane and carbon dioxide directly, by *acetotrophic* archaea. These acetotrophs are archaea in the order Methanosarcinales, and are a major part of the communities of microorganisms that produce biogas.

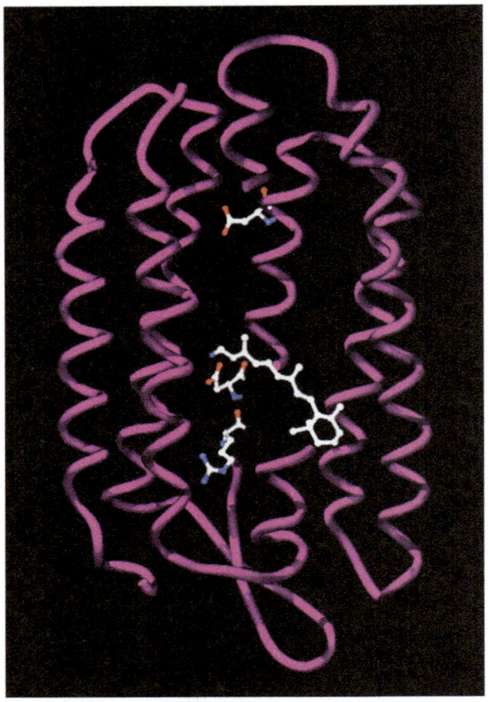

Bacteriorhodopsin from *Halobacterium salinarum*. The retinol cofactor and residues involved in proton transfer are shown as ball-and-stick models.

Other archaea use CO2 in the atmosphere as a source of carbon, in a process called carbon fixation (they are autotrophs). This process involves either a highly modified form of the Calvin cycle or a recently discovered metabolic pathway called the 3-hydroxypropionate/4-hydroxybutyrate cycle. The Crenarchaeota also use the reverse Krebs cycle while the Euryarchaeota also use the reductive acetyl-CoA pathway. Carbon–fixation is powered by inorganic energy sources. No known archaea carry out photosynthesis. Archaeal energy sources are extremely diverse, and range from the oxidation of ammonia by the Nitrosopumilales to the oxidation of hydrogen sulfide or elemental sulfur by species of *Sulfolobus*, using either oxygen or metal ions as electron acceptors.

Phototrophic archaea use light to produce chemical energy in the form of ATP. In the Halobacteria, light-activated ion pumps like bacteriorhodopsin and halorhodopsin generate ion gradients by pumping ions out of the cell across the plasma membrane. The energy stored in these electrochemical gradients is then converted into ATP by ATP synthase. This process is a form of photophosphorylation. The ability of these light-driven pumps to move ions across membranes depends on light-driven changes in the structure of a retinol cofactor buried in the center of the protein.

Genetics

Archaea usually have a single circular chromosome, with as many as 5,751,492 base pairs in *Methanosarcina acetivorans*, the largest known archaeal genome. The tiny 490,885 base-pair genome of *Nanoarchaeum equitans* is one-tenth of this size and the smallest archaeal genome known; it is estimated to contain only 537 protein-encoding genes. Smaller independent pieces of DNA, called *plasmids*, are also found in archaea. Plasmids may be transferred between cells by physical contact, in a process that may be similar to bacterial conjugation.

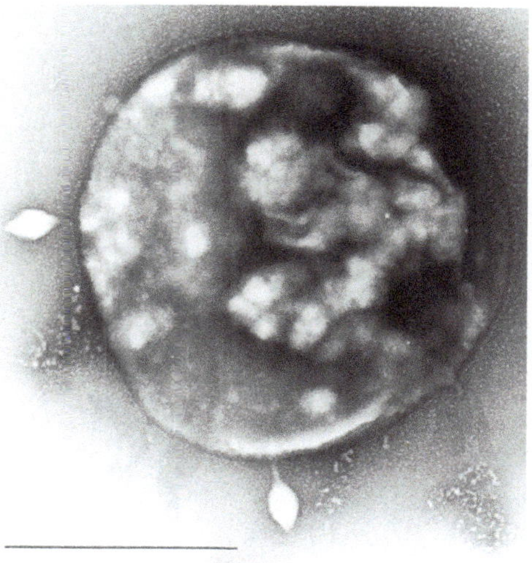

Sulfolobus infected with the DNA virus STSV1. Bar is 1 micrometer.

Archaea can be infected by double-stranded DNA viruses that are unrelated to any other form of virus and have a variety of unusual shapes, including bottles, hooked rods, or teardrops. These viruses have been studied in most detail in thermophilics, particularly the orders Sulfolobales and Thermoproteales. Two groups of single-stranded DNA viruses that infect archaea have been recently isolated. One group is exemplified by the *Halorubrum pleomorphic virus 1* ("Pleolipoviridae") infecting halophilic archaea and the other one by the Aeropyrum coil-shaped virus ("Spiraviridae") infecting a hyperthermophilic (optimal growth at 90–95 °C) host. Notably, the latter virus has the largest currently reported ssDNA genome. Defenses against these viruses may involve RNA interference from repetitive DNA sequences that are related to the genes of the viruses.

Archaea are genetically distinct from bacteria and eukaryotes, with up to 15% of the proteins encoded by any one archaeal genome being unique to the domain, although most of these unique genes have no known function. Of the remainder of the unique proteins that have an identified function, most belong to the Euryarchaea and are involved in methanogenesis. The proteins that archaea, bacteria and eukaryotes share form a common core of cell function, relating mostly to transcription, translation, and nucleotide metabolism. Other characteristic archaeal features are the organization of genes of related function—such as enzymes that catalyze steps in the same metabolic pathway into novel operons, and large differences in tRNA genes and their aminoacyl tRNA synthetases.

Transcription in archaea more closely resembles eukaryotic than bacterial transcription, with the archaeal RNA polymerase being very close to its equivalent in eukaryotes; while archaeal translation shows signs of both bacterial and eukaryal equivalents. Although archaea only have one type of RNA polymerase, its structure and function in transcription seems to be close to that of the eukaryotic RNA polymerase II, with similar protein assemblies (the general transcription factors) directing the binding of the RNA polymerase to a gene's promoter. However, other archaeal transcription factors are closer to those found in bacteria. Post-transcriptional modification is simpler than in eukaryotes, since most archaeal genes lack introns, although there are many introns in their transfer RNA and ribosomal RNA genes, and introns may occur in a few protein-encoding genes.

Gene Transfer and Genetic Exchange

Halobacterium volcanii, an extreme halophilic archaeon, forms cytoplasmic bridges between cells that appear to be used for transfer of DNA from one cell to another in either direction.

When the hyperthermophilic archaea *Sulfolobus solfataricus* and *Sulfolobus acidocaldarius* are exposed to the DNA damaging agents UV irradiation, bleomycin or mitomycin C, species-specific cellular aggregation is induced. Aggregation in *S. solfataricus* could not be induced by other physical stressors, such as pH or temperature shift, suggesting that aggregation is induced specifically by DNA damage. Ajon et al. showed that UV-induced cellular aggregation mediates chromosomal marker exchange with high frequency in *S. acidocaldarius*. Recombination rates exceeded those of uninduced cultures by up to three orders of magnitude. Frols et al. and Ajon et al. hypothesized that cellular aggregation enhances species specific DNA transfer between *Sulfolobus* cells in order to provide increased repair of damaged DNA by means of homologous recombination. This response may be a primitive form of sexual interaction similar to the more well-studied bacterial transformation systems that are also associated with species specific DNA transfer between cells leading to homologous recombinational repair of DNA damage.

Reproduction

Archaea reproduce asexually by binary or multiple fission, fragmentation, or budding; meiosis does not occur, so if a species of archaea exists in more than one form, all have the same genetic material. Cell division is controlled in a cell cycle; after the cell's chromosome is replicated and the two daughter chromosomes separate, the cell divides. In the genus *Sulfolobus*, the cycle has characteristics that are similar to both bacterial and eukaryotic systems. The chromosomes replicate from multiple starting-points (origins of replication) using DNA polymerases that resemble the equivalent eukaryotic enzymes.

In euryarchaea the cell division protein FtsZ, which forms a contracting ring around the cell, and the components of the septum that is constructed across the center of the cell, are similar to their bacterial equivalents. In cren- and thaumarchaea, however, the cell division machinery Cdv fulfills a similar role. This machinery is related to the eukaryotic ESCRT-III machinery which, while best known for its role in cell sorting, also has been seen to fulfill a role in separation between divided cell, suggesting an ancestral role in cell division.

Both bacteria and eukaryotes, but not archaea, make spores. Some species of Haloarchaea undergo phenotypic switching and grow as several different cell types, including thick-walled structures that are resistant to osmotic shock and allow the archaea to survive in water at low salt concentrations, but these are not reproductive structures and may instead help them reach new habitats.

Ecology

Habitats

Archaea that grow in the hot water of the Morning Glory Hot Spring in Yellowstone National Park produce a bright colour

Archaea exist in a broad range of habitats, and as a major part of global ecosystems, may represent about 20% of microbial cells in the oceans. The first-discovered archaeans were extremophiles. Indeed, some archaea survive high temperatures, often above 100 °C (212 °F), as found in geysers, black smokers, and oil wells. Other common habitats include very cold habitats and highly saline, acidic, or alkaline water. However, archaea include mesophiles that grow in mild conditions, in swamps and marshland, sewage, the oceans, the intestinal tract of animals, and soils.

Extremophile archaea are members of four main physiological groups. These are the halophiles, thermophiles, alkaliphiles, and acidophiles. These groups are not comprehensive or phylum-specific, nor are they mutually exclusive, since some archaea belong to several groups. Nonetheless, they are a useful starting point for classification.

Halophiles, including the genus *Halobacterium*, live in extremely saline environments such as salt lakes and outnumber their bacterial counterparts at salinities greater than 20–25%. Thermophiles grow best at temperatures above 45 °C (113 °F), in places such as hot springs; *hyperthermophilic* archaea grow optimally at temperatures greater than 80 °C (176 °F). The archaeal *Methanopyrus kandleri* Strain 116 can even reproduce at 122 °C (252 °F), the highest recorded temperature of any organism.

Other archaea exist in very acidic or alkaline conditions. For example, one of the most extreme archaean acidophiles is *Picrophilus torridus*, which grows at pH 0, which is equivalent to thriving in 1.2 molar sulfuric acid.

This resistance to extreme environments has made archaea the focus of speculation about the possible properties of extraterrestrial life. Some extremophile habitats are not dissimilar to those on Mars, leading to the suggestion that viable microbes could be transferred between planets in meteorites.

Recently, several studies have shown that archaea exist not only in mesophilic and thermophilic environments but are also present, sometimes in high numbers, at low temperatures as well. For example, archaea are common in cold oceanic environments such as polar seas. Even more significant are the large numbers of archaea found throughout the world's oceans in non-extreme habitats among the plankton community (as part of the picoplankton). Although these archaea can be present in extremely high numbers (up to 40% of the microbial biomass), almost none of these species have been isolated and studied in pure culture. Consequently, our understanding of the role of archaea in ocean ecology is rudimentary, so their full influence on global biogeochemical cycles remains largely unexplored. Some marine Crenarchaeota are capable of nitrification, suggesting these organisms may affect the oceanic nitrogen cycle, although these oceanic Crenarchaeota may also use other sources of energy. Vast numbers of archaea are also found in the sediments that cover the sea floor, with these organisms making up the majority of living cells at depths over 1 meter below the ocean bottom. It has been demonstrated that in all oceanic surface sediments (from 1000- to 10,000-m water depth), the impact of viral infection is higher on archaea than on bacteria and virus-induced lysis of archaea accounts for up to one-third of the total microbial biomass killed, resulting in the release of ~0.3 to 0.5 gigatons of carbon per year globally.

Role in Chemical Cycling

Archaea recycle elements such as carbon, nitrogen and sulfur through their various habitats. Although these activities are vital for normal ecosystem function, archaea can also contribute to human-made changes, and even cause pollution.

Archaea carry out many steps in the nitrogen cycle. This includes both reactions that remove nitrogen from ecosystems (such as nitrate-based respiration and denitrification) as well as processes that introduce nitrogen (such as nitrate assimilation and nitrogen fixation). Researchers recently discovered archaeal involvement in ammonia oxidation reactions. These reactions are particularly important in the oceans. The archaea also appear crucial for ammonia oxidation in soils. They produce nitrite, which other microbes then oxidize to nitrate. Plants and other organisms consume the latter.

In the sulfur cycle, archaea that grow by oxidizing sulfur compounds release this element from rocks, making it available to other organisms. However, the archaea that do this, such as *Sulfolobus*, produce sulfuric acid as a waste product, and the growth of these organisms in abandoned mines can contribute to acid mine drainage and other environmental damage.

In the carbon cycle, methanogen archaea remove hydrogen and play an important role in the decay of organic matter by the populations of microorganisms that act as decomposers in anaerobic ecosystems, such as sediments, marshes and sewage-treatment works.

Global methane levels in 2011 had increased by a factor of 2.5 since pre-industrial times: from 722 ppb to 1800 ppb, the highest value in at least 800,000 years. Methane has an anthropogenic global warming potential (AGWP) of 29, which means that it's 29 times stronger in heat-trapping than carbon dioxide is, over a 100-year time scale.

Interactions with Other Organisms

Methanogenic archaea form a symbiosis with termites.

The well-characterized interactions between archaea and other organisms are either mutual or commensal. There are no clear examples of known archaeal pathogens or parasites. However, some species of methanogens have been suggested to be involved in infections in the mouth, and *Nanoarchaeum equitans* may be a parasite of another species of archaea, since it only survives and reproduces within the cells of the Crenarchaeon *Ignicoccus hospitalis*, and appears to offer no benefit to its host. In contrast, Archaeal Richmond Mine Acidophilic Nanoorganisms (ARMAN) occasionally connect with other archaeal cells in acid mine drainage biofilms. The nature of this relationship is unknown. However, it is distinct from that of *Nanarchaeaum–Ignicoccus* in that the ultrasmall ARMAN cells are usually seen independent of the Thermoplasmatales cells.

Mutualism

One well-understood example of mutualism is the interaction between protozoa and methanogenic archaea in the digestive tracts of animals that digest cellulose, such as ruminants and termites. In these anaerobic environments, protozoa break down plant cellulose to obtain energy. This process releases hydrogen as a waste product, but high levels of hydrogen reduce energy production. When methanogens convert hydrogen to methane, protozoa benefit from more energy.

In anaerobic protozoa, such as *Plagiopyla frontata*, archaea reside inside the protozoa and consume hydrogen produced in their hydrogenosomes. Archaea also associate with larger organisms.

For example, the marine archaean *Cenarchaeum symbiosum* lives within (is an endosymbiont of) the sponge *Axinella mexicana*.

Commensalism

Archaea can also be commensals, benefiting from an association without helping or harming the other organism. For example, the methanogen *Methanobrevibacter smithii* is by far the most common archaean in the human flora, making up about one in ten of all the prokaryotes in the human gut. In termites and in humans, these methanogens may in fact be mutualists, interacting with other microbes in the gut to aid digestion. Archaean communities also associate with a range of other organisms, such as on the surface of corals, and in the region of soil that surrounds plant roots (the rhizosphere).

Significance in Technology and Industry

Extremophile archaea, particularly those resistant either to heat or to extremes of acidity and alkalinity, are a source of enzymes that function under these harsh conditions. These enzymes have found many uses. For example, thermostable DNA polymerases, such as the Pfu DNA polymerase from *Pyrococcus furiosus*, revolutionized molecular biology by allowing the polymerase chain reaction to be used in research as a simple and rapid technique for cloning DNA. In industry, amylases, galactosidases and pullulanases in other species of *Pyrococcus* that function at over 100 °C (212 °F) allow food processing at high temperatures, such as the production of low lactose milk and whey. Enzymes from these thermophilic archaea also tend to be very stable in organic solvents, allowing their use in environmentally friendly processes in green chemistry that synthesize organic compounds. This stability makes them easier to use in structural biology. Consequently, the counterparts of bacterial or eukaryotic enzymes from extremophile archaea are often used in structural studies.

In contrast to the range of applications of archaean enzymes, the use of the organisms themselves in biotechnology is less developed. Methanogenic archaea are a vital part of sewage treatment, since they are part of the community of microorganisms that carry out anaerobic digestion and produce biogas. In mineral processing, acidophilic archaea display promise for the extraction of metals from ores, including gold, cobalt and copper.

Archaea host a new class of potentially useful antibiotics. A few of these archaeocins have been characterized, but hundreds more are believed to exist, especially within *Haloarchaea* and *Sulfolobus*. These compounds differ in structure from bacterial antibiotics, so they may have novel modes of action. In addition, they may allow the creation of new selectable markers for use in archaeal molecular biology.

Bacteria

Bacteria constitute a large domain of prokaryotic microorganisms. Typically a few micrometres in length, bacteria have a number of shapes, ranging from spheres to rods and spirals. Bacteria were among the first life forms to appear on Earth, and are present in most of its habitats. Bacteria inhabit soil, water, acidic hot springs, radioactive waste, and the deep portions of Earth's crust. Bacteria also live in symbiotic and parasitic relationships with plants and animals. Most bacteria have not been characterised, and only about half of the bacterial phyla have species that can be grown in the laboratory. The study of bacteria is known as bacteriology, a branch of microbiology.

There are typically 40 million bacterial cells in a gram of soil and a million bacterial cells in a millilitre of fresh water. There are approximately 5×10^{30} bacteria on Earth, forming a biomass which exceeds that of all plants and animals. Bacteria are vital in many stages of the nutrient cycle by recycling nutrients such as the fixation of nitrogen from the atmosphere. The nutrient cycle includes the decomposition of dead bodies and bacteria are responsible for the putrefaction stage in this process. In the biological communities surrounding hydrothermal vents and cold seeps, extremophile bacteria provide the nutrients needed to sustain life by converting dissolved compounds, such as hydrogen sulphide and methane, to energy. In March 2013, data reported by researchers in October 2012, was published. It was suggested that bacteria thrive in the Mariana Trench, which with a depth of up to 11 kilometres is the deepest known part of the oceans. Other researchers reported related studies that microbes thrive inside rocks up to 580 metres below the sea floor under 2.6 kilometres of ocean off the coast of the northwestern United States. According to one of the researchers, "You can find microbes everywhere—they're extremely adaptable to conditions, and survive wherever they are."

There are approximately ten times as many bacterial cells in the human microbiota as there are human cells in the body, with their largest number being in the gut flora, and a large number on the skin. The vast majority of the bacteria in the body are rendered harmless by the protective effects of the immune system, though many are beneficial particularly in the gut flora. However several species of bacteria are pathogenic and cause infectious diseases, including cholera, syphilis, anthrax, leprosy, and bubonic plague. The most common fatal bacterial diseases are respiratory infections, with tuberculosis alone killing about 2 million people per year, mostly in sub-Saharan Africa. In developed countries, antibiotics are used to treat bacterial infections and are also used in farming, making antibiotic resistance a growing problem. In industry, bacteria are important in sewage treatment and the breakdown of oil spills, the production of cheese and yogurt through fermentation, and the recovery of gold, palladium, copper and other metals in the mining sector, as well as in biotechnology, and the manufacture of antibiotics and other chemicals.

Once regarded as plants constituting the class *Schizomycetes*, bacteria are now classified as prokaryotes. Unlike cells of animals and other eukaryotes, bacterial cells do not contain a nucleus and rarely harbour membrane-bound organelles. Although the term *bacteria* traditionally included all prokaryotes, the scientific classification changed after the discovery in the 1990s that prokaryotes consist of two very different groups of organisms that evolved from an ancient common ancestor. These evolutionary domains are called *Bacteria* and *Archaea*.

Origin and Early Evolution

The ancestors of modern bacteria were unicellular microorganisms that were the first forms of life to appear on Earth, about 4 billion years ago. For about 3 billion years, most organisms were microscopic, and bacteria and archaea were the dominant forms of life. In 2008, fossils of macroorganisms were discovered and named as the Francevillian biota. Although bacterial fossils exist, such as stromatolites, their lack of distinctive morphology prevents them from being used to examine the history of bacterial evolution, or to date the time of origin of a particular bacterial species. However, gene sequences can be used to reconstruct the bacterial phylogeny, and these studies indicate that bacteria diverged first from the archaeal/eukaryotic lineage. Bacteria were also involved in the second great evolutionary divergence, that of the archaea and eukaryotes. Here, eukaryotes resulted from the entering of ancient bacteria into endosymbiotic associations with the ancestors of eukaryotic cells, which were themselves possibly related to the Archaea. This involved the engulfment by proto-eukaryotic cells of alphaproteobacterial symbionts to form either mitochondria or hydrogenosomes, which are still found in all known Eukarya (sometimes in highly reduced form, e.g. in ancient "amitochondrial" protozoa). Later on, some eukaryotes that already contained mitochondria also engulfed cyanobacterial-like organisms. This led to the formation of chloroplasts in algae and plants. There are also some algae that originated from even later endosymbiotic events. Here, eukaryotes engulfed a eukaryotic algae that developed into a "second-generation" plastid. This is known as secondary endosymbiosis.

In March 2017, researchers reported evidence of possibly the oldest forms of life on Earth. Putative fossilized microorganisms were discovered in hydrothermal vent precipitates in the Nuvvuagittuq Belt of Quebec, Canada, that may have lived as early as 4.280 billion years ago, not long after the oceans formed 4.4 billion years ago, and not long after the formation of the Earth 4.54 billion years ago. According to biologist Stephen Blair Hedges, "If life arose relatively quickly on Earth ... then it could be common in the universe."

Morphology

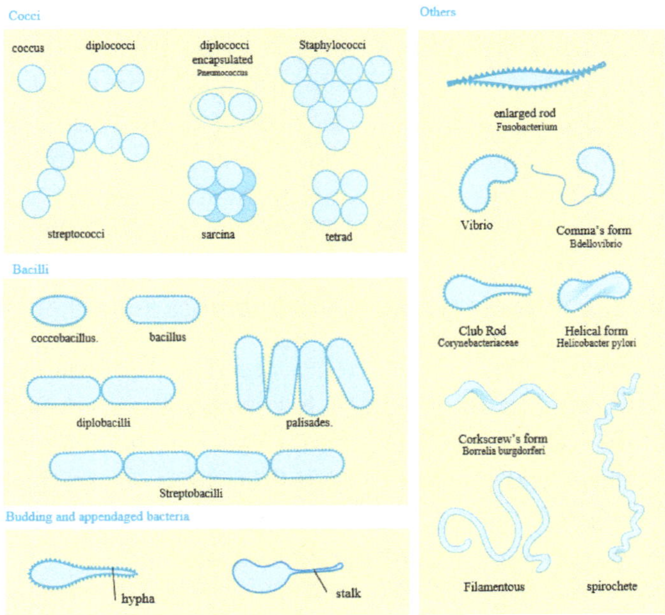

Bacteria display many cell morphologies and arrangements

Bacteria display a wide diversity of shapes and sizes, called *morphologies*. Bacterial cells are about one-tenth the size of eukaryotic cells and are typically 0.5–5.0 micrometres in length. However, a few species are visible to the unaided eye—for example, *Thiomargarita namibiensis* is up to half a millimetre long and *Epulopiscium fishelsoni* reaches 0.7 mm. Among the smallest bacteria are members of the genus *Mycoplasma*, which measure only 0.3 micrometres, as small as the largest viruses. Some bacteria may be even smaller, but these ultramicrobacteria are not well-studied.

Most bacterial species are either spherical, called *cocci* (*sing*. coccus, from Greek *kókkos*, grain, seed), or rod-shaped, called *bacilli* (*sing*. bacillus, from Latin *baculus*, stick). Elongation is associated with swimming. Some bacteria, called *vibrio*, are shaped like slightly curved rods or comma-shaped; others can be spiral-shaped, called *spirilla*, or tightly coiled, called *spirochaetes*. A small number of species even have tetrahedral or cuboidal shapes. More recently, some bacteria were discovered deep under Earth's crust that grow as branching filamentous types with a star-shaped cross-section. The large surface area to volume ratio of this morphology may give these bacteria an advantage in nutrient-poor environments. This wide variety of shapes is determined by the bacterial cell wall and cytoskeleton, and is important because it can influence the ability of bacteria to acquire nutrients, attach to surfaces, swim through liquids and escape predators.

A biofilm of thermophilic bacteria in the outflow of Mickey Hot Springs, Oregon, approximately 20 mm thick.

Many bacterial species exist simply as single cells, others associate in characteristic patterns: *Neisseria* form diploids (pairs), *Streptococcus* form chains, and *Staphylococcus* group together in "bunch of grapes" clusters. Bacteria can also be elongated to form filaments, for example the Actinobacteria. Filamentous bacteria are often surrounded by a sheath that contains many individual cells. Certain types, such as species of the genus *Nocardia*, even form complex, branched filaments, similar in appearance to fungal mycelia.

Bacteria often attach to surfaces and form dense aggregations in bacterial mats called biofilms. These biofilms can range from a few micrometres in thickness to up to half a metre in depth, and may contain multiple species of bacteria, protists and archaea. Bacteria living in biofilms display a complex arrangement of cells and extracellular components, forming secondary structures, such as microcolonies, through which there are networks of channels to enable better diffusion of nutrients. In natural environments, such as soil or the surfaces of plants, the majority of bacteria are bound to surfaces in biofilms. Biofilms are also important in medicine, as these structures are often present during chronic bacterial infections or in infections of implanted medical devices, and bacteria protected within biofilms are much harder to kill than individual isolated bacteria.

Even more complex morphological changes are sometimes possible. For example, when starved of amino acids, Myxobacteria detect surrounding cells in a process known as quorum sensing, migrate towards each other, and aggregate to form fruiting bodies up to 500 micrometres long and containing approximately 100,000 bacterial cells. In these fruiting bodies, the bacteria perform separate tasks; this type of cooperation is a simple type of multicellular organisation. For example, about one in 10 cells migrate to the top of these fruiting bodies and differentiate into a specialised dormant state called myxospores, which are more resistant to drying and other adverse environmental conditions than are ordinary cells.

Cellular Structure

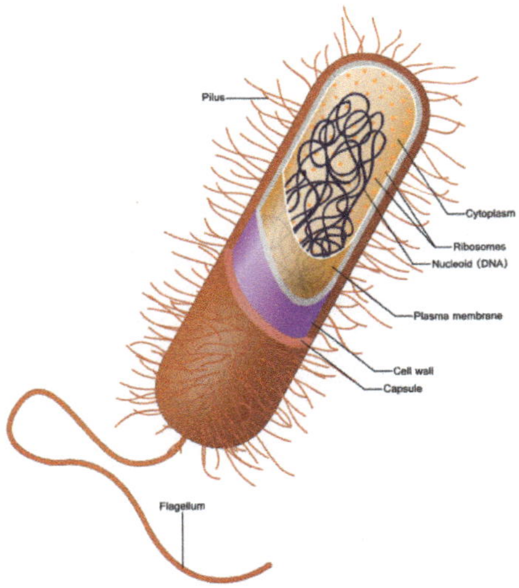

Structure and contents of a typical gram-positive bacterial cell (seen by the fact that only *one* cell membrane is present).

Intracellular Structures

The bacterial cell is surrounded by a cell membrane (also known as a lipid, cytoplasmic or plasma membrane). This membrane encloses the contents of the cell and acts as a barrier to hold nutrients, proteins and other essential components of the *cytoplasm* within the cell. As they are prokaryotes, bacteria do not usually have membrane-bound organelles in their cytoplasm, and thus contain few large intracellular structures. They lack a true nucleus, mitochondria, chloroplasts and the other organelles present in eukaryotic cells. Bacteria were once seen as simple bags of cytoplasm, but structures such as the *prokaryotic cytoskeleton* and the localisation of proteins to specific locations within the cytoplasm that give bacteria some complexity have been discovered. These subcellular levels of organisation have been called "bacterial hyperstructures".

Bacterial microcompartments, such as carboxysomes, provide a further level of organisation; they are compartments within bacteria that are surrounded by polyhedral protein shells, rather than by lipid membranes. These "polyhedral organelles" localise and compartmentalise bacterial metabolism, a function performed by the membrane-bound organelles in eukaryotes.

Many important biochemical reactions, such as energy generation, use concentration gradients across membranes. The general lack of internal membranes in bacteria means reactions such as electron transport occur across the cell membrane between the cytoplasm and the *periplasmic space*. However, in many photosynthetic bacteria the plasma membrane is highly folded and fills most of the cell with layers of light-gathering membrane. These light-gathering complexes may even form lipid-enclosed structures called chlorosomes in green sulfur bacteria. Other proteins import nutrients across the cell membrane, or expel undesired molecules from the cytoplasm.

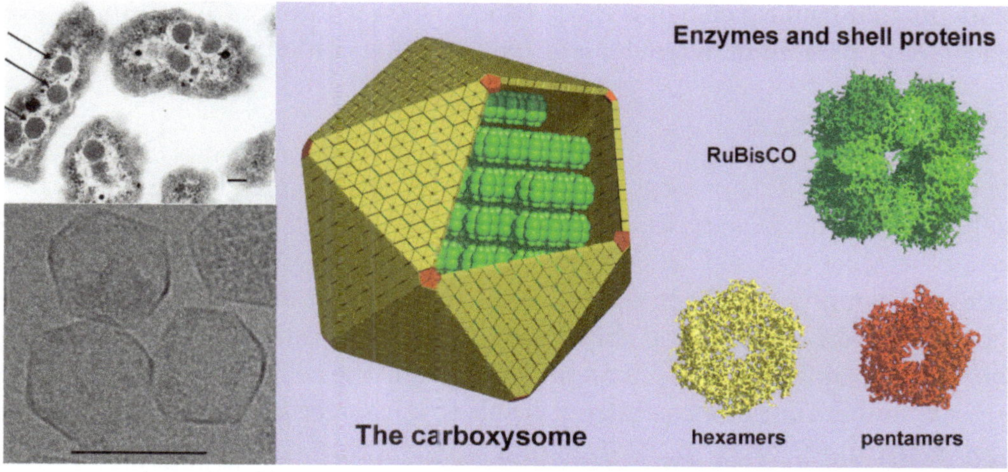

Carboxysomes are protein-enclosed bacterial organelles. Top left is an electron microscope image of carboxysomes in *Halothiobacillus neapolitanus*, below is an image of purified carboxysomes. On the right is a model of their structure. Scale bars are 100 nm.

Bacteria do not have a membrane-bound nucleus, and their genetic material is typically a single circular bacterial chromosome of DNA located in the cytoplasm in an irregularly shaped body called the *nucleoid*. The nucleoid contains the chromosome with its associated proteins and RNA. The phylum Planctomycetes and candidate phylum Poribacteria may be exceptions to the general absence of internal membranes in bacteria, because they appear to have a double membrane around their nucleoids and contain other membrane-bound cellular structures. Like all living organisms, bacteria contain *ribosomes*, often grouped in chains called polyribosomes, for the production of proteins, but the structure of the bacterial ribosome is different from that of eukaryotes and Archaea. Bacterial ribosomes have a sedimentation rate of 70S (measured in Svedberg units): their subunits have rates of 30S and 50S. Some antibiotics bind specifically to 70S ribosomes and inhibit bacterial protein synthesis. Those antibiotics kill bacteria without affecting the larger 80S ribosomes of eukaryotic cells and without harming the host.

Some bacteria produce intracellular nutrient storage granules for later use, such as glycogen, polyphosphate, sulfur or polyhydroxyalkanoates. Certain bacterial species, such as the photosynthetic Cyanobacteria, produce internal gas vesicles, which they use to regulate their buoyancy—allowing them to move up or down into water layers with different light intensities and nutrient levels. *Intracellular membranes* called *chromatophores* are also found in membranes of phototrophic bacteria. Used primarily for photosynthesis, they contain bacteriochlorophyll pigments and carotenoids. An early idea was that bacteria might contain membrane folds termed mesosomes, but these were later shown to be artefacts produced by the chemicals used to prepare the cells for electron microscopy. *Inclusions* are considered to be nonliving components of the cell that do not possess metabolic activity and are not bounded by membranes. The most common inclusions are glycogen, lipid droplets, crystals, and pigments. *Volutin granules* are cytoplasmic inclusions of complexed inorganic polyphosphate. These granules are called *metachromatic granules* due to their displaying the metachromatic effect; they appear red or blue when stained with the blue dyes methylene blue or toluidine blue. *Gas vacuoles*, which are freely permeable to gas, are membrane-bound vesicles present in some species of *Cyanobacteria*. They allow the bacteria to control their buoyancy. *Microcompartments* are widespread, membrane-bound organelles that are made of a protein shell that surrounds and encloses various enzymes. *Carboxysomes* are bacterial microcompartments that contain enzymes involved in carbon fixation. *Magnetosomes* are bacterial microcompartments, present in magnetotactic bacteria, that contain magnetic crystals.

Extracellular Structures

In most bacteria, a *cell wall* is present on the outside of the cell membrane. The cell membrane and cell wall comprise the *cell envelope.* A common bacterial cell wall material is *peptidoglycan* (called "murein" in older sources), which is made from polysaccharide chains cross-linked by peptides containing D-amino acids. Bacterial cell walls are different from the cell walls of plants and fungi, which are made of cellulose and chitin, respectively. The cell wall of bacteria is also distinct from that of Archaea, which do not contain peptidoglycan. The cell wall is essential to the survival of many bacteria, and the antibiotic penicillin is able to kill bacteria by inhibiting a step in the synthesis of peptidoglycan.

There are broadly speaking two different types of cell wall in bacteria, a thick one in the gram-positives and a thinner one in the gram-negatives. The names originate from the reaction of cells to the Gram stain, a long-standing test for the classification of bacterial species.

Gram-positive bacteria possess a thick cell wall containing many layers of peptidoglycan and *teichoic acids.* In contrast, *gram-negative bacteria* have a relatively thin cell wall consisting of a few layers of peptidoglycan surrounded by a second lipid membrane containing *lipopolysaccharides* and lipoproteins. Lipopolysaccharides, also called *endotoxins,* are composed of polysaccharides and *lipid A* that is responsible for much of the toxicity of gram-negative bacteria. Most bacteria have the gram-negative cell wall, and only the Firmicutes and Actinobacteria have the alternative gram-positive arrangement. These two groups were previously known as the low G+C and high G+C gram-positive bacteria, respectively. These differences in structure can produce differences in antibiotic susceptibility; for instance, vancomycin can kill only gram-positive bacteria and is ineffective against gram-negative pathogens, such as *Haemophilus influenzae* or *Pseudomonas aeruginosa.* If the bacterial cell wall is entirely removed, it is called a *protoplast,* whereas if it is partially removed, it is called a *spheroplast.* β-Lactam antibiotics, such as penicillin, inhibit the formation of peptidoglycan cross-links in the bacterial cell wall. The enzyme lysozyme, found in human tears, also digests the cell wall of bacteria and is the body's main defence against eye infections.

Acid-fast bacteria, such as *Mycobacteria,* are resistant to decolorisation by acids during staining procedures. The high mycolic acid content of *Mycobacteria* is responsible for the staining pattern of poor absorption followed by high retention. The most common staining technique used to identify acid-fast bacteria is the Ziehl-Neelsen stain or acid-fast stain, in which the acid-fast bacilli are stained bright-red and stand out clearly against a blue background. *L-form bacteria* are strains of bacteria that lack cell walls. The main pathogenic bacteria in this class is *Mycoplasma* (not to be confused with *Mycobacteria*).

In many bacteria, an *S-layer* of rigidly arrayed protein molecules covers the outside of the cell. This layer provides chemical and physical protection for the cell surface and can act as a macromolecular diffusion barrier. S-layers have diverse but mostly poorly understood functions, but are known to act as virulence factors in *Campylobacter* and contain surface enzymes in *Bacillus stearothermophilus.*

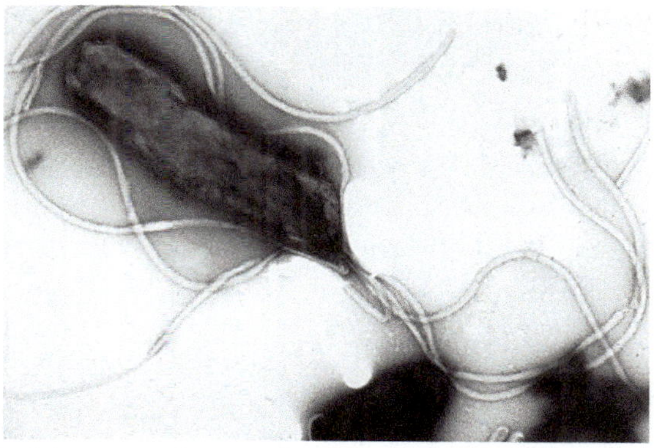

Helicobacter pylori electron micrograph, showing multiple flagella on the cell surface

Flagella are rigid protein structures, about 20 nanometres in diameter and up to 20 micrometres in length, that are used for motility. Flagella are driven by the energy released by the transfer of ions down an electrochemical gradient across the cell membrane.

Fimbriae (sometimes called "attachment pili") are fine filaments of protein, usually 2–10 nanometres in diameter and up to several micrometres in length. They are distributed over the surface of the cell, and resemble fine hairs when seen under the electron microscope. Fimbriae are believed to be involved in attachment to solid surfaces or to other cells, and are essential for the virulence of some bacterial pathogens. *Pili* (*sing*. pilus) are cellular appendages, slightly larger than fimbriae, that can transfer genetic material between bacterial cells in a process called conjugation where they are called *conjugation pili* or "sex pili". They can also generate movement where they are called *type IV pili*.

Glycocalyx are produced by many bacteria to surround their cells, and vary in structural complexity: ranging from a disorganised *slime layer* of extra-cellular polymer to a highly structured *capsule*. These structures can protect cells from engulfment by eukaryotic cells such as macrophages (part of the human immune system). They can also act as antigens and be involved in cell recognition, as well as aiding attachment to surfaces and the formation of biofilms.

The assembly of these extracellular structures is dependent on bacterial secretion systems. These transfer proteins from the cytoplasm into the periplasm or into the environment around the cell. Many types of secretion systems are known and these structures are often essential for the virulence of pathogens, so are intensively studied.

Endospores

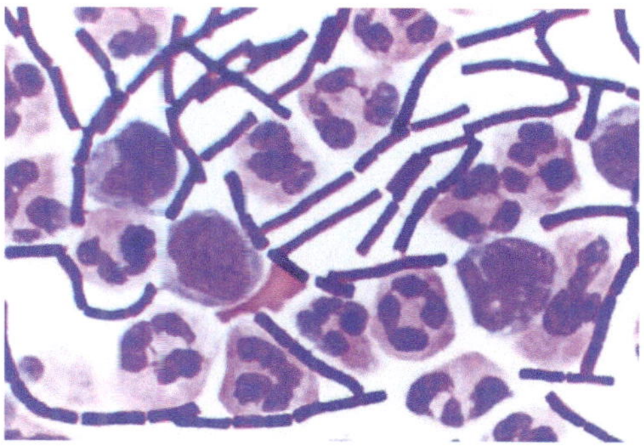

Bacillus anthracis (stained purple) growing in cerebrospinal fluid

Certain genera of gram-positive bacteria, such as *Bacillus*, *Clostridium*, *Sporohalobacter*, *Anaerobacter*, and *Heliobacterium*, can form highly resistant, dormant structures called *endospores*. In almost all cases, one endospore is formed and this is not a reproductive process, although *Anaerobacter* can make up to seven endospores in a single cell. Endospores have a central core of cytoplasm containing DNA and ribosomes surrounded by a cortex layer and protected by an impermeable and rigid coat. Dipicolinic acid is a chemical compound that composes 5% to 15% of the dry weight of bacterial spores. It is implicated as responsible for the heat resistance of the endospore.

Endospores show no detectable metabolism and can survive extreme physical and chemical stresses, such as high levels of UV light, gamma radiation, detergents, disinfectants, heat, freezing, pressure, and desiccation. In this dormant state, these organisms may remain viable for millions of years, and endospores even allow bacteria to survive exposure to the vacuum and radiation in space. According to scientist Dr. Steinn Sigurdsson, "There are viable bacterial spores that have been found that are 40 million years old on Earth—and we know they're very hardened to radiation." Endospore-forming bacteria can also cause disease: for example, anthrax can be contracted by the inhalation of *Bacillus anthracis* endospores, and contamination of deep puncture wounds with *Clostridium tetani* endospores causes tetanus.

Metabolism

Bacteria exhibit an extremely wide variety of metabolic types. The distribution of metabolic traits within a group of bacteria has traditionally been used to define their taxonomy, but these traits often do not correspond with modern genetic classifications. Bacterial metabolism is classified into nutritional groups on the basis of three major criteria: the kind of energy used for growth, the source of carbon, and the electron donors used for growth. An additional criterion of respiratory microorganisms are the electron acceptors used for aerobic or anaerobic respiration.

Nutritional types in bacterial metabolism			
Nutritional type	**Source of energy**	**Source of carbon**	**Examples**
Phototrophs	Sunlight	Organic compounds (photoheterotrophs) or carbon fixation (photoautotrophs)	Cyanobacteria, Green sulfur bacteria, Chloroflexi, or Purple bacteria
Lithotrophs	Inorganic compounds	Organic compounds (lithoheterotrophs) or carbon fixation (lithoautotrophs)	Thermodesulfobacteria, *Hydrogenophilaceae*, or Nitrospirae
Organotrophs	Organic compounds	Organic compounds (chemoheterotrophs) or carbon fixation (chemoautotrophs)	*Bacillus*, *Clostridium* or *Enterobacteriaceae*

Carbon metabolism in bacteria is either *heterotrophic*, where organic carbon compounds are used as carbon sources, or *autotrophic*, meaning that cellular carbon is obtained by fixing carbon dioxide. Heterotrophic bacteria include parasitic types. Typical autotrophic bacteria are phototrophic cyanobacteria, green sulfur-bacteria and some purple bacteria, but also many chemolithotrophic species, such as nitrifying or sulfur-oxidising bacteria. Energy metabolism of bacteria is either based on *phototrophy*, the use of light through photosynthesis, or based on *chemotrophy*, the use of chemical substances for energy, which are mostly oxidised at the expense of oxygen or alternative electron acceptors (aerobic/anaerobic respiration).

Filaments of photosynthetic cyanobacteria

Bacteria are further divided into *lithotrophs* that use inorganic electron donors and *organotrophs* that use organic compounds as electron donors. Chemotrophic organisms use the respective electron donors for energy conservation (by aerobic/anaerobic respiration or fermentation) and biosynthetic reactions (e.g., carbon dioxide fixation), whereas phototrophic organisms use them only for biosynthetic purposes. Respiratory organisms use chemical compounds as a source of energy by taking electrons from the reduced substrate and transferring them to a terminal electron acceptor in a redox reaction. This reaction releases energy that can be used to synthesise ATP and drive metabolism. In *aerobic organisms*, oxygen is used as the electron acceptor. In *anaerobic organisms* other inorganic compounds, such as nitrate, sulfate or carbon dioxide are used as electron acceptors. This leads to the ecologically important processes of denitrification, sulfate reduction, and acetogenesis, respectively.

Another way of life of chemotrophs in the absence of possible electron acceptors is fermentation, wherein the electrons taken from the reduced substrates are transferred to oxidised intermediates to generate reduced fermentation products (e.g., lactate, ethanol, hydrogen, butyric acid). Fermentation is possible, because the energy content of the substrates is higher than that of the products, which allows the organisms to synthesise ATP and drive their metabolism.

These processes are also important in biological responses to pollution; for example, sulfate-reducing bacteria are largely responsible for the production of the highly toxic forms of mercury (methyl- and dimethylmercury) in the environment. Non-respiratory anaerobes use fermentation to generate energy and reducing power, secreting metabolic by-products (such as ethanol in brewing) as waste. Facultative anaerobes can switch between fermentation and different terminal electron acceptors depending on the environmental conditions in which they find themselves.

Lithotrophic bacteria can use inorganic compounds as a source of energy. Common inorganic electron donors are hydrogen, carbon monoxide, ammonia (leading to nitrification), ferrous iron and other reduced metal ions, and several reduced sulfur compounds. In unusual circumstances, the gas methane can be used by methanotrophic bacteria as both a source of electrons and a substrate for carbon anabolism. In both aerobic phototrophy and chemolithotrophy, oxygen is used as a terminal electron acceptor, whereas under anaerobic conditions inorganic compounds are used instead. Most lithotrophic organisms are autotrophic, whereas organotrophic organisms are heterotrophic.

In addition to fixing carbon dioxide in photosynthesis, some bacteria also fix nitrogen gas using the enzyme nitrogenase. This environmentally important trait can be found in most bacteria of the metabolic types listed above.

Regardless of the type of metabolic process they employ, the majority of bacteria are able to take in raw materials only in the form of relatively small molecules, which enter the cell by diffusion or through molecular channels in cell membranes. The Planctomycetes are the exception (as they are in possessing membranes around their nuclear material). It has recently been shown that *Gemmata obscuriglobus* is able to take in large molecules via a process that in some ways resembles endocytosis, the process used by eukaryotic cells to engulf external items.

Growth and Reproduction

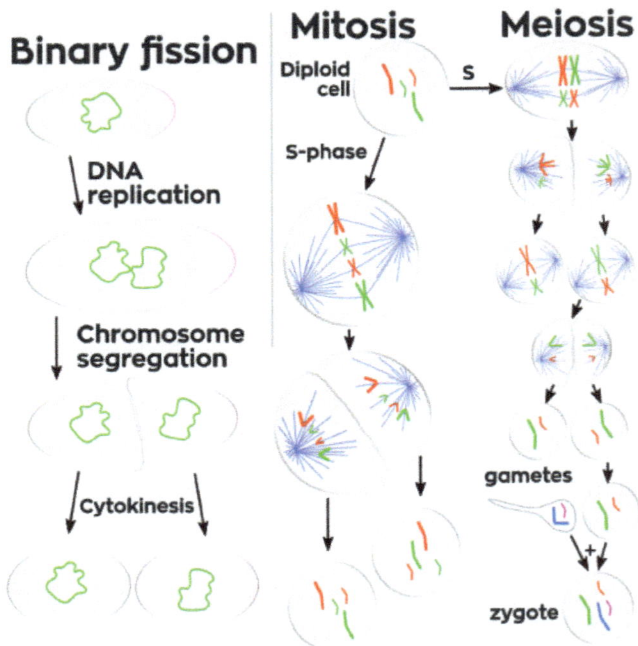

Many bacteria reproduce through binary fission, which is compared to mitosis and meiosis in this image.

Unlike in multicellular organisms, increases in cell size (cell growth) and reproduction by cell division are tightly linked in unicellular organisms. Bacteria grow to a fixed size and then reproduce through *binary fission*, a form of asexual reproduction. Under optimal conditions, bacteria can grow and divide extremely rapidly, and bacterial populations can double as quickly as every 9.8 minutes. In cell division, two identical clone daughter cells are produced. Some bacteria, while still reproducing asexually, form more complex reproductive structures that help disperse the newly formed daughter cells. Examples include fruiting body formation by *Myxobacteria* and aerial hyphae formation by *Streptomyces*, or budding. Budding involves a cell forming a protrusion that breaks away and produces a daughter cell.

A colony of *Escherichia coli*

In the laboratory, bacteria are usually grown using solid or liquid media. Solid *growth media*, such as agar plates, are used to isolate pure cultures of a bacterial strain. However, liquid growth media are used when measurement of growth or large volumes of cells are required. Growth in stirred liquid media occurs as an even cell suspension, making the cultures easy to divide and transfer, although isolating single bacteria from liquid media is difficult. The use of selective media (media with specific nutrients added or deficient, or with antibiotics added) can help identify specific organisms.

Most laboratory techniques for growing bacteria use high levels of nutrients to produce large amounts of cells cheaply and quickly. However, in natural environments, nutrients are limited, meaning that bacteria cannot continue to reproduce indefinitely. This nutrient limitation has led the evolution of different growth strategies. Some organisms can grow extremely rapidly when nutrients become available, such as the formation of algal (and cyanobacterial) blooms that often occur in lakes during the summer. Other organisms have adaptations to harsh environments, such as the production of multiple antibiotics by *Streptomyces* that inhibit the growth of competing microorganisms. In nature, many organisms live in communities (e.g., biofilms) that may allow for increased supply of nutrients and protection from environmental stresses. These relationships can be essential for growth of a particular organism or group of organisms (syntrophy).

Bacterial growth follows four phases. When a population of bacteria first enter a high-nutrient environment that allows growth, the cells need to adapt to their new environment. The first phase of growth is the *lag phase*, a period of slow growth when the cells are adapting to the high-nutrient environment and preparing for fast growth. The lag phase has high biosynthesis rates, as proteins necessary for rapid growth are produced. The second phase of growth is the *log phase*, also known as the *logarithmic or exponential phase*. The log phase is marked by rapid exponential growth. The rate at which cells grow during this phase is known as the *growth rate* (k), and the time it takes the cells to double is known as the *generation time* (g). During log phase, nutrients are metabolised at maximum speed until one of the nutrients is depleted and starts limiting growth. The third phase of growth is the *stationary phase* and is caused by depleted nutrients. The cells reduce their metabolic activity and consume non-essential cellular proteins. The stationary phase is a transition from rapid growth to a stress response state and there is increased expression of genes involved in DNA repair, antioxidant metabolism and nutrient transport. The final phase is the *death phase* where the bacteria run out of nutrients and die.

Genomes

The genomes of thousands of bacterial species have been sequenced, with at least 9,000 sequences completed and more than 42,000 left as "permanent" drafts (as of Sep 2016).

Most bacteria have a single circular chromosome that can range in size from only 160,000 base pairs in the endosymbiotic bacteria *Candidatus Carsonella ruddii*, to 12,200,000 base pairs in the soil-dwelling bacteria *Sorangium cellulosum*. The genes in bacterial genomes are usually a single continuous stretch of DNA and although several different types of introns do exist in bacteria, these are much rarer than in eukaryotes. Some bacteria, including the Spirochaetes of the genus *Borrelia* are a notable exception to this arrangement. *Borrelia burgdorferi*, the cause of Lyme disease, contains a single linear chromosome and several linear and circular plasmids.

Plasmids are small extra-chromosomal DNAs that may contain genes for antibiotic resistance or virulence factors. Plasmids replicate independently of chromosomes, so it is possible that plasmids could be lost in bacterial cell division. Against this possibility is the fact that a single bacterium can contain hundreds of copies of a single plasmid.

Genetics

Bacteria, as asexual organisms, inherit identical copies of their parent's genes (i.e., they are clonal). However, all bacteria can evolve by selection on changes to their genetic material DNA caused by genetic recombination or mutations. Mutations come from errors made during the replication of DNA or from exposure to mutagens. Mutation rates vary widely among different species of bacteria and even among different clones of a single species of bacteria. Genetic changes in bacterial genomes come from either random mutation during replication or "stress-directed mutation", where genes involved in a particular growth-limiting process have an increased mutation rate.

DNA Transfer

Some bacteria also transfer genetic material between cells. This can occur in three main ways. First, bacteria can take up exogenous DNA from their environment, in a process called *transformation*. Genes can also be transferred by the process of *transduction*, when the integration of a bacteriophage introduces foreign DNA into the chromosome. The third method of gene transfer is *conjugation*, whereby DNA is transferred through direct cell contact.

Transduction of bacterial genes by bacteriophage appears to be a consequence of infrequent errors during intracellular assembly of virus particles, rather than a bacterial adaptation. Conjugation, in the much-studied E. coli system is determined by plasmid genes, and is an adaptation for transferring copies of the plasmid from one bacterial host to another. It is seldom that a conjugative plasmid integrates into the host bacterial chromosome, and subsequently transfers part of the host bacterial DNA to another bacterium. Plasmid-mediated transfer of host bacterial DNA also appears to be an accidental process rather than a bacterial adaptation.

Transformation, unlike transduction or conjugation, depends on numerous bacterial gene products that specifically interact to perform this complex process, and thus transformation is clearly a bacterial adaptation for DNA transfer. In order for a bacterium to bind, take up and recombine donor DNA into its own chromosome, it must first enter a special physiological state termed competence. In *Bacillus subtilis*, about 40 genes are required for the development of competence. The length of DNA transferred during *B. subtilis* transformation can be between a third of a chromosome up to the whole chromosome. Transformation appears to be common among bacterial species, and thus far at least 60 species are known to have the natural ability to become competent for transformation. The development of competence in nature is usually associated with stressful environmental conditions, and seems to be an adaptation for facilitating repair of DNA damage in recipient cells.

In ordinary circumstances, transduction, conjugation, and transformation involve transfer of DNA between individual bacteria of the same species, but occasionally transfer may occur between individuals of different bacterial species and this may have significant consequences, such as the transfer of antibiotic resistance. In such cases, gene acquisition from other bacteria or the environment is called *horizontal gene transfer* and may be common under natural conditions. Gene transfer is particularly important in antibiotic resistance as it allows the rapid transfer of resistance genes between different pathogens.

Bacteriophages

Bacteriophages are viruses that infect bacteria. Many types of bacteriophage exist, some simply infect and lyse their host bacteria, while others insert into the bacterial chromosome. A bacteriophage can contain genes that contribute to its host's phenotype: for example, in the evolution of *Escherichia coli* O157:H7 and *Clostridium botulinum*, the toxin genes in an integrated phage converted a harmless ancestral bacterium into a lethal pathogen. Bacteria resist phage infection through restriction modification systems that degrade foreign DNA, and a system that uses CRISPR sequences to retain fragments of the genomes of phage that the bacteria have come into contact with in the past, which allows them to block virus replication through a form of RNA interference. This CRISPR system provides bacteria with acquired immunity to infection.

Behaviour

Secretion

Bacteria frequently secrete chemicals into their environment in order to modify it favourably. The secretions are often proteins and may act as enzymes that digest some form of food in the environment.

Bioluminescence

A few bacteria have chemical systems that generate light. This bioluminescence often occurs in bacteria that live in association with fish, and the light probably serves to attract fish or other large animals.

Multicellularity

Bacteria often function as multicellular aggregates known as biofilms, exchanging a variety of molecular signals for inter-cell communication, and engaging in coordinated multicellular behaviour.

The communal benefits of multicellular cooperation include a cellular division of labour, accessing resources that cannot effectively be used by single cells, collectively defending against antagonists, and optimising population survival by differentiating into distinct cell types. For example, bacteria in biofilms can have more than 500 times increased resistance to antibacterial agents than individual "planktonic" bacteria of the same species.

One type of inter-cellular communication by a molecular signal is called quorum sensing, which serves the purpose of determining whether there is a local population density that is sufficiently high that it is productive to invest in processes that are only successful if large numbers of similar organisms behave similarly, as in excreting digestive enzymes or emitting light.

Quorum sensing allows bacteria to coordinate gene expression, and enables them to produce, release and detect autoinducers or pheromones which accumulate with the growth in cell population.

Movement

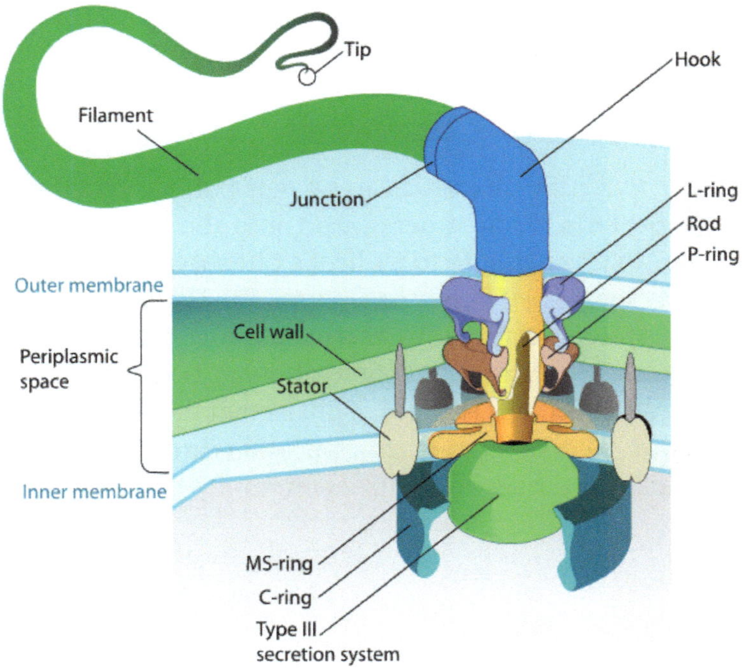

Flagellum of gram-negative bacteria. The base drives the rotation of the hook and filament.

Many bacteria can move using a variety of mechanisms: flagella are used for swimming through fluids; bacterial gliding and twitching motility move bacteria across surfaces; and changes of buoyancy allow vertical motion.

Swimming bacteria frequently move near 10 body lengths per second and a few as fast as 100. This makes them at least as fast as fish, on a relative scale.

In bacterial gliding and twitching motility, bacteria use their *type IV pili* as a grappling hook, repeatedly extending it, anchoring it and then retracting it with remarkable force (>80 pN).

> Our observations redefine twitching motility as a rapid, highly organized mechanism of bacterial translocation by which Pseudomonas aeruginosa can disperse itself over large areas to colonize new territories. It is also now clear, both morphologically and genetically, that twitching motility and social gliding motility, such as occurs in Myxococcus xanthus, are essentially the same process.
>
> —*Semmler, Whitchurch & Mattick (1999)*

Flagella are semi-rigid cylindrical structures that are rotated and function much like the propeller on a ship. Objects as small as bacteria operate a low Reynolds number and cylindrical forms are more efficient than the flat, paddle-like, forms appropriate at human-size scale.

Bacterial species differ in the number and arrangement of flagella on their surface; some have a single flagellum (*monotrichous*), a flagellum at each end (*amphitrichous*), clusters of flagella at the poles of the cell (*lophotrichous*), while others have flagella distributed over the entire surface of the cell (*peritrichous*). The bacterial flagella is the best-understood motility structure in any organism and is made of about 20 proteins, with approximately another 30 proteins required for its regulation and assembly. The flagellum is a rotating structure driven by a reversible motor at the base that uses the electrochemical gradient across the membrane for power. This motor drives the motion of the filament, which acts as a propeller.

Many bacteria (such as *E. coli*) have two distinct modes of movement: forward movement (swimming) and tumbling. The tumbling allows them to reorient and makes their movement a three-dimensional random walk. The flagella of a unique group of bacteria, the spirochaetes, are found between two membranes in the periplasmic space. They have a distinctive helical body that twists about as it moves.

Motile bacteria are attracted or repelled by certain stimuli in behaviours called taxes: these include chemotaxis, phototaxis, energy taxis, and magnetotaxis. In one peculiar group, the myxobacteria, individual bacteria move together to form waves of cells that then differentiate to form fruiting bodies containing spores. The myxobacteria move only when on solid surfaces, unlike *E. coli*, which is motile in liquid or solid media.

Several *Listeria* and *Shigella* species move inside host cells by usurping the cytoskeleton, which is normally used to move organelles inside the cell. By promoting actin polymerisation at one pole of their cells, they can form a kind of tail that pushes them through the host cell's cytoplasm.

Classification and identification

Classification seeks to describe the diversity of bacterial species by naming and grouping organisms based on similarities. Bacteria can be classified on the basis of cell structure, cellular metabolism or on differences in cell components, such as DNA, fatty acids, pigments, antigens and quinones. While these schemes allowed the identification and classification of bacterial strains, it was unclear whether these differences represented variation between distinct species or between strains of the same species. This uncertainty was due to the lack of distinctive structures in most bacteria, as well as lateral gene transfer between unrelated species. Due to lateral gene transfer, some closely related bacteria can have very different morphologies and metabolisms. To overcome this uncertainty, modern bacterial classification emphasises molecular systematics, using genetic techniques such as guanine cytosine ratio determination, genome-genome hybridisation, as well as sequencing genes that have not undergone extensive lateral gene transfer, such as the rRNA gene. Classification of bacteria is determined by publication in the International Journal of Systematic Bacteriology, and Bergey's Manual of Systematic Bacteriology. The International Committee on Systematic Bacteriology (ICSB) maintains international rules for the naming of bacteria and taxonomic categories and for the ranking of them in the International Code of Nomenclature of Bacteria.

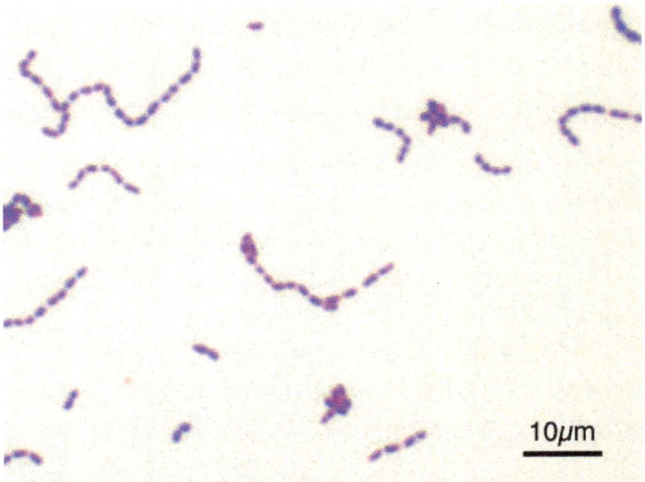

Streptococcus mutans visualised with a Gram stain

The term "bacteria" was traditionally applied to all microscopic, single-cell prokaryotes. However, molecular systematics showed prokaryotic life to consist of two separate domains, originally called *Eubacteria* and *Archaebacteria*, but now called *Bacteria* and *Archaea* that evolved independently from an ancient common ancestor. The archaea and eukaryotes are more closely related to each other than either is to the bacteria. These two domains, along with Eukarya, are the basis of the three-domain system, which is currently the most widely used classification system in microbiolology. However, due to the relatively recent introduction of molecular systematics and a rapid increase in the number of genome sequences that are available, bacterial classification remains a changing and expanding field. For example, a few biologists argue that the Archaea and Eukaryotes evolved from gram-positive bacteria.

The identification of bacteria in the laboratory is particularly relevant in medicine, where the correct treatment is determined by the bacterial species causing an infection. Consequently, the need to identify human pathogens was a major impetus for the development of techniques to identify bacteria.

The *Gram stain*, developed in 1884 by Hans Christian Gram, characterises bacteria based on the structural characteristics of their cell walls. The thick layers of peptidoglycan in the "gram-positive" cell wall stain purple, while the thin "gram-negative" cell wall appears pink. By combining morphology and Gram-staining, most bacteria can be classified as belonging to one of four groups (gram-positive cocci, gram-positive bacilli, gram-negative cocci and gram-negative bacilli). Some organisms are best identified by stains other than the Gram stain, particularly mycobacteria or *No-*

cardia, which show acid-fastness on Ziehl–Neelsen or similar stains. Other organisms may need to be identified by their growth in special media, or by other techniques, such as serology.

Culture techniques are designed to promote the growth and identify particular bacteria, while restricting the growth of the other bacteria in the sample. Often these techniques are designed for specific specimens; for example, a sputum sample will be treated to identify organisms that cause pneumonia, while stool specimens are cultured on selective media to identify organisms that cause diarrhoea, while preventing growth of non-pathogenic bacteria. Specimens that are normally sterile, such as blood, urine or spinal fluid, are cultured under conditions designed to grow all possible organisms. Once a pathogenic organism has been isolated, it can be further characterised by its morphology, growth patterns (such as aerobic or anaerobic growth), patterns of hemolysis, and staining.

As with bacterial classification, identification of bacteria is increasingly using molecular methods. Diagnostics using DNA-based tools, such as polymerase chain reaction, are increasingly popular due to their specificity and speed, compared to culture-based methods. These methods also allow the detection and identification of "viable but nonculturable" cells that are metabolically active but non-dividing. However, even using these improved methods, the total number of bacterial species is not known and cannot even be estimated with any certainty. Following present classification, there are a little less than 9,300 known species of prokaryotes, which includes bacteria and archaea; but attempts to estimate the true number of bacterial diversity have ranged from 10^7 to 10^9 total species—and even these diverse estimates may be off by many orders of magnitude.

Interactions with other Organisms

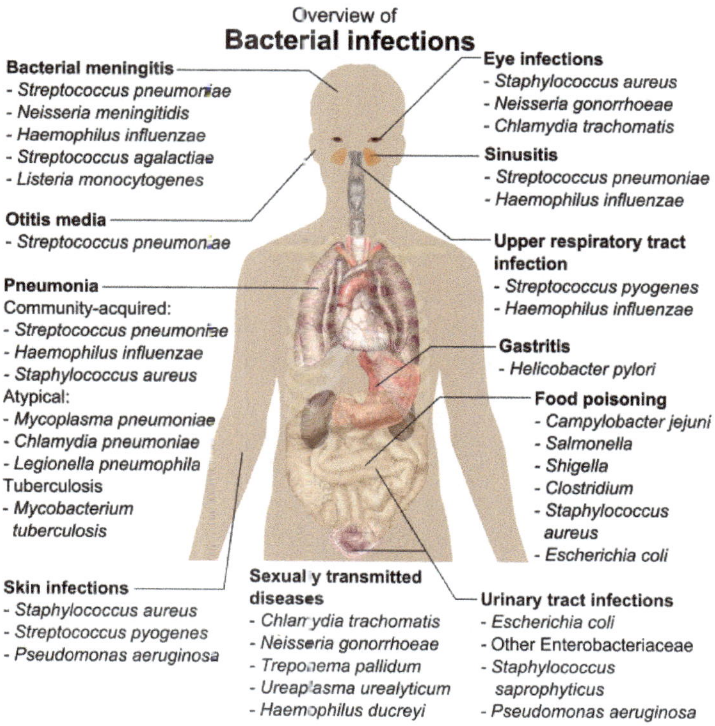

Overview of bacterial infections and main species involved.

Despite their apparent simplicity, bacteria can form complex associations with other organisms. These symbiotic associations can be divided into parasitism, mutualism and commensalism. Due to their small size, commensal bacteria are ubiquitous and grow on animals and plants exactly as they will grow on any other surface. However, their growth can be increased by warmth and sweat, and large populations of these organisms in humans are the cause of body odour.

Predators

Some species of bacteria kill and then consume other microorganisms, these species are called *predatory bacteria*. These include organisms such as *Myxococcus xanthus*, which forms swarms of cells that kill and digest any bacteria they encounter. Other bacterial predators either attach to their prey in order to digest them and absorb nutrients, such as *Vampirovibrio chlorellavorus*, or invade another cell and multiply inside the cytosol, such as *Daptobacter*. These predatory bacteria are thought to have evolved from saprophages that consumed dead microorganisms, through adaptations that allowed them to entrap and kill other organisms.

Mutualists

Certain bacteria form close spatial associations that are essential for their survival. One such mutualistic association, called interspecies hydrogen transfer, occurs between clusters of anaerobic bacteria that consume organic acids, such as butyric acid or propionic acid, and produce hydrogen, and methanogenic Archaea that consume hydrogen. The bacteria in this association are unable to consume the organic acids as this reaction produces hydrogen that accumulates in their surroundings. Only the intimate association with the hydrogen-consuming Archaea keeps the hydrogen concentration low enough to allow the bacteria to grow.

In soil, microorganisms that reside in the rhizosphere (a zone that includes the root surface and the soil that adheres to the root after gentle shaking) carry out nitrogen fixation, converting nitrogen gas to nitrogenous compounds. This serves to provide an easily absorbable form of nitrogen for many plants, which cannot fix nitrogen themselves. Many other bacteria are found as symbionts in humans and other organisms. For example, the presence of over 1,000 bacterial species in the normal human gut flora of the intestines can contribute to gut immunity, synthesise vitamins, such as folic acid, vitamin K and biotin, convert sugars to lactic acid, as well as fermenting complex undigestible carbohydrates. The presence of this gut flora also inhibits the growth of potentially pathogenic bacteria (usually through competitive exclusion) and these beneficial bacteria are consequently sold as probiotic dietary supplements.

Pathogens

If bacteria form a parasitic association with other organisms, they are classed as pathogens. Pathogenic bacteria are a major cause of human death and disease and cause infections such as tetanus, typhoid fever, diphtheria, syphilis, cholera, foodborne illness, leprosy and tuberculosis. A pathogenic cause for a known medical disease may only be discovered many years after, as was the case with *Helicobacter pylori* and peptic ulcer disease. Bacterial diseases are also important in agriculture, with bacteria causing leaf spot, fire blight and wilts in plants, as well as Johne's disease, mastitis, salmonella and anthrax in farm animals.

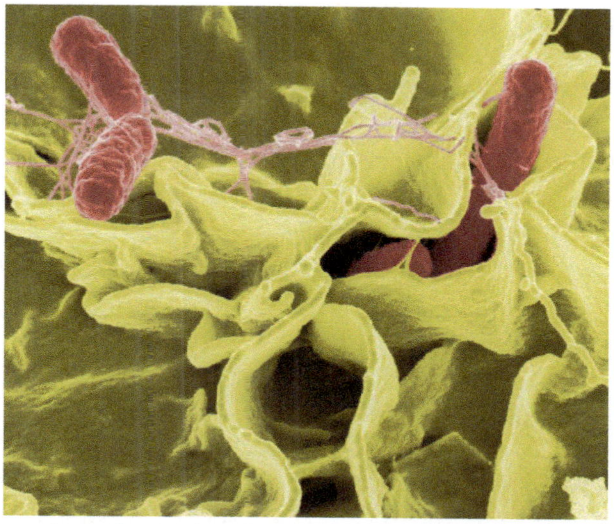

Colour-enhanced scanning electron micrograph showing *Salmonella typhimurium* (red) invading cultured human cells

Each species of pathogen has a characteristic spectrum of interactions with its human hosts. Some organisms, such as *Staphylococcus* or *Streptococcus*, can cause skin infections, pneumonia, meningitis and even overwhelming sepsis, a systemic inflammatory response producing shock, massive vasodilation and death. Yet these organisms are also part of the normal human flora and usually exist on the skin or in the nose without causing any disease at all. Other organisms invariably cause disease in humans, such as the Rickettsia, which are obligate intracellular parasites able to grow and reproduce only within the cells of other organisms. One species of Rickettsia causes typhus, while another causes Rocky Mountain spotted fever. *Chlamydia*, another phylum of obligate intracellular parasites, contains species that can cause pneumonia, or urinary tract infection and may be involved in coronary heart disease. Finally, some species, such as *Pseudomonas aeruginosa*, *Burkholderia cenocepacia*, and *Mycobacterium avium*, are opportunistic pathogens and cause disease mainly in people suffering from immunosuppression or cystic fibrosis.

Bacterial infections may be treated with antibiotics, which are classified as bacteriocidal if they kill bacteria, or bacteriostatic if they just prevent bacterial growth. There are many types of antibiotics and each class inhibits a process that is different in the pathogen from that found in the host. An example of how antibiotics produce selective toxicity are chloramphenicol and puromycin, which inhibit the bacterial ribosome, but not the structurally different eukaryotic ribosome. Antibiotics are used both in treating human disease and in intensive farming to promote animal growth, where they may be contributing to the rapid development of antibiotic resistance in bacterial populations. Infections can be prevented by antiseptic measures such as sterilising the skin prior to piercing it with the needle of a syringe, and by proper care of indwelling catheters. Surgical and dental instruments are also sterilised to prevent contamination by bacteria. Disinfectants such as bleach are used to kill bacteria or other pathogens on surfaces to prevent contamination and further reduce the risk of infection.

Significance in Technology and Industry

Bacteria, often lactic acid bacteria, such as *Lactobacillus* and *Lactococcus*, in combination with yeasts and moulds, have been used for thousands of years in the preparation of fermented foods, such as cheese, pickles, soy sauce, sauerkraut, vinegar, wine and yogurt.

The ability of bacteria to degrade a variety of organic compounds is remarkable and has been used in waste processing and bioremediation. Bacteria capable of digesting the hydrocarbons in petroleum are often used to clean up oil spills. Fertiliser was added to some of the beaches in Prince William Sound in an attempt to promote the growth of these naturally occurring bacteria after the 1989 *Exxon Valdez* oil spill. These efforts were effective on beaches that were not too thickly covered in oil. Bacteria are also used for the bioremediation of industrial toxic wastes. In the chemical industry, bacteria are most important in the production of enantiomerically pure chemicals for use as pharmaceuticals or agrichemicals.

Bacteria can also be used in the place of pesticides in the biological pest control. This commonly involves *Bacillus thuringiensis* (also called BT), a gram-positive, soil dwelling bacterium. Subspecies of this bacteria are used as a Lepidopteran-specific insecticides under trade names such as Dipel and Thuricide. Because of their specificity, these pesticides are regarded as environmentally friendly, with little or no effect on humans, wildlife, pollinators and most other beneficial insects.

Because of their ability to quickly grow and the relative ease with which they can be manipulated, bacteria are the workhorses for the fields of molecular biology, genetics and biochemistry. By making mutations in bacterial DNA and examining the resulting phenotypes, scientists can determine the function of genes, enzymes and metabolic pathways in bacteria, then apply this knowledge to more complex organisms. This aim of understanding the biochemistry of a cell reaches its most complex expression in the synthesis of huge amounts of enzyme kinetic and gene expression data into mathematical models of entire organisms. This is achievable in some well-studied bacteria, with models of *Escherichia coli* metabolism now being produced and tested. This understanding of bacterial metabolism and genetics allows the use of biotechnology to bioengineer bacteria for the production of therapeutic proteins, such as insulin, growth factors, or antibodies.

Because of their importance for research in general, samples of bacterial strains are isolated and preserved in Biological Resource Centers. This ensures the availability of the strain to scientists worldwide.

History of Bacteriology

Bacteria were first observed by the Dutch microscopist Antonie van Leeuwenhoek in 1676, using a single-lens microscope of his own design. He then published his observations in a series of letters to the Royal Society of London. Bacteria were Leeuwenhoek's most remarkable microscopic discovery. They were just at the limit of what his simple lenses could make out and, in one of the most striking hiatuses in the history of science, no one else would see them again for over a century. Only then were his by-then-largely-forgotten observations of bacteria—as opposed to his famous "animalcules" (spermatozoa)—taken seriously.

Christian Gottfried Ehrenberg introduced the word "bacterium" in 1828. In fact, his *Bacterium* was a genus that contained non-spore-forming rod-shaped bacteria, as opposed to *Bacillus*, a genus of spore-forming rod-shaped bacteria defined by Ehrenberg in 1835.

Louis Pasteur demonstrated in 1859 that the growth of microorganisms causes the fermentation process, and that this growth is not due to spontaneous generation. (Yeasts and moulds, commonly associated with fermentation, are not bacteria, but rather fungi.) Along with his contemporary Robert Koch, Pasteur was an early advocate of the germ theory of disease.

Robert Koch, a pioneer in medical microbiology, worked on cholera, anthrax and tuberculosis. In his research into tuberculosis Koch finally proved the germ theory, for which he received a Nobel Prize in 1905. In *Koch's postulates*, he set out criteria to test if an organism is the cause of a disease, and these postulates are still used today.

Though it was known in the nineteenth century that bacteria are the cause of many diseases, no effective antibacterial treatments were available. In 1910, Paul Ehrlich developed the first antibiotic, by changing dyes that selectively stained *Treponema pallidum*—the spirochaete that causes syphilis—into compounds that selectively killed the pathogen. Ehrlich had been awarded a 1908 Nobel Prize for his work on immunology, and pioneered the use of stains to detect and identify bacteria, with his work being the basis of the Gram stain and the Ziehl–Neelsen stain.

A major step forward in the study of bacteria came in 1977 when Carl Woese recognised that archaea have a separate line of evolutionary descent from bacteria. This new phylogenetic taxonomy depended on the sequencing of 16S ribosomal RNA, and divided prokaryotes into two evolutionary domains, as part of the three-domain system.

Algae

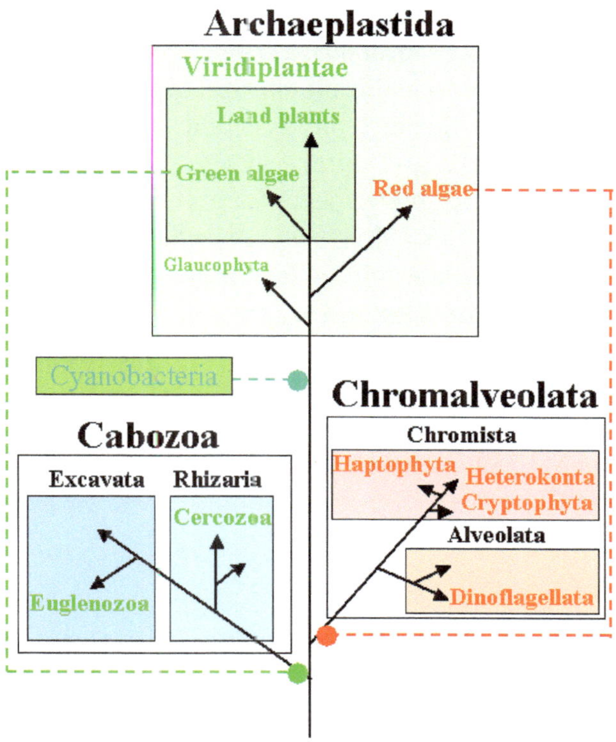

The lineage of algae according to Thomas Cavalier-Smith. The exact number and placement of endosymbiotic events is currently unknown, so this diagram can be taken only as a general guide. It represents the most parsimonious way of explaining the three types of endosymbiotic origins of plastids. These types include the endosymbiotic events of cyanobacteria, red algae and green algae, leading to the hypothesis of the supergroups Archaeplastida, Chromalveolata and Cabozoa respectively. Endosymbiotic events are noted with dotted lines.

Algae is an informal term for a large, diverse group of photosynthetic organisms which are not necessarily closely related, and is thus polyphyletic. Included organisms range from unicellular genera, such as *Chlorella* and the diatoms, to multicellular forms, such as the giant kelp, a large brown alga which may grow up to 50 m in length. Most are aquatic and autotrophic and lack many of the distinct cell and tissue types, such as stomata, xylem, and phloem, which are found in land plants. The largest and most complex marine algae are called seaweeds, while the most complex freshwater forms are the Charophyta, a division of green algae which includes, for example, *Spirogyra* and the stoneworts.

No definition of algae is generally accepted. One definition is that algae "have chlorophyll as their primary photosynthetic pigment and lack a sterile covering of cells around their reproductive cells". Some authors exclude all prokaryotes thus do not consider cyanobacteria (blue-green algae) as algae.

Algae constitute a polyphyletic group since they do not include a common ancestor, and although their plastids seem to have a single origin, from cyanobacteria, they were acquired in different ways. Green algae are examples of algae that have primary chloroplasts derived from endosymbiotic cyanobacteria. Diatoms and brown algae are examples of algae with secondary chloroplasts derived from an endosymbiotic red alga.

Algae exhibit a wide range of reproductive strategies, from simple asexual cell division to complex forms of sexual reproduction.

Algae lack the various structures that characterize land plants, such as the phyllids (leaf-like structures) of bryophytes, rhizoids in nonvascular plants, and the roots, leaves, and other organs found in tracheophytes (vascular plants). Most are phototrophic, although some are mixotrophic, deriving energy both from photosynthesis and uptake of organic carbon either by osmotrophy, myzotrophy, or phagotrophy. Some unicellular species of green algae, many golden algae, euglenids, dinoflagellates, and other algae have become heterotrophs (also called colorless or apochlorotic algae), sometimes parasitic, relying entirely on external energy sources and have limited or no photosynthetic apparatus. Some other heterotrophic organisms, such as the apicomplexans, are also derived from cells whose ancestors possessed plastids, but are not traditionally considered as algae. Algae have photosynthetic machinery ultimately derived from cyanobacteria that produce oxygen as a by-product of photosynthesis, unlike other photosynthetic bacteria such as purple and green sulfur bacteria. Fossilized filamentous algae from the Vindhya basin have been dated back to 1.6 to 1.7 billion years ago.

Etymology and Study

The singular *alga* is the Latin word for "seaweed" and retains that meaning in English. The etymology is obscure. Although some speculate that it is related to Latin *algēre*, "be cold", no reason is known to associate seaweed with temperature. A more likely source is *alliga*, "binding, entwining".

The Ancient Greek word for seaweed (*fūkos* or *phykos*), which could mean either the seaweed (probably red algae) or a red dye derived from it. The Latinization, *fūcus*, meant primarily the cosmetic rouge. The etymology is uncertain, but a strong candidate has long been some word related to the Biblical, "paint" (if not that word itself), a cosmetic eye-shadow used by the ancient

Egyptians and other inhabitants of the eastern Mediterranean. It could be any color: black, red, green, or blue.

Accordingly, the modern study of marine and freshwater algae is called either phycology or algology, depending on whether the Greek or Latin root is used. The name *Fucus* appears in a number of taxa.

Classification

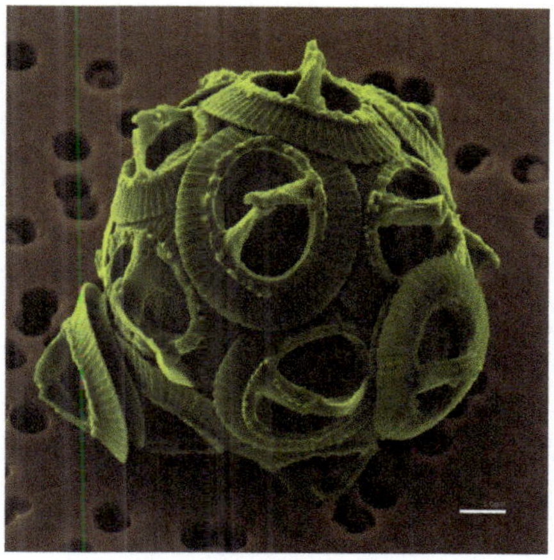

False-color scanning electron micrograph of the unicellular coccolithophore *Gephyrocapsa oceanica*

Most algae contain chloroplasts that are similar in structure to cyanobacteria. Chloroplasts contain circular DNA like that in cyanobacteria and presumably represent reduced endosymbiotic cyanobacteria. However, the exact origin of the chloroplasts is different among separate lineages of algae, reflecting their acquisition during different endosymbiotic events. Their lineage relationships are shown in the figure. Many of these groups contain some members that are no longer photosynthetic. Some retain plastids, but not chloroplasts, while others have lost plastids entirely.

Phylogeny based on plastid not nucleocytoplasmic genealogy:

HISTORIA

FVCORVM

AVCTORE

SAMVEL GOTTLIEB GMELIN,

MED. DOCT. ACADEM. IMPER. PETROPOL. BOTA-
NICES PROFESSORE ET MEMBRO ORDINARIO

PETROPOLI

EX TYPOGRAPHIA ACADEMIAE SCIENTIARVM

CIↃ IↃCC LXVIII.

Title page of Gmelin's *Historia Fucorum*, dated 1768

Linnaeus, in *Species Plantarum* (1753), the starting point for modern botanical nomenclature, recognized 14 genera of algae, of which only four are currently considered among algae. In *Systema Naturae*, Linnaeus described the genera *Volvox* and *Corallina*, and a species of *Acetabularia* (as *Madrepora*), among the animals.

In 1768, Samuel Gottlieb Gmelin (1744–1774) published the *Historia Fucorum*, the first work dedicated to marine algae and the first book on marine biology to use the then new binomial nomenclature of Linnaeus. It included elaborate illustrations of seaweed and marine algae on folded leaves.

W. H. Harvey (1811—1866) and Lamouroux (1813) were the first to divide macroscopic algae into four divisions based on their pigmentation. This is the first use of a biochemical criterion in plant systematics. Harvey's four divisions are: red algae (Rhodospermae), brown algae (Melanospermae), green algae (Chlorospermae), and Diatomaceae.

At this time, microscopic algae were discovered and reported by a different group of workers (e.g., O. F. Müller and Ehrenberg) studying the Infusoria (microscopic organisms). Unlike macroalgae, which were clearly viewed as plants, microalgae were frequently considered animals because they are often motile. Even the nonmotile (coccoid) microalgae were sometimes merely seen as stages of the lifecycle of plants, macroalgae, or animals.

Although used as a taxonomic category in some pre-Darwinian classifications, e.g., Linnaeus (1753), de Jussieu (1789), Horaninow (1843), Agassiz (1859), Wilson & Cassin (1864), in further classifications, the "algae" are seen as an artificial, polyphyletic group.

Throughout the 20th century, most classifications treated the following groups as divisions or classes of algae: cyanophytes, rhodophytes, chrysophytes, xanthophytes, bacillariophytes, phaeophytes, pyrrhophytes (cryptophytes and dinophytes), euglenophytes, and chlorophytes. Later, many new groups were discovered (e.g., Bolidophyceae), and others were splintered from older groups: charophytes and glaucophytes (from chlorophytes), many heterokontophytes (e.g., synurophytes from chrysophytes, or eustigmatophytes from xanthophytes), haptophytes (from chrysophytes), and chlorarachniophytes (from xanthophytes).

With the abandonment of plant-animal dichotomous classification, most groups of algae (sometimes all) were included in Protista, later also abandoned in favour of Eukaryota. However, as a legacy of the older plant life scheme, some groups that were also treated as protozoans in the past still have duplicated classifications.

Some parasitic algae (e.g., the green algae *Prototheca* and *Helicosporidium*, parasites of metazoans, or *Cephaleuros*, parasites of plants) were originally classified as fungi, sporozoans, or protistans of *incertae sedis*, while others (e.g., the green algae *Phyllosiphon* and *Rhodochytrium*, parasites of plants, or the red algae *Pterocladiophila* and *Gelidiocolax mammillatus*, parasites of other red algae, or the dinoflagellates *Oodinium*, parasites of fish) had their relationship with algae conjectured early. In other cases, some groups were originally characterized as parasitic algae (e.g., *Chlorochytrium*), but later were seen as endophytic algae. Some filamentous bacteria (e.g., *Beggiatoa*) were originally seen as algae. Furthermore, groups like the apicomplexans are also parasites derived from ancestors that possessed plastids, but are not included in any group traditionally seen as algae.

Relationship to Land Plants

The first land plants probably evolved from shallow freshwater charophyte algae much like *Chara* almost 500 million years ago. These probably had an isomorphic alternation of generations and were probably filamentous. Fossils of isolated land plant spores suggest land plants may have been around as long as 475 million years ago.

Morphology

The kelp forest exhibit at the Monterey Bay Aquarium: A three-dimensional, multicellular thallus

A range of algal morphologies is exhibited, and convergence of features in unrelated groups is common. The only groups to exhibit three-dimensional multicellular thalli are the reds and browns, and some chlorophytes. Apical growth is constrained to subsets of these groups: the florideophyte reds, various browns, and the charophytes. The form of charophytes is quite different from those of reds and browns, because they have distinct nodes, separated by internode 'stems'; whorls of branches reminiscent of the horsetails occur at the nodes. Conceptacles are another polyphyletic trait; they appear in the coralline algae and the Hildenbrandiales, as well as the browns.

Most of the simpler algae are unicellular flagellates or amoeboids, but colonial and nonmotile forms have developed independently among several of the groups. Some of the more common organizational levels, more than one of which may occur in the lifecycle of a species, are

- Colonial: small, regular groups of motile cells
- Capsoid: individual non-motile cells embedded in mucilage
- Coccoid: individual non-motile cells with cell walls
- Palmelloid: nonmotile cells embedded in mucilage

- Filamentous: a string of nonmotile cells connected together, sometimes branching
- Parenchymatous: cells forming a thallus with partial differentiation of tissues

In three lines, even higher levels of organization have been reached, with full tissue differentiation. These are the brown algae,—some of which may reach 50 m in length (kelps)—the red algae, and the green algae. The most complex forms are found among the green algae, in a lineage that eventually led to the higher land plants. The point where these nonalgal plants begin and algae stop is usually taken to be the presence of reproductive organs with protective cell layers, a characteristic not found in the other algal groups.

Physiology

Many algae, particularly members of the Characeae, have served as model experimental organisms to understand the mechanisms of the water permeability of membranes, osmoregulation, turgor regulation, salt tolerance, cytoplasmic streaming, and the generation of action potentials.

Phytohormones are found not only in higher plants, but in algae, too.

Symbiotic Algae

Some species of algae form symbiotic relationships with other organisms. In these symbioses, the algae supply photosynthates (organic substances) to the host organism providing protection to the algal cells. The host organism derives some or all of its energy requirements from the algae. Examples are:

Lichens

Rock lichens in Ireland

Lichens are defined by the International Association for Lichenology to be "an association of a fungus and a photosynthetic symbiont resulting in a stable vegetative body having a specific structure." The fungi, or mycobionts, are mainly from the Ascomycota with a few from the Basidiomycota. They are not found alone in nature; but when they began to associate is not known. One mycobiont associates with the same phycobiont species, rarely two, from the green algae, except that alternatively, the mycobiont may associate with a species of cyanobacteria (hence "photobiont" is the more accurate term). A photobiont may be associated with many different mycobionts or may live independently; accordingly, lichens are named and classified as fungal species. The association is termed a morphogenesis because the lichen has a form and capabilities not possessed by the symbiont species alone (they can be experimentally isolated). The photobiont possibly triggers otherwise latent genes in the mycobiont.

Coral Reefs

Floridian coral reef

Coral reefs are accumulated from the calcareous exoskeletons of marine invertebrates of the order Scleractinia (stony corals). These animals metabolize sugar and oxygen to obtain energy for their cell-building processes, including secretion of the exoskeleton, with water and carbon dioxide as byproducts. Dinoflagellates (algal protists) are often endosymbionts in the cells of the coral-forming marine invertebrates, where they accelerate host-cell metabolism by generating immediately available sugar and oxygen through photosynthesis using incident light and the carbon dioxide produced by the host. Reef-building stony corals (hermatypic corals) require endosymbiotic algae from the genus *Symbiodinium* to be in a healthy condition. The loss of *Symbiodinium* from the host is known as coral bleaching, a condition which leads to the deterioration of a reef.

Sea Sponges

Green algae live close to the surface of some sponges, for example, breadcrumb sponge (*Halichondria panicea*). The alga is thus protected from predators; the sponge is provided with oxygen and sugars which can account for 50 to 80% of sponge growth in some species.

Lifecycle

Rhodophyta, Chlorophyta, and Heterokontophyta, the three main algal divisions, have lifecycles which show considerable variation and complexity. In general, an asexual phase exists where the seaweed's cells are diploid, a sexual phase where the cells are haploid, followed by fusion of the male and female gametes. Asexual reproduction permits efficient population increases, but less variation is possible. Commonly, in sexual reproduction of unicellular and colonial algae, two specialized, sexually compatible, haploid gametes make physical contact and fuse to form a zygote. To ensure a successful mating, the development and release of gametes is highly synchronized and regulated; pheromones may play a key role in these processes. Sexual reproduction allows for more variation and provides the benefit of efficient recombinational repair of DNA damages during meiosis, a key stage of the sexual cycle. However, sexual reproduction is more costly than asexual reproduction. Meiosis has been shown to occur in many different species of algae.

Numbers

Algae on coastal rocks at Shihtiping in Taiwan

The *Algal Collection of the US National Herbarium* (located in the National Museum of Natural History) consists of approximately 320,500 dried specimens, which, although not exhaustive (no exhaustive collection exists), gives an idea of the order of magnitude of the number of algal species (that number remains unknown). Estimates vary widely. For example, according to one standard textbook, in the British Isles the *UK Biodiversity Steering Group Report* estimated there to be 20,000 algal species in the UK. Another checklist reports only about 5,000 species. Regarding the difference of about 15,000 species, the text concludes: "It will require many detailed field surveys before it is possible to provide a reliable estimate of the total number of species ..."

Regional and group estimates have been made, as well:

- 5,000–5,500 species of red algae worldwide
- "some 1,300 in Australian Seas"
- 400 seaweed species for the western coastline of South Africa, and 212 species from the coast of KwaZulu-Natal. Some of these are duplicates, as the range extends across both coasts, and the total recorded is probably about 500 species. Most of these are listed in List of seaweeds of South Africa. These exclude phytoplankton and crustose corallines.
- 669 marine species from California (US)
- 642 in the check-list of Britain and Ireland

and so on, but lacking any scientific basis or reliable sources, these numbers have no more credibility than the British ones mentioned above. Most estimates also omit microscopic algae, such as phytoplankton.

The most recent estimate suggests 72,500 algal species worldwide.

Distribution

The distribution of algal species has been fairly well studied since the founding of phytogeography in the mid-19th century. Algae spread mainly by the dispersal of spores analogously to the dispersal of Plantae by seeds and spores. This dispersal can be accomplished by air, water, or other organisms. Due to this, spores can be found in a variety of environments: fresh and marine waters, air, soil, and in or on other organisms. Whether a spore is to grow into an organism depends on the combination of the species and the environmental conditions where the spore lands.

The spores of freshwater algae are dispersed mainly by running water and wind, as well as by living carriers. However, not all bodies of water can carry all species of algae, as the chemical composition of certain water bodies limits the algae that can survive within them. Marine spores are often spread by ocean currents. Ocean water presents many vastly different habitats based on temperature and nutrient availability, resulting in phytogeographic zones, regions, and provinces.

To some degree, the distribution of algae is subject to floristic discontinuities caused by geographical features, such as Antarctica, long distances of ocean or general land masses. It is, therefore, possible to identify species occurring by locality, such as "Pacific algae" or "North Sea algae". When they occur out of their localities, hypothesizing a transport mechanism is usually possible, such as the hulls of ships. For example, *Ulva reticulata* and *U. fasciata* travelled from the mainland to Hawaii in this manner.

Mapping is possible for select species only: "there are many valid examples of confined distribution patterns." For example, *Clathromorphum* is an arctic genus and is not mapped far south of there. However, scientists regard the overall data as insufficient due to the "difficulties of undertaking such studies."

Ecology

Phytoplankton, Lake Chuzenji

Algae are prominent in bodies of water, common in terrestrial environments, and are found in unusual environments, such as on snow and ice. Seaweeds grow mostly in shallow marine waters, under 100 m (330 ft) deep; however, some such as Navicula pennata have been recorded to a depth of 360 m (1,180 ft).

The various sorts of algae play significant roles in aquatic ecology. Microscopic forms that live suspended in the water column (phytoplankton) provide the food base for most marine food chains. In very high densities (algal blooms), these algae may discolor the water and outcompete, poison, or asphyxiate other life forms.

Algae can be used as indicator organisms to monitor pollution in various aquatic systems. In many cases, algal metabolism is sensitive to various pollutants. Due to this, the species composition of algal populations may shift in the presence of chemical pollutants. To detect these changes, algae can be sampled from the environment and maintained in laboratories with relative ease.

On the basis of their habitat, algae can be categorized as: aquatic (planktonic, benthic, marine, freshwater, lentic, lotic), terrestrial, aerial (subareial), lithophytic, halophytic (or euryhaline), psammon, thermophilic, cryophilic, epibiont (epiphytic, epizoic), endosymbiont (endophytic, endozoic), parasitic, calcifilic or lichenic (phycobiont).

Cultural Associations

In classical Chinese, the word is used both for "algae" and (in the modest tradition of the imperial scholars) for "literary talent". The third island in Kunming Lake beside the Summer Palace in Beijing is known as the Zaojian Tang Dao, which thus simultaneously means "Island of the Algae-Viewing Hall" and "Island of the Hall for Reflecting on Literary Talent".

Uses

Agar

Harvesting algae

Agar, a gelatinous substance derived from red algae, has a number of commercial uses. It is a good medium on which to grow bacteria and fungi, as most microorganisms cannot digest agar.

Alginates

Alginic acid, or alginate, is extracted from brown algae. Its uses range from gelling agents in food, to medical dressings. Alginic acid also has been used in the field of biotechnology as a biocompatible medium for cell encapsulation and cell immobilization. Molecular cuisine is also a user of the substance for its gelling properties, by which it becomes a delivery vehicle for flavours.

Between 100,000 and 170,000 wet tons of *Macrocystis* are harvested annually in New Mexico for alginate extraction and abalone feed.

Energy Source

To be competitive and independent from fluctuating support from (local) policy on the long run, biofuels should equal or beat the cost level of fossil fuels. Here, algae-based fuels hold great promise, directly related to the potential to produce more biomass per unit area in a year than any other form of biomass. The break-even point for algae-based biofuels is estimated to occur by 2025.

Fertilizer

Seaweed-fertilized gardens on Inisheer

For centuries, seaweed has been used as a fertilizer; George Owen of Henllys writing in the 16th century referring to drift weed in South Wales:

"...This kind of ore they often gather and lay on great heapes, where it heteth and rotteth, and will have a strong and loathsome smell; when being so rotten they cast on the land, as they do their muck, and thereof springeth good corn, especially barley ... After spring-tydes or great rigs of the sea, they fetch it in sacks on horse backes, and carie the same three, four, or five miles, and cast it on the lande, which doth very much better the ground for corn and grass..."

Today, algae are used by humans in many ways; for example, as fertilizers, soil conditioners, and livestock feed. Aquatic and microscopic species are cultured in clear tanks or ponds and are either harvested or used to treat effluents pumped through the ponds. Algaculture on a large scale is an important type of aquaculture in some places. Maerl is commonly used as a soil conditioner.

Nutrition

Dulse, a type of food

Naturally growing seaweeds are an important source of food, especially in Asia. They provide many vitamins including: A, B_1, B_2, B_6, niacin, and C, and are rich in iodine, potassium, iron, magnesium, and calcium. In addition, commercially cultivated microalgae, including both algae and cyanobacteria, are marketed as nutritional supplements, such as spirulina, *Chlorella* and the vitamin-C supplement from *Dunaliella*, high in beta-carotene.

Algae are national foods of many nations: China consumes more than 70 species, including *fat choy*, a cyanobacterium considered a vegetable; Japan, over 20 species; Ireland, dulse; Chile, cochayuyo. Laver is used to make "laver bread" in Wales, where it is known as *bara lawr*; in Korea, *gim*; in Japan, *nori* and *aonori*. It is also used along the west coast of North America from California to British Columbia, in Hawaii and by the Māori of New Zealand. Sea lettuce and badderlocks are salad ingredients in Scotland, Ireland, Greenland, and Iceland.

The oils from some algae have high levels of unsaturated fatty acids. For example, *Parietochloris incisa* is very high in arachidonic acid, where it reaches up to 47% of the triglyceride pool. Some varieties of algae favored by vegetarianism and veganism contain the long-chain, essential omega-3 fatty acids, docosahexaenoic acid (DHA) and eicosapentaenoic acid (EPA). Fish oil contains the omega-3 fatty acids, but the original source is algae (microalgae in particular), which are eaten by marine life such as copepods and are passed up the food chain. Algae have emerged in recent years as a popular source of omega-3 fatty acids for vegetarians who cannot get long-chain EPA and DHA from other vegetarian sources such as flaxseed oil, which only contains the short-chain alpha-linolenic acid (ALA).

Pollution Control

- Sewage can be treated with algae, reducing the use of large amounts of toxic chemicals that would otherwise be needed.
- Algae can be used to capture fertilizers in runoff from farms. When subsequently harvested, the enriched algae can be used as fertilizer.

- Aquaria and ponds can be filtered using algae, which absorb nutrients from the water in a device called an algae scrubber, also known as an algae turf scrubber.

Agricultural Research Service scientists found that 60–90% of nitrogen runoff and 70–100% of phosphorus runoff can be captured from manure effluents using a horizontal algae scrubber, also called an algal turf scrubber (ATS). Scientists developed the ATS, which consists of shallow, 100-foot raceways of nylon netting where algae colonies can form, and studied its efficacy for three years. They found that algae can readily be used to reduce the nutrient runoff from agricultural fields and increase the quality of water flowing into rivers, streams, and oceans. Researchers collected and dried the nutrient-rich algae from the ATS and studied its potential as an organic fertilizer. They found that cucumber and corn seedlings grew just as well using ATS organic fertilizer as they did with commercial fertilizers. Algae scrubbers, using bubbling upflow or vertical waterfall versions, are now also being used to filter aquaria and ponds.

Bioremediation

The alga *Stichococcus bacillaris* has been seen to colonize silicone resins used at archaeological sites; biodegrading the synthetic substance.

Pigments

The natural pigments (carotenoids and chlorophylls) produced by algae can be used as alternatives to chemical dyes and coloring agents. The presence of some individual algal pigments, together with specific pigment concentration ratios, are taxon-specific: analysis of their concentrations with various analytical methods, particularly high-performance liquid chromatography, can therefore offer deep insight into the taxonomic composition and relative abundance of natural alga populations in sea water samples.

Brown Algae

The Phaeophyceae or brown algae (singular: alga), are a large group of mostly marine multicellular algae, including many seaweeds of colder Northern Hemisphere waters. They play an important role in marine environments, both as food and for the habitats they form. For instance, *Macrocystis*, a kelp of the order Laminariales, may reach 60 m (200 ft) in length and forms prominent underwater forests. Another example is *Sargassum*, which creates unique habitats in the tropical waters of the Sargasso Sea. Many brown algae, such as members of the order Fucales, commonly grow along rocky seashores. Some members of the class, such as kelp, are used as food for humans.

Worldwide, about 1500–2000 species of brown algae are known. Some species, such as *Ascophyllum nodosum*, are of sufficient commercial importance that they have become subjects of extensive research in their own right.

Brown algae belong to a very large group, the Heterokontophyta, a group of eukaryotic organisms distinguished most prominently by having chloroplasts surrounded by four membranes, suggesting an origin from a symbiotic relationship between a basal eukaryote and another eukaryotic organism. Most brown algae contain the pigment fucoxanthin, which is responsible for the distinctive greenish-brown color that gives them their name. Brown algae are unique among heterokonts in developing into multicellular forms with differentiated tissues, but they reproduce by means of flagellated spores and gametes that closely resemble cells of other heterokonts. Genetic studies show their closest relatives to be the yellow-green algae.

Morphology

Brown algae exist in a wide range of sizes and forms. The smallest members of the group grow as tiny, feathery tufts of threadlike cells no more than a few centimeters long. Some species have a stage in their life cycle that consists of only a few cells, making the entire alga microscopic. Other groups of brown algae grow to much larger sizes. The rockweeds and leathery kelps are often the most conspicuous algae in their habitats. Kelps can range in size from the two-foot-tall sea palm *Postelsia* to the giant kelp *Macrocystis pyrifera*, which grows to over 45 m (150 ft) long and is the largest of all the algae. In form, the brown algae range from small crusts or cushions to leafy free-floating mats formed by species of *Sargassum*. They may consist of delicate felt-like strands of cells, as in *Ectocarpus*, or of foot-long flattened branches resembling a fan, as in *Padina*.

Regardless of size or form, two visible features set the Phaeophyceae apart from all other algae. First, members of the group possess a characteristic color that ranges from an olive green to various shades of brown. The particular shade depends upon the amount of fucoxanthin present in the alga. Second, all brown algae are multicellular. There are no known species that exist as single cells or as colonies of cells, and the brown algae are the only major group of seaweeds that does not include such forms. However, this may be the result of classification rather than a consequence of evolution, as all the groups hypothesized to be the closest relatives of the browns include single-celled or colonial forms.

Visible Structures

Two specimens of *Laminaria hyperborea*, each showing the rootlike holdfast at lower left, a divided blade at upper right, and a stemlike stipe connecting the blade to the holdfast.

Whatever their form, the body of all brown algae is termed a *thallus*, indicating that it lacks the complex xylem and phloem of vascular plants. This does not mean that brown algae completely lack specialized structures. But, because some botanists define "true" stems, leaves, and roots by the presence of these tissues, their absence in the brown algae means that the stem-like and leaf-like structures found in some groups of brown algae must be described using different terminology. Although not all brown algae are structurally complex, those that are typically possess one or more characteristic parts.

A *holdfast* is a rootlike structure present at the base of the alga. Like a root system in plants, a holdfast serves to anchor the alga in place on the *substrate* where it grows, and thus prevents the alga from being carried away by the current. Unlike a root system, the holdfast generally does not serve as the primary organ for water uptake, nor does it take in nutrients from the substrate. The overall physical appearance of the holdfast differs among various brown algae and among various substrates. It may be heavily branched, or it may be cup-like in appearance. A single alga typically has just one holdfast, although some species have more than one stipe growing from their holdfast.

A *stipe* is a stalk or stemlike structure present in an alga. It may grow as a short structure near the base of the alga (as in *Laminaria*), or it may develop into a large, complex structure running throughout the algal body (as in *Sargassum* or *Macrocystis*). In the most structurally differentiated brown algae (such as *Fucus*), the tissues within the stipe are divided into three distinct layers or regions. These regions include a central pith, a surrounding cortex, and an outer epidermis, each of which has an analog in the stem of a vascular plant. In some brown algae, the pith region includes a core of elongated cells that resemble the phloem of vascular plants both in structure and function. In others (such as *Nereocystis*), the center of the stipe is hollow and filled with gas that serves to keep that part of the alga buoyant. The stipe may be relatively flexible and elastic in species like *Macrocystis pyrifera* that grow in strong currents, or may be more rigid in species like *Postelsia palmaeformis* that are exposed to the atmosphere at low tide.

Many algae have a flattened portion that may resemble a leaf, and this is termed a *blade*, *lamina*, or *frond*. The name *blade* is most often applied to a single undivided structure, while *frond* may be applied to all or most of an algal body that is flattened, but this distinction is not universally applied. The name *lamina* refers to that portion of a structurally differentiated alga that is flattened. It may be a single or a divided structure, and may be spread over a substantial portion of the alga. In rockweeds, for example, the lamina is a broad wing of tissue that runs continuously along both sides of a branched *midrib*. The midrib and lamina together constitute almost all of a rockweed, so that the lamina is spread throughout the alga rather than existing as a localized portion of it.

Species like *Fucus vesiculosus* produce numerous gas-filled pneumatocysts (air bladders) to increase buoyancy.

In some brown algae, there is a single lamina or blade, while in others there may be many separate blades. Even in those species that initially produce a single blade, the structure may tear with rough currents or as part of maturation to form additional blades. These blades may be attached directly to the stipe, to a holdfast with no stipe present, or there may be an air bladder between the stipe and blade. The surface of the lamina or blade may be smooth or wrinkled; its tissues may be thin and flexible or thick and leathery. In species like *Egregia menziesii*, this characteristic may change depending upon the turbulence of the waters in which it grows. In other species, the surface of the blade is coated with slime to discourage the attachment of epiphytes or to deter herbivores. Blades are also often the parts of the alga that bear the reproductive structures.

Gas-filled floats called *pneumatocysts* provide buoyancy in many kelps and members of the Fucales. These bladder-like structures occur in or near the *lamina*, so that it is held nearer the water surface and thus receives more light for photosynthesis. Pneumatocysts are most often spherical or ellipsoidal, but can vary in shape among different species. Species such as *Nereocystis luetkeana* and *Pelagophycus porra* bear a single large pneumatocyst between the top of the stipe and the base of the blades. In contrast, the giant kelp *Macrocystis pyrifera* bears many blades along its stipe, with a pneumatocyst at the base of each blade where it attaches to the main stipe. Species of *Sargassum* also bear many blades and pneumatocysts, but both kinds of structures are attached separately to the stipe by short stalks. In species of *Fucus*, the pneumatocysts develop within the lamina itself, either as discrete spherical bladders or as elongated gas-filled regions that take the outline of the lamina in which they develop.

Growth

Growth in *Dictyota dichotoma* occurs at each frond tip, where new cells are produced.

The brown algae include the largest and fastest growing of seaweeds. Fronds of *Macrocystis* may grow as much as 50 centimetres (20 in) per day, and the stipes can grow 6 centimetres (2.4 in) in a single day.

Growth in most brown algae occurs at the tips of structures as a result of divisions in a single *apical cell* or in a row of such cells. As this apical cell divides, the new cells that it produces develop into all the tissues of the alga. Branchings and other lateral structures appear when the apical cell divides to produce two new apical cells. However, a few groups (such as *Ectocarpus*) grow by a diffuse, unlocalized production of new cells that can occur anywhere on the thallus.

Tissue Organization

The simplest browns are filamentous—that is, their cells are elongate and have septa cutting across their width. They branch by getting wider at their tip, and then dividing the widening.

Aside from filamentous forms, there are two main types of tissue organization in the brown algae: pseudoparenchymatous (haplostichous) and parenchymatous (polystichous). The fronds may be multiaxial or monoaxial.

The cell wall consists of two layers; the inner layer bears the strength, and consists of cellulose; the outer wall layer is mainly algin, and is gummy when wet but becomes hard and brittle when it dries out.

Evolutionary History

Genetic and ultrastructural evidence place the Phaeophyceae among the heterokonts (Stramenopiles), a large assemblage of organisms that includes both photosynthetic members with plastids (such as the diatoms) as well as non-photosynthetic groups (such as the slime nets and water molds). Although some heterokont relatives of the brown algae lack plastids in their cells, scientists believe this is a result of evolutionary loss of that organelle in those groups rather than independent acquisition by the several photosynthetic members. Thus, all heterokonts are believed to descend from a single heterotrophic ancestor that became photosynthetic when it acquired plastids through endosymbiosis of another unicellular eukaryote.

The closest relatives of the brown algae include unicellular and filamentous species, but no unicellular species of brown algae are known. However, most scientists assume that the Phaeophyceae evolved from unicellular ancestors. DNA sequence comparison also suggests that the brown algae evolved from the filamentous Phaeothamniophyceae, Xanthophyceae, or the Chrysophyceae between 150 and 200 million years ago. In many ways, the evolution of the brown algae parallels that of the green algae and red algae, as all three groups possess complex multicellular species with an alternation of generations. Analysis of 5S rRNA sequences reveals much smaller evolutionary distances among genera of the brown algae than among genera of red or green algae, which suggests that the brown algae have diversified much more recently than the other two groups.

Fossils

The occurrence of Phaeophyceae as fossils is rare due to their generally soft-bodied nature, and scientists continue to debate the identification of some finds. Part of the problem with identification lies in the convergent evolution of morphologies between many brown and red algae. Most fossils of soft-tissue algae preserve only a flattened outline, without the microscopic features that permit the major groups of multicellular algae to be reliably distinguished. Among the brown algae, only species of the genus *Padina* deposit significant quantities of minerals in or around their cell walls. Other algal groups, such as the red algae and green algae, have a number of calcareous members. Because of this, they are more likely to leave evidence in the fossil record than the soft bodies of most brown algae and more often can be precisely classified.

Fossils comparable in morphology to brown algae are known from strata as old as the Upper Ordovician, but the taxonomic affinity of these impression fossils is far from certain. Claims that earlier

Ediacaran fossils are brown algae have since been dismissed. While many carbonaceous fossils have been described from the Precambrian, they are typically preserved as flattened outlines or fragments measuring only millimeters long. Because these fossils lack features diagnostic for identification at even the highest level, they are assigned to fossil form taxa according to their shape and other gross morphological features. A number of Devonian fossils termed *fucoids*, from their resemblance in outline to species in the genus *Fucus*, have proven to be inorganic rather than true fossils. The Devonian megafossil *Prototaxites*, which consists of masses of filaments grouped into trunk-like axes, has been considered a possible brown alga. However, modern research favors reinterpretation of this fossil as a terrestrial fungus or fungal-like organism. Likewise, the fossil *Protosalvinia* was once considered a possible brown alga, but is now thought to be an early land plant.

A number of Paleozoic fossils have been tentatively classified with the brown algae, although most have also been compared to known red algae species. *Phascolophyllaphycus* possesses numerous elongate, inflated blades attached to a stipe. It is the most abundant of algal fossils found in a collection made from Carboniferous strata in Illinois. Each hollow blade bears up to eight pneumatocysts at its base, and the stipes appear to have been hollow and inflated as well. This combination of characteristics is similar to certain modern genera in the order Laminariales (kelps). Several fossils of *Drydenia* and a single specimen of *Hungerfordia* from the Upper Devonian of New York have also been compared to both brown and red algae. Fossils of *Drydenia* consist of an elliptical blade attached to a branching filamentous holdfast, not unlike some species of *Laminaria*, *Porphyra*, or *Gigartina*. The single known specimen of *Hungerfordia* branches dichotomously into lobes and resembles genera like *Chondrus* and *Fucus* or *Dictyota*.

The earliest known fossils that can be assigned reliably to the Phaeophyceae come from Miocene diatomite deposits of the Monterey Formation in California. Several soft-bodied brown macroalgae, such as *Julescraneia*, have been found.

Life Cycle

Sexual reproduction may be isogamous, oogamous, or anisogamous. There is evidence of sex chromosomes. Union of gametes may take place in water or within the oogonium (oogamous species). The life cycle shows great variability from one group to another.

In *Laminaria*, there is a conspicuous diploid generation. Meiosis takes place within sporangia, before the spores are released. As they are haploid, with sex chromosomes, there are equal numbers of male and female spores. With the exception of the Fucales, all brown algae have a life cycle with an alternation between haploid and diploid forms.

Ecology

Brown algae have adapted to a wide variety of marine ecological niches including the tidal splash zone, rock pools, the whole intertidal zone and relatively deep near shore waters. They are an important constituent of some brackish water ecosystems, and four species are restricted to life in fresh water. A large number of Phaeophyceae are intertidal or upper littoral, and they are predominantly cool and cold water organisms that benefit from nutrients in up welling cold water currents and inflows from land; *Sargassum* being a prominent exception to this generalisation.

Brown algae growing in brackish waters are almost solely asexual.

Saccharina latissima on a beach.

Chemistry

Algal group	δ¹³C range
HCO_3-using red algae	−22.5‰ to −9.6‰
CO_2-using red algae	−34.5‰ to −29.9‰
Brown algae	−20.8‰ to −10.5‰
Green algae	−20.3‰ to −8.8‰

Brown algae have a $\delta^{13}C$ value in the range of −20.8‰ to −10.5‰, in contrast with red algae and greens. This reflects their different metabolic pathways.

They have cellulose walls with alginic acid and also contain the polysaccharide fucoidan in the amorphous sections of their cell walls. A few species (of *Padina*) calcify with aragonite needles.

The photosynthetic system of brown algae is made of a P700 complex containing chlorophyll a. Their plastids also contain chlorophyll c and carotenoids (the most widespread of those being fucoxanthin).

Brown algae produce a specific type of tannin called phlorotannins in higher amounts than red algae do.

Importance and Uses

The brown algae include a number of edible seaweeds, and a number of brown algae contains alginic acid that is extracted commercially and used in a large number of products. Sargachromanol G, found in brown algae, may be anti-inflammatory.

Diatom

Diatoms are a major group of algae, and are among the most common types of phytoplankton. Diatoms are unicellular, although they can form colonies in the shape of filaments or ribbons (e.g.

Fragilaria), fans (e.g. *Meridion*), zigzags (e.g. *Tabellaria*), or stars (e.g. *Asterionella*). The first diatom formally described in scientific literature, the colonial *Bacillaria paradoxa*, was found in 1783 by Danish naturalist Otto Friedrich Müller. Diatoms are producers within the food chain. A unique feature of diatom cells is that they are enclosed within a cell wall made of silica (hydrated silicon dioxide) called a frustule. These frustules show a wide diversity in form, but are usually almost bilaterally symmetrical, hence the group name. The symmetry is not perfect since one of the valves is slightly larger than the other, allowing one valve to fit inside the edge of the other. Fossil evidence suggests that they originated during, or before, the early Jurassic period. Only male gametes of centric diatoms are capable of movement by means of flagella. Diatom communities are a popular tool for monitoring environmental conditions, past and present, and are commonly used in studies of water quality.

General Biology

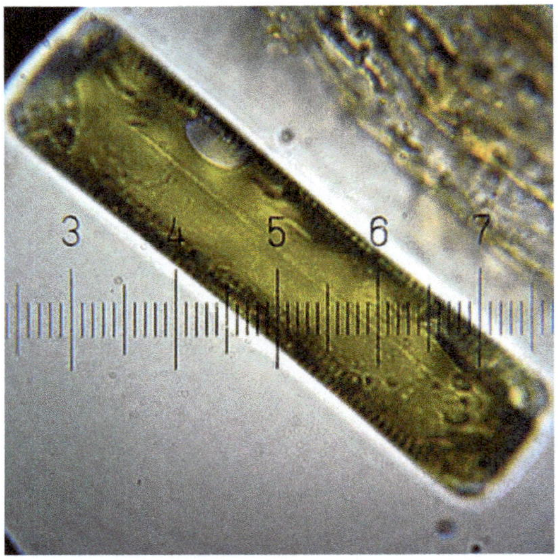

A diatom. Numbered graduations are 10 micrometres apart

More than 200 genera of living diatoms are known, with an estimated 100,000 extant species. Diatoms are a widespread group and can be found in the oceans, in fresh water, in soils, and on damp surfaces. They dominate over phytoplankton in nutrient-rich coastal waters and during oceanic spring blooms since they can divide more rapidly than phytoplankton. Most live pelagically in open water, although some live as surface films at the water-sediment interface (benthic), or even under damp atmospheric conditions. They are especially important in oceans, where they contribute an estimated 45% of the total oceanic primary production of organic material. Spatial distribution of marine phytoplankton species is restricted both horizontally and vertically. Though most are microscopic, some species of diatoms can reach up to 2 mm in length.

Diatoms belong to a large group called the heterokonts, including both autotrophs (*e.g.*, golden algae, kelp) and heterotrophs (*e.g.*, water moulds). Their yellowish-brown chloroplasts are typical of heterokonts, having four membranes and containing pigments such as the carotenoid fucoxanthin. Individuals usually lack flagella, but they are present in male gametes of the centric diatoms and have the usual heterokont structure, except they lack the hairs (mastigonemes) characteristic in other groups. Most diatoms are nonmotile, as their relatively dense cell walls cause them to readily sink. Planktonic forms in open water usually rely on turbulent mixing of the upper layers by the wind to keep them suspended in sunlit surface waters. Some species actively regulate their buoyancy with intracellular lipids to counter sinking.

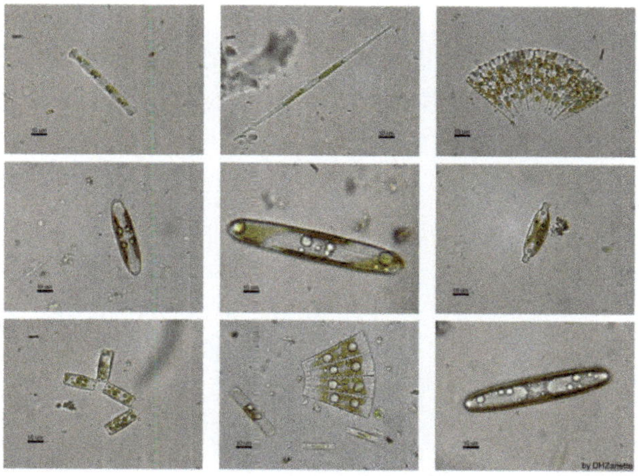

Several species of freshwater diatoms

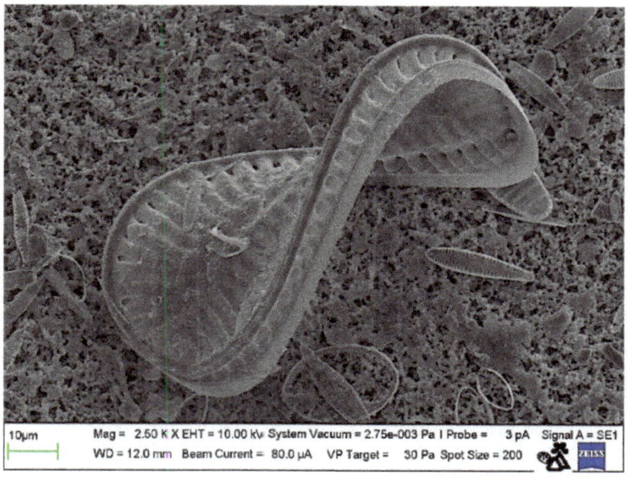

Scanning electron microscope image of the benthic diatom *Surirella spiralis*

A feature of diatoms is the urea cycle, which links them evolutionarily to animals. This was discovered in research carried out by Andrew Allen, Chris Bowler and colleagues. Their findings, published in 2011, that diatoms have a functioning urea cycle was highly significant, since prior to this, the urea cycle was thought to have originated with the metazoans which appeared several hundreds of millions of years after the diatoms. Their study showed that while diatoms and animals use the urea cycle for different ends, they are seen to be evolutionally linked in such a way that animals and plants are not.

Diatom cells are contained within a unique silica cell wall known as a frustule made up of two valves called thecae, that typically overlap one another. The biogenic silica composing the cell wall is synthesised intracellularly by the polymerisation of silicic acid monomers. This material is then extruded to the cell exterior and added to the wall. In most species, when a diatom divides to produce two daughter cells, each cell keeps one of the two halves and grows a smaller half within it. As a result, after each division cycle, the average size of diatom cells in the population gets smaller.

Once such cells reach a certain minimum size, rather than simply divide, they reverse this decline by forming an auxospore. This expands in size to give rise to a much larger cell, which then returns to size-diminishing divisions. Auxospore production is almost always linked to meiosis and sexual reproduction.

Decomposition and decay of diatoms leads to organic and inorganic (in the form of silicates) sediment, the inorganic component of which can lead to a method of analyzing past marine environments by corings of ocean floors or bay muds, since the inorganic matter is embedded in deposition of clays and silts and forms a permanent geological record of such marine strata.

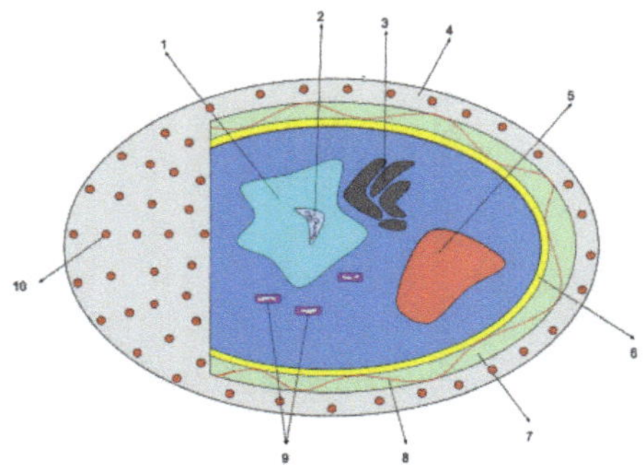

A single diatom at first look may not seem to be a cell, but diatoms are eukaryotic organisms with common organelles such as a nucleus, mitochondria, and golgi complex.

1) Nucleus; holds the genetic material

2) Nucleolus; Location of the chromosomes

3) Golgi complex; modifies proteins and sends them out of the cell

4) Cell Wall; Outer membrane of the cell

5) Pyrenoid; center of carbon fixation

6) Chromatophore; pigment carrying membrane structure

7) Vacuoles; vesicle of a cell that contains fluid bound by a membrane

8) Cytoplasmic strands; hold the nucleus

9) Mitochondria; creates ATP (energy) for the cell

10) Valve/Striae; allows nutrients and waste in and out of the cell

The study of diatoms is a branch of phycology, and phycologists specializing in diatoms are called diatomists.

Classification

Diatomaceous earth as viewed under bright field illumination on a light microscope. Diatomaceous earth is a soft, siliceous, sedimentary rock made up of the cell walls of diatoms and readily crumbles to a fine powder. This sample consists of a mixture of centric (radially symmetric) and pennate (bilaterally symmetric) diatoms. This image of diatomaceous earth particles in water is at a scale of 6.236 pixels/µm, the entire image covers a region of approximately 1.13 by 0.69 mm.

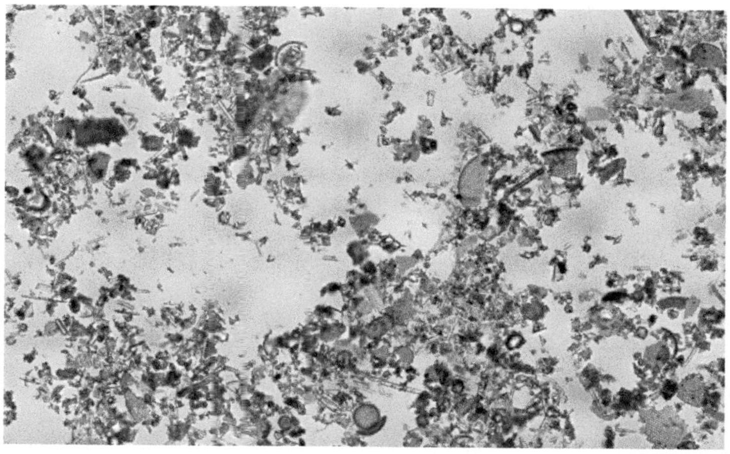

The classification of heterokonts is still unsettled, and they may be treated as a division (or phylum), kingdom, or something in-between. Accordingly, groups like the diatoms may be ranked anywhere from class (usually called Diatomophyceae or Bacillariophyceae) to division (usually called Bacillariophyta), with corresponding changes in the ranks of their subgroups.

Diatoms are traditionally divided into two main groups that are primarily distinguished by frustule structures: the centrics and the pennates. Centric diatoms are radially symmetrically and are composed of upper and lower valves (*epitheca* and *hypotheca*); each consisting of a valve and a girdle band that can easily slide underneath each other and expand to increase cell content over the diatoms progression. The Centric diatom's cytoplasm is located along the inner surface of the shell and provides a hollow lining around the large vacuole located in the center of the cell. This large, central vacuole is filled by a fluid known as "Cell sap" which is similar to seawater but varies with specific ion content. The cytoplasmic layer is home to several organelles, like the chloroplasts and mitochondria. Before the Centric begins to expand, its nucleus can be found at the center of one of the valves and will begin to move towards the center of the cytoplasmic layer before division is complete. This type of diatom can be found with a variety of shapes and sizes which heavily depends on how which axis the shell extends from and if spines are added to the Centrics. Unlike Centric diatoms, Pennate diatoms are bilaterally not radially symmetric. Each one of their valves have openings that are slits along the raphes and their shells are typically elongated parallel to these raphes. They generate cell movement through cytoplasm that streams along the raphes, always moving along solid surfaces.

Due to the difference noted in pennate diatoms, of the presence or absence of a raphe (a longitudinal groove in the valve), a more recent classification by Round, Crawford & Mann (1990) divides the diatoms (as Bacillarophyta) into three classes and a number of orders:

- Class Coscinodiscophyceae Round & R.M.Crawford: centric diatoms
    - Anaulales Round & R.M.Crawford
    - Arachnoidiscales Round

- Asterolamprales Round
- Aulacoseirales R.M.Crawford
- Biddulphiales
- Chaetocerotales Round & R.M.Crawfor
- Chrysanthemodiscales Round
- Corethrales Round & R.M.Crawford
- Coscinodiscales Round
- Cymatosirales Round & R.M.Crawford
- Ethmodiscales Round
- Hemiaulales Round & R.M.Crawford
- Leptocylindrales Round & R.M.Crawford
- Lithodesmiales
- Melosirales R.M.Crawford
- Orthoseirales R.M.Crawford
- Paraliales R.M.Crawford
- Rhizosoleniales
- Stictocyclales Round
- Stictodiscales Round & R.M.Crawford
- Thalassiosirales
- Triceratiales Round & R.M.Crawford

- Class Fragilariophyceae F.E.Round: pennate diatoms without a raphe (araphids)
    - Ardissoneales F.E.Round
    - Climacospheniales Round
    - Cyclophorales Round & R.M.Crawford
    - Fragilariales P.C.Silva
    - Licmophorales Round
    - Protoraphidales Round
    - Rhabdonematales Round & R.M.Crawford

- Rhaphoneidales Round
- Striatellales F.E.Round
- Tabellariales Round
- Thalassionematales Round
- Toxariales Round

- Class Bacillariophyceae Haeckel, 1878, *emend.* D.G.Mann: pennate diatoms with a raphe (raphids)
    - Achnanthales P.C.Silva
    - Bacillariales Hendey
    - Cymbellales D.G.Mann
    - Dictyoneidales D.G.Mann
    - Eunotiales P.C.Silva
    - Lyrellales D.G.Mann
    - Mastogloiales D.G.Mann
    - Naviculales Bessey
    - Rhopalodiales D.G.Mann
    - Surirellales D.G.Mann
    - Thalassiophysales D.G.Mann

It is probable there will be further revisions as understanding of their relationships increases. Medlin & Kaczmarska (2004) propose the following classification for the diatoms:

- Bacillaryophyta
    - Coscinodiscophytina
        - Coscinodiscophyceae ('radial centrics')
    - Bacillariophytina
        - Mediophyceae ('polar centrics')
        - Bacillariophyceae (pennate diatoms)

Diatoms generally range in size from 2 to 200μm, and build intricate hard but porous cell walls (called frustules or tests) composed primarily of silica. This siliceous wall can be highly patterned with a variety of pores, ribs, minute spines, marginal ridges and elevations; all of which can be used to delineate genera and species. The cell itself consists of two halves, each containing an essentially flat plate, or valve and marginal connecting, or girdle band. One half, the hypotheca, is slightly smaller than the other half, the epitheca. Diatom morphology varies. Although the shape of the cell is typically circular, some cells may be triangular, square, or elliptical. Their distinguishing feature is a hard mineral shell or frustule composed of opal (hydrated, polymerized silicic acid).

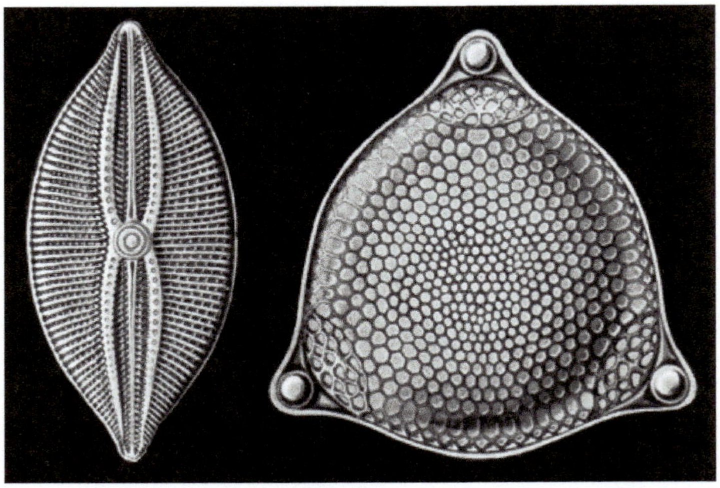

Selections from Ernst Haeckel's 1904 *Kunstformen der Natur* (Art Forms of Nature), showing pennate (left) and centric (right) frustules.

Cells are solitary or united into colonies of various kinds, which may be linked by siliceous structures; mucilage pads, stalks or tubes; amorphous masses of mucilage, or by threads of chitin, (polysaccharide) which are secreted through strutted processes of the cell. Major pigments of diatoms are chlorophylls a and c, beta-carotene, fucoxanthin, diatoxanthin and diadinoxanthin. Diatoms are mainly photosynthetic. A few, however, are obligate heterotrophs, while others can live heterotrophically in the absence of light, provided an appropriate organic carbon source is available. Storage products are chrysolaminarin and lipids.

Hoek *et al.* (1995) also provide a comprehensive coverage of diatom taxonomy.

Ecology

Planktonic diatoms in freshwater and marine environments typically exhibit a "boom and bust" (or "bloom and bust") lifestyle. When conditions in the upper mixed layer (nutrients and light) are favourable (as at the spring), their competitive edge and rapid growth rate enables them to dominate phytoplankton communities ("boom" or "bloom"). As such they are often classed as opportunistic r-strategists (*i.e.* those organisms whose ecology is defined by a high growth rate, *r*).

The freshwater diatom *Didymosphenia geminata*, commonly known as *Didymo* causes severe environmental degradation in water-courses where it blooms, producing large quantities of a brown jelly-like material called "brown snot" or "rock snot". This diatom is native to Europe and is an invasive species both in the antipodes and in parts of North America. The problem is most frequently recorded from Australia and New Zealand.

When conditions turn unfavourable, usually upon depletion of nutrients, diatom cells typically increase in sinking rate and exit the upper mixed layer ("bust"). This sinking is induced by either a loss of buoyancy control, the synthesis of mucilage that sticks diatoms cells together, or the production of heavy resting spores. Sinking out of the upper mixed layer removes diatoms from conditions unfavourable to growth, including grazer populations and higher temperatures (which would otherwise increase cell metabolism). Cells reaching deeper water or the shallow seafloor can then rest until conditions become more favourable again. In the open ocean, many sinking cells are lost to the deep, but refuge populations can persist near the thermocline.

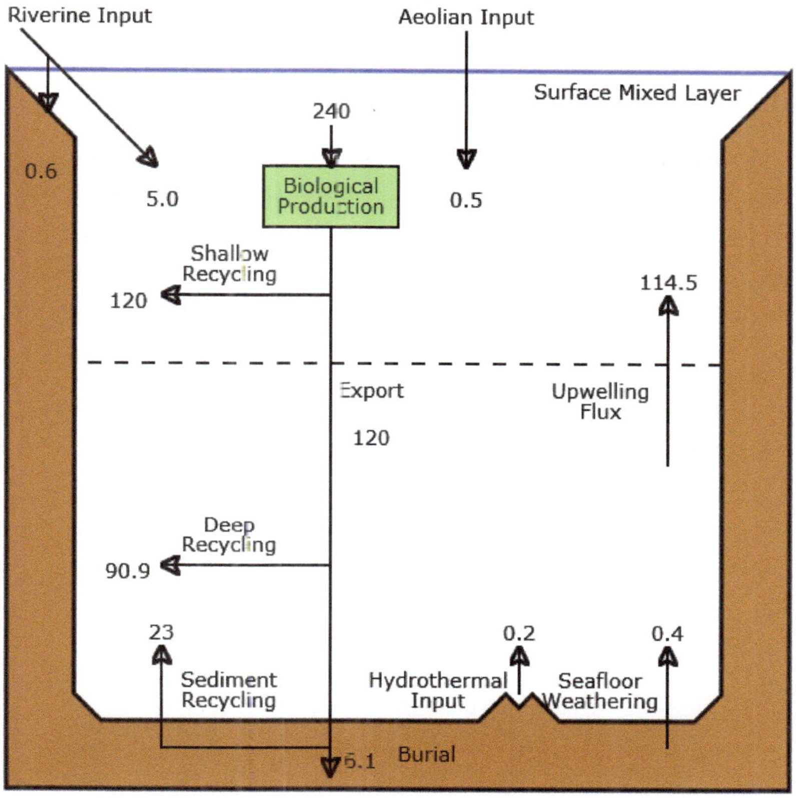

A budget of the ocean's silicon cycle

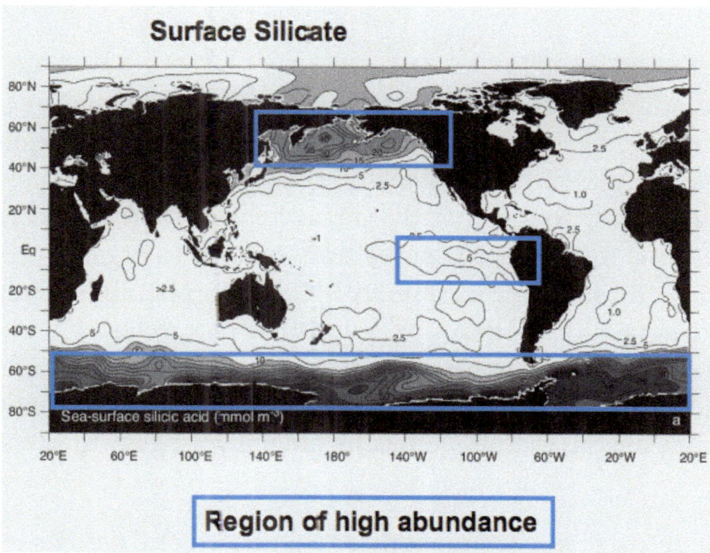

Regions of high abundance of Diatoms in the ocean

Ultimately, diatom cells in these resting populations re-enter the upper mixed layer when vertical mixing entrains them. In most circumstances, this mixing also replenishes nutrients in the upper mixed layer, setting the scene for the next round of diatom blooms. In the open ocean (away from areas of continuous upwelling), this cycle of bloom, bust, then return to pre-bloom conditions typically occurs over an annual cycle, with diatoms only being prevalent during the spring and early summer. In some locations, however, an autumn bloom may occur, caused by the breakdown of summer stratification and the entrainment of nutrients while light levels are still sufficient for growth. Since vertical mixing is increasing, and light levels are falling as winter approaches, these blooms are smaller and shorter-lived than their spring equivalents.

In the open ocean, the diatom (spring) bloom is typically ended by a shortage of silicon. Unlike other minerals, the requirement for silicon is unique to diatoms and it is not regenerated in the plankton ecosystem as efficiently as, for instance, nitrogen or phosphorus nutrients. This can be seen in maps of surface nutrient concentrations – as nutrients decline along gradients, silicon is usually the first to be exhausted (followed normally by nitrogen then phosphorus).

Because of this bloom-and-bust cycle, diatoms are believed to play a disproportionately important role in the export of carbon from oceanic surface waters. Significantly, they also play a key role in the regulation of the biogeochemical cycle of silicon in the modern ocean.

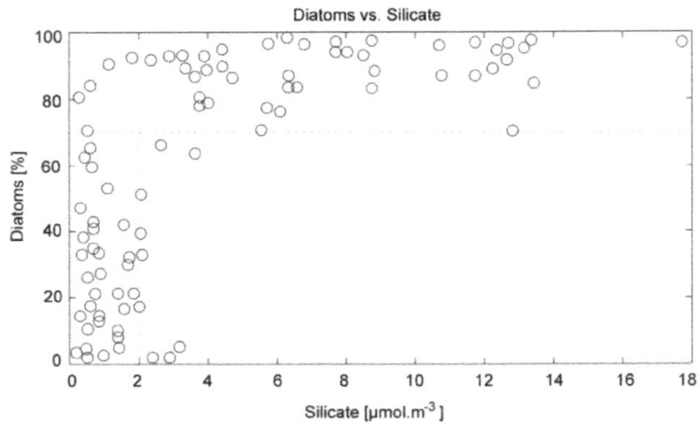

Egge & Aksnes (1992) figure.

The use of silicon by diatoms is believed by many researchers to be the key to their ecological success. In a now classic study, Egge and Aksnes (1992) found that diatom dominance of mesocosm communities was directly related to the availability of silicic acid — when concentrations were greater than 2 μmol m^{-3}, they found that diatoms typically represented more than 70% of the phytoplankton community. Raven (1983) noted that, relative to organic cell walls, silica frustules require less energy to synthesize (approximately 8% of a comparable organic wall), potentially a significant saving on the overall cell energy budget. Other researchers have suggested that the biogenic silica in diatom cell walls acts as an effective pH buffering agent, facilitating the conversion of bicarbonate to dissolved CO_2 (which is more readily assimilated). Notwithstanding the possible advantages conferred by silicon, diatoms typically have higher growth rates than other algae of a corresponding size.

Diatoms occur in virtually every environment that contains water. Diatoms inhabit not only oceans, seas, lakes, and streams, but also soil and wetlands.

Life-cycle

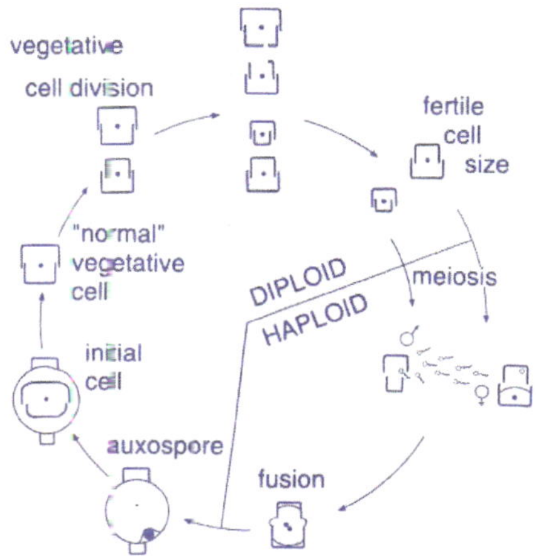

Sexual reproduction of a centric diatom (oogamy)

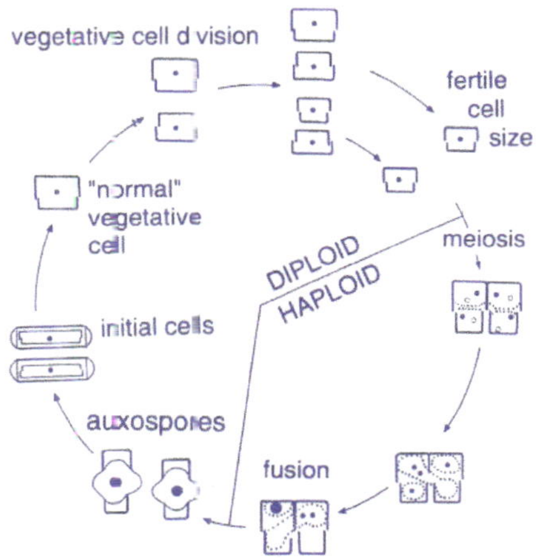

Sexual reproduction of a pinnate diatom (morphological isogamy, physiological anisogamy)

Reproduction among these organisms is primarily asexual by binary fission, with each daughter cell receiving one of the parent cell's two frustules (or theca). This is used by each daughter cell as the larger frustule (or epitheca) into which a second, small frustule (or hypotheca) is constructed.

This form of division results in a size reduction of the daughter cell that received the smaller frustule from the parent and therefore the average cell size of a diatom population decreases, until the cells are about one-third their maximum size. It has been observed, however, that certain taxa have the ability to divide without causing a reduction in cell size. Nonetheless, in order to restore the cell size of a diatom population for those that do endure size reduction, sexual reproduction and auxospore formation must occur. Vegetative cells of diatoms are diploid (2N) and so meiosis can take place, producing male and female gametes which then fuse to form the zygote. The zygote sheds its silica theca and grows into a large sphere covered by an organic membrane, the auxospore. A new diatom cell of maximum size, the initial cell, forms within the auxospore thus beginning a new generation. Resting spores may also be formed as a response to unfavourable environmental conditions with germination occurring when conditions improve.

Diatoms are mostly non-motile; however, sperm found in some species can be flagellated, though motility is usually limited to a gliding motion. In centric diatoms, the small male gametes have one flagellum while the female gametes are large and non-motile (oogamous). Conversely, in pennate diatoms both gametes lack flagella (isoogamous). Certain araphid species, that is pennate diatoms without a raphe, have been documented as anisogamous and are, therefore, considered to represent a transitional stage between centric and raphid pennate diatoms, diatoms with a raphe.

Evolutionary History

Heterokont chloroplasts appear to derive from those of red algae, rather than directly from prokaryotes as occurred in plants. This suggests they had a more recent origin than many other algae. However, fossil evidence is scant, and only with the evolution of the diatoms themselves do the heterokonts make a serious impression on the fossil record.

The earliest known fossil diatoms date from the early Jurassic (~185 Ma ago), although the molecular clock and sedimentary evidence suggests an earlier origin. It has been suggested that their origin may be related to the end-Permian mass extinction (~250 Ma), after which many marine niches were opened. The gap between this event and the time that fossil diatoms first appear may indicate a period when diatoms were unsilicified and their evolution was cryptic. Since the advent of silicification, diatoms have made a significant impression on the fossil record, with major deposits of fossil diatoms found as far back as the early Cretaceous, and some rocks (diatomaceous earth, diatomite, kieselguhr) being composed almost entirely of them.

Although the diatoms may have existed since the Triassic, the timing of their ascendancy and "take-over" of the silicon cycle occurred more recently. Prior to the Phanerozoic (before 544 Ma), it is believed that microbial or inorganic processes weakly regulated the ocean's silicon cycle. Subsequently, the cycle appears dominated (and more strongly regulated) by the radiolarians and siliceous sponges, the former as zooplankton, the latter as sedentary filter-feeders primarily on the continental shelves. Within the last 100 My, it is thought that the silicon cycle has come under even tighter control, and that this derives from the ecological ascendancy of the diatoms.

However, the precise timing of the "take-over" remains unclear, and different authors have conflicting interpretations of the fossil record. Some evidence, such as the displacement of siliceous sponges from the shelves, suggests that this takeover began in the Cretaceous (146 Ma to 65 Ma), while evidence from radiolarians suggests "take-over" did not begin until the Cenozoic (65 Ma to present). The expansion of grassland biomes and the evolutionary radiation of grasses during the Miocene is believed to have increased the flux of soluble silicon to the oceans, and it has been argued that this promoted the diatoms during the Cenozoic era. Recent work suggests that diatom success is decoupled from the evolution of grasses, although both diatom and grassland diversity increased strongly from the middle Miocene. Diatom diversity over the Cenozoic has been very sensitive to global temperature, particularly to the equator-pole temperature gradient. Warmer oceans, particularly warmer polar regions, have in the past been shown to have had substantially lower diatom diversity. Future warm oceans with enhanced polar warming, as projected in global-warming scenarios, could thus in theory result in a significant loss of diatom diversity, although from current knowledge it is impossible to say if this would occur rapidly or only over many tens of thousands of years.

Fossil Record

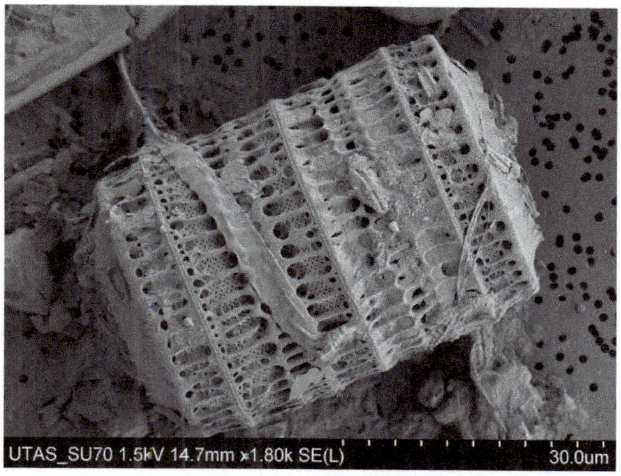

Paralia sulcata diatom imaged using the University of Tasmania scanning electron microscope.

The fossil record of diatoms has largely been established through the recovery of their siliceous frustules in marine and non-marine sediments. Although diatoms have both a marine and non-marine stratigraphic record, diatom biostratigraphy, which is based on time-constrained evolutionary originations and extinctions of unique taxa, is only well developed and widely applicable in marine systems. The duration of diatom species ranges have been documented through the study of ocean cores and rock sequences exposed on land. Where diatom biozones are well established and calibrated to the geomagnetic polarity time scale (e.g., Southern Ocean, North Pacific, eastern equatorial Pacific), diatom-based age estimates may be resolved to within <100,000 years, although typical age resolution for Cenozoic diatom assemblages is several hundred thousand years.

The Cretaceous record of diatoms is limited, but recent studies reveal a progressive diversification of diatom types. The Cretaceous–Paleogene extinction event, which in the oceans dramatically affected organisms with calcareous skeletons, appears to have had relatively little impact on diatom evolution.

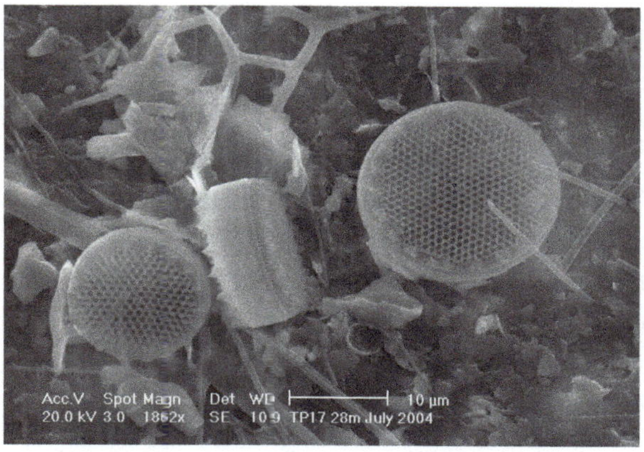

Diatoms *Chaetoceros* spp. on a membrane filter (pore size 0.4 μm) imaged using the Hellenic Centre for Marine Research scanning electron microscope. Thermaikos Gulf, Greece, 2004.

Although no mass extinctions of marine diatoms have been observed during the Cenozoic, times of relatively rapid evolutionary turnover in marine diatom assemblages occurred near the Paleocene–Eocene boundary and at the Eocene–Oligocene boundary. Further turnover of assemblages took place at various times between the middle Miocene and late Pliocene, in response to progressive cooling of polar regions and the development of more endemic diatom assemblages. A global trend toward more delicate diatom frustules has been noted from the Oligocene to the Quaternary. This coincides with an increasingly more vigorous circulation of the ocean's surface and deep waters brought about by increasing latitudinal thermal gradients at the onset of major ice sheet expansion on Antarctica and progressive cooling through the Neogene and Quaternary towards a bipolar glaciated world. This drove the diatoms into uptaking silica more competitively (i.e., to use less silica in formation of their frustules). Increased mixing of the oceans renews silica and other nutrients necessary for diatom growth in surface waters, especially in regions of coastal and oceanic upwelling.

Collection

Living diatoms are often found clinging in great numbers to filamentous algae, or forming gelatinous masses on various submerged plants. *Cladophora* is frequently covered with *Cocconeis*, an elliptically shaped diatom; *Vaucheria* is often covered with small forms. Diatoms are frequently present as a brown, slippery coating on submerged stones and sticks, and may be seen to "stream" with river current.

The surface mud of a pond, ditch, or lagoon will almost always yield some diatoms. They can be made to emerge by filling a jar with water and mud, wrapping it in black paper and letting direct sunlight fall on the surface of the water. Within a day, the diatoms will come to the top in a scum and can be isolated.

Since diatoms form an important part of the food of molluscs, tunicates, and fishes, the alimentary tracts of these animals often yield forms that are not easily secured in other ways. Marine diatoms can be collected by direct water sampling, though benthic forms can be secured by scraping barnacles, oyster shells, and other shells.

Expressed Sequence Tagging

The first insights into the properties of the *Phaeodactylum tricornutum* gene repertoire were described using 1,000 expressed sequence tags (ESTs). Subsequently, the number of ESTs was extended to 12,000 and the diatom EST database was constructed for functional analyses. These sequences have been used to make a comparative analysis between *P. tricornutum* and the putative complete proteomes from the green alga *Chlamydomonas reinhardtii*, the red alga *Cyanidioschyzon merolae*, and the diatom *Thalassiosira pseudonana*. The diatom EST database now consists of over 200,000 ESTs from *P. tricornutum* (16 libraries) and *T. pseudonana* (7 libraries) cells grown in a range of different conditions, many of which correspond to different abiotic stresses .

Genome Sequencing

The entire genomes of the centric diatom, *Thalassiosira pseudonana* (32.4 Mb), and the pennate diatom, *Phaeodactylum tricornutum* (27.4 Mb), have been sequenced. Comparisons of the two fully sequenced diatom genomes finds that the *P. tricornutum* genome includes fewer genes (10,402 opposed to 11,776) than *T. pseudonana* and no major synteny (gene order) could be detected between the two genomes. *T. pseudonana* genes show an average of ~1.52 introns per gene as opposed to 0.79 in *P. tricornutum*, suggesting recent widespread intron gain in the centric diatom. Despite relatively recent evolutionary divergence (90 million years), the extent of molecular divergence between centrics and pennates indicates rapid evolutionary rates within the Bacillariophyceae compared to other eukaryotic groups. Comparative genomics also established that a specific class of transposable elements, the Diatom Copia-like retrotransposons (or CoDis), has been significantly amplified in the *P. tricornutum* genome with respect to *T. pseudonana*, constituting 5.8 and 1% of the respective genomes.

Importantly, diatom genomics brought much information about the extent and dynamics of the endosymbiotic gene transfer (EGT) process. Comparison of the *T. pseudonana* proteins with homologs in other organisms suggested that hundreds have their closest homologs in the Plantae lineage. EGT towards diatom genomes can be illustrated by the fact that the *T. pseudonana* genome encodes six proteins which are most closely related to genes encoded by the *Guillardia theta* (cryptomonad) nucleomorph genome. Four of these genes are also found in red algal plastid genomes, thus demonstrating successive EGT from red algal plastid to red algal nucleus (nucleomorph) to heterokont host nucleus. More recent phylogenomic analyses of diatom proteomes provided evidence for a prasinophyte-like endosymbiont in the common ancestor of chromalveolates as supported by the fact the 70% of diatom genes of Plantae origin are of green lineage provenance and that such genes are also found in the genome of other stramenopiles. Therefore, it was proposed that chromalveolates are the product of serial secondary endosymbiosis first with a green algae, followed by a second one with a red algae that conserved the genomic footprints of the previous but displaced the green plastid. However, phylogenomic analyses of diatom proteomes and chromalveolate evolutionary history will likely take advantage of complementary genomic data from under-sequenced lineages such as red algae.

In addition to EGT, horizontal gene transfer (HGT) can occur independently of an endosymbiotic event. The publication of the *P. tricornutum* genome reported that at least 587 *P. tricornutum* genes appear to be most closely related to bacterial genes, accounting for more than 5% of the *P. tricornutum* proteome. About half of these are also found in the *T. pseudonana* genome, attesting their ancient incorporation in the diatom lineage.

Photonic Properties

Diatoms are oftern referred as "jewels of the sea" or "living opals" thanks to their photonic crystal properties. It is not clear what is the biological significant of such structural coloration but it is speculated that it varies from communication, camouflage, and thermal exchange to UV protection.

Forensic Research

The main goal of diatom analysis in forensics is to differentiate a death by submersion from a post-mortem immersion of a body in water. Laboratory tests may reveal the presence of diatoms in the body. Since the silica-based skeletons of diatoms do not readily decay, they can sometimes be detected even in heavily decomposed bodies. As they do not occur naturally in the body, if laboratory tests show diatoms in the corpse that are of the same species found in the water where the body was recovered, then it may be good evidence of drowning as the cause of death. The blend of diatom species found in a corpse may be the same or different from the surrounding water, indicating whether the victim drowned in the same site in which the body was found.

Nanotechnology Research

The deposition of silica by diatoms may also prove to be of utility to nanotechnology. Diatom cells repeatedly and reliably manufacture valves of various shapes and sizes, potentially allowing diatoms to manufacture micro- or nano-scale structures which may be of use in a range of devices, including: optical systems; semiconductor nanolithography; and even vehicles for drug delivery. With an appropriate artificial selection procedure, diatoms that produce valves of particular shapes and sizes might be evolved for cultivation in chemostat cultures to mass-produce nanoscale components. It has also been proposed that diatoms could be used as a component of solar cells by substituting photosensitive titanium dioxide for the silicon dioxide that diatoms normally use to create their cell walls. Diatom biofuel producing solar panels have also been proposed.

Microbial Degradation

Certain species of bacteria in oceans and lakes can accelerate the rate of dissolution of silica in dead and living diatoms by using hydrolytic enzymes to break down the organic algal material.

Fungus

A fungus is any member of the group of eukaryotic organisms that includes microorganisms such as yeasts and molds, as well as the more familiar mushrooms. These organisms are classified as a kingdom, Fungi, which is separate from the other eukaryotic life kingdoms of plants and animals.

A characteristic that places fungi in a different kingdom from plants, bacteria and some protists, is chitin in their cell walls. Similar to animals, fungi are heterotrophs; they acquire their food by absorbing dissolved molecules, typically by secreting digestive enzymes into their environment. Fungi do not photosynthesise. Growth is their means of mobility, except for spores (a few of which are flagellated), which may travel through the air or water. Fungi are the principal decomposers in ecological systems. These and other differences place fungi in a single group of related organisms, named the *Eumycota* (*true fungi* or *Eumycetes*), which share a common ancestor (form a *monophyletic group*), an interpretation that is also strongly supported by molecular phylogenetics. This fungal group is distinct from the structurally similar myxomycetes (slime molds) and oomycetes (water molds). The discipline of biology devoted to the study of fungi is known as mycology (from the Greek μύκης, mukēs, meaning "fungus"). In the past, mycology was regarded as a branch of botany, although it is now known fungi are genetically more closely related to animals than to plants.

Abundant worldwide, most fungi are inconspicuous because of the small size of their structures, and their cryptic lifestyles in soil or on dead matter. Fungi include symbionts of plants, animals, or other fungi and also parasites. They may become noticeable when fruiting, either as mushrooms or as molds. Fungi perform an essential role in the decomposition of organic matter and have fundamental roles in nutrient cycling and exchange in the environment. They have long been used as a direct source of human food, in the form of mushrooms and truffles; as a leavening agent for bread; and in the fermentation of various food products, such as wine, beer, and soy sauce. Since the 1940s, fungi have been used for the production of antibiotics, and, more recently, various enzymes produced by fungi are used industrially and in detergents. Fungi are also used as biological pesticides to control weeds, plant diseases and insect pests. Many species produce bioactive compounds called mycotoxins, such as alkaloids and polyketides, that are toxic to animals including humans. The fruiting structures of a few species contain psychotropic compounds and are consumed recreationally or in traditional spiritual ceremonies. Fungi can break down manufactured materials and buildings, and become significant pathogens of humans and other animals. Losses of crops due to fungal diseases (e.g., rice blast disease) or food spoilage can have a large impact on human food supplies and local economies.

The fungus kingdom encompasses an enormous diversity of taxa with varied ecologies, life cycle strategies, and morphologies ranging from unicellular aquatic chytrids to large mushrooms. However, little is known of the true biodiversity of Kingdom Fungi, which has been estimated at 1.5 million to 5 million species, with about 5% of these having been formally classified. Ever since the pioneering 18th and 19th century taxonomical works of Carl Linnaeus, Christian Hendrik Persoon, and Elias Magnus Fries, fungi have been classified according to their morphology (e.g., characteristics such as spore color or microscopic features) or physiology. Advances in molecular genetics have opened the way for DNA analysis to be incorporated into taxonomy, which has sometimes challenged the historical groupings based on morphology and other traits. Phylogenetic studies published in the last decade have helped reshape the classification within Kingdom Fungi, which is divided into one subkingdom, seven phyla, and ten subphyla.

Etymology

The English word *fungus* is directly adopted from the Latin *fungus* (mushroom), used in the writings of Horace and Pliny. This in turn is derived from the Greek word *sphongos* ("sponge"), which refers to the macroscopic structures and morphology of mushrooms and molds; the root is also used in other languages, such as the German *Schwamm* ("sponge") and *Schimmel* ("mold"). The use of the word *mycology*, which is derived from the Greek *mykes* ("mushroom") and *logos* ("discourse"), to denote the scientific study of fungi is thought to have originated in 1836 with English naturalist Miles Joseph Berkeley's publication *The English Flora of Sir James Edward Smith, Vol. 5*. A group of all the fungi present in a particular area or geographic region is known as mycobiota (plural noun, no singular), e.g., "the mycobiota of Ireland".

Characteristics

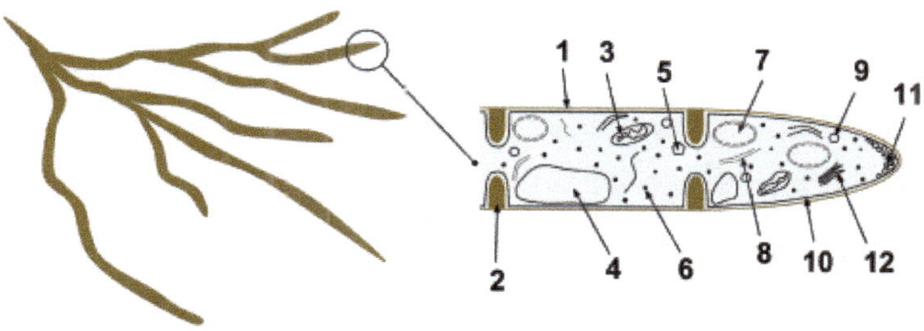

Fungal hyphae cells

Fungal hyphae cells:

1. Hyphal wall
2. Septum
3. Mitochondrion
4. Vacuole
5. Ergosterol crystal
6. Ribosome
7. Nucleus
8. Endoplasmic reticulum
9. Lipid body
10. Plasma membrane
11. Spitzenkörper
12. Golgi apparatus

Before the introduction of molecular methods for phylogenetic analysis, taxonomists considered fungi to be members of the plant kingdom because of similarities in lifestyle: both fungi and plants are mainly immobile, and have similarities in general morphology and growth habitat. Like plants, fungi often grow in soil and, in the case of mushrooms, form conspicuous fruit bodies, which sometimes resemble plants such as mosses. The fungi are now considered a separate kingdom, distinct from both plants and animals, from which they appear to have diverged around one billion years ago. Some morphological, biochemical, and genetic features are shared with other organisms, while others are unique to the fungi, clearly separating them from the other kingdoms:

Shared features:

- With other eukaryotes: Fungal cells contain membrane-bound nuclei with chromosomes that contain DNA with noncoding regions called introns and coding regions called exons. Fungi have membrane-bound cytoplasmic organelles such as mitochondria, sterol-containing membranes, and ribosomes of the 80S type. They have a characteristic range of soluble carbohydrates and storage compounds, including sugar alcohols (e.g., mannitol), disaccharides, (e.g., trehalose), and polysaccharides (e.g., glycogen, which is also found in animals).
- With animals: Fungi lack chloroplasts and are heterotrophic organisms and so require preformed organic compounds as energy sources.
- With plants: Fungi have a cell wall and vacuoles. They reproduce by both sexual and asexual means, and like basal plant groups (such as ferns and mosses) produce spores. Similar to mosses and algae, fungi typically have haploid nuclei.

- With euglenoids and bacteria: Higher fungi, euglenoids, and some bacteria produce the amino acid L-lysine in specific biosynthesis steps, called the α-aminoadipate pathway.
- The cells of most fungi grow as tubular, elongated, and thread-like (filamentous) structures called hyphae, which may contain multiple nuclei and extend by growing at their tips. Each tip contains a set of aggregated vesicles—cellular structures consisting of proteins, lipids, and other organic molecules—called the Spitzenkörper. Both fungi and oomycetes grow as filamentous hyphal cells. In contrast, similar-looking organisms, such as filamentous green algae, grow by repeated cell division within a chain of cells. There are also single-celled fungi (yeasts) that do not form hyphae, and some fungi have both hyphal and yeast forms.
- In common with some plant and animal species, more than 70 fungal species display bioluminescence.

Unique features:

- Some species grow as unicellular yeasts that reproduce by budding or binary fission. Dimorphic fungi can switch between a yeast phase and a hyphal phase in response to environmental conditions.
- The fungal cell wall is composed of glucans and chitin; while glucans are also found in plants and chitin in the exoskeleton of arthropods, fungi are the only organisms that combine these two structural molecules in their cell wall. Unlike those of plants and oomycetes, fungal cell walls do not contain cellulose.

Omphalotus nidiformis, a bioluminescent mushroom

Most fungi lack an efficient system for the long-distance transport of water and nutrients, such as the xylem and phloem in many plants. To overcome this limitation, some fungi, such as *Armillaria*, form rhizomorphs, which resemble and perform functions similar to the roots of plants. As eukaryotes, fungi possess a biosynthetic pathway for producing terpenes that uses mevalonic acid and pyrophosphate as chemical building blocks. Plants and some other organisms have an additional terpene biosynthesis pathway in their chloroplasts, a structure fungi and animals do not have. Fungi produce several secondary metabolites that are similar or identical in structure to those made by plants. Many of the plant and fungal enzymes that make these compounds differ from each other in sequence and other characteristics, which indicates separate origins and convergent evolution of these enzymes in the fungi and plants.

Diversity

Bracket fungi on a tree stump

Fungi have a worldwide distribution, and grow in a wide range of habitats, including extreme environments such as deserts or areas with high salt concentrations or ionizing radiation, as well as in deep sea sediments. Some can survive the intense UV and cosmic radiation encountered during space travel. Most grow in terrestrial environments, though several species live partly or solely in aquatic habitats, such as the chytrid fungus *Batrachochytrium dendrobatidis*, a parasite that has been responsible for a worldwide decline in amphibian populations. This organism spends part of its life cycle as a motile zoospore, enabling it to propel itself through water and enter its amphibian host. Other examples of aquatic fungi include those living in hydrothermal areas of the ocean.

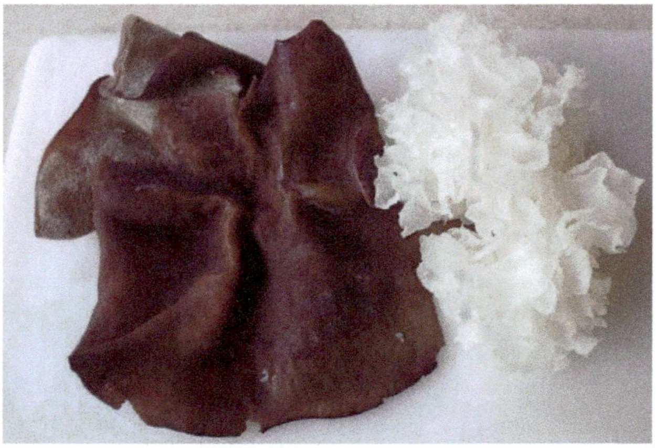

Two types of edible fungi

Around 100,000 species of fungi have been formally described by taxonomists, but the global biodiversity of the fungus kingdom is not fully understood. On the basis of observations of the ratio of the number of fungal species to the number of plant species in selected environments, the fungal kingdom has been estimated to contain about 1.5 million species. A recent (2011) estimate suggests there may be over 5 million species. In mycology, species have historically been distinguished by a variety of methods and concepts. Classification based on morphological characteristics, such as the size and shape of spores or fruiting structures, has traditionally dominated fungal taxonomy. Species may also be distinguished by their biochemical and physiological characteristics, such as their ability to metabolize certain biochemicals, or their reaction to chemical tests. The biological species concept discriminates species based on their ability to mate. The application of molecular tools, such as DNA sequencing and phylogenetic analysis, to study diversity has greatly enhanced the resolution and added robustness to estimates of genetic diversity within various taxonomic groups.

Mycology

Mycology is the branch of biology concerned with the systematic study of fungi, including their genetic and biochemical properties, their taxonomy, and their use to humans as a source of medicine, food, and psychotropic substances consumed for religious purposes, as well as their dangers, such as poisoning or infection. The field of phytopathology, the study of plant diseases, is closely related because many plant pathogens are fungi.

In 1729, Pier A. Micheli first published descriptions of fungi.

The use of fungi by humans dates back to prehistory; Ötzi the Iceman, a well-preserved mummy of a 5,300-year-old Neolithic man found frozen in the Austrian Alps, carried two species of polypore mushrooms that may have been used as tinder (*Fomes fomentarius*), or for medicinal purposes (*Piptoporus betulinus*). Ancient peoples have used fungi as food sources–often unknowingly–for millennia, in the preparation of leavened bread and fermented juices. Some of the oldest written records contain references to the destruction of crops that were probably caused by pathogenic fungi.

History

Mycology is a relatively new science that became systematic after the development of the microscope in the 17th century. Although fungal spores were first observed by Giambattista della Porta in 1588, the seminal work in the development of mycology is considered to be the publication of Pier Antonio Micheli's 1729 work *Nova plantarum genera*. Micheli not only observed spores but also showed that, under the proper conditions, they could be induced into growing into the same species of fungi from which they originated. Extending the use of the binomial system of nomenclature introduced by Carl Linnaeus in his *Species plantarum* (1753), the Dutch Christian Hendrik Persoon (1761–1836) established the first classification of mushrooms with such skill so as to be considered a founder of modern mycology. Later, Elias Magnus Fries (1794–1878) further elaborated the classification of fungi, using spore color and various microscopic characteristics, methods still used by taxonomists today. Other notable early contributors to mycology in the 17th–19th and early 20th centuries include Miles Joseph Berkeley, August Carl Joseph Corda, Anton de Bary, the brothers Louis René and Charles Tulasne, Arthur H. R. Buller, Curtis G. Lloyd, and Pier Andrea Saccardo. The 20th century has seen a modernization of mycology that has come from advances in biochemistry, genetics, molecular biology, and biotechnology. The use of DNA sequencing technologies and phylogenetic analysis has provided new insights into fungal relationships and biodiversity, and has challenged traditional morphology-based groupings in fungal taxonomy.

Morphology

Microscopic Structures

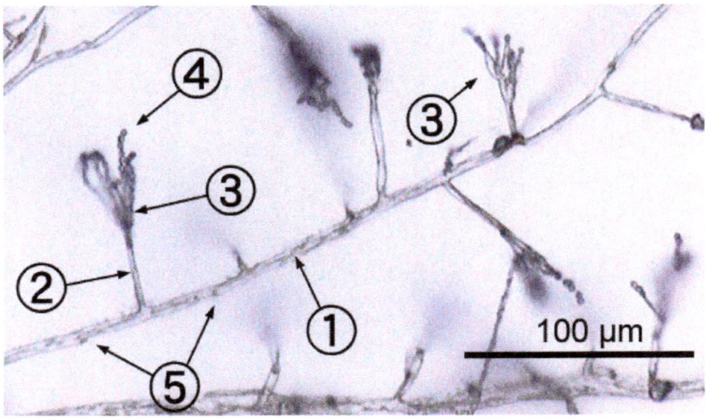

An environmental isolate of *Penicillium*

1. Hypha
2. Conidiophore
3. Phialide
4. Conidia
5. Septa

Most fungi grow as hyphae, which are cylindrical, thread-like structures 2–10 μm in diameter and up to several centimeters in length. Hyphae grow at their tips (apices); new hyphae are typically formed by emergence of new tips along existing hyphae by a process called *branching*, or occasionally growing hyphal tips fork, giving rise to two parallel-growing hyphae. Hyphae also sometimes fuse when they come into contact, a process called hyphal fusion (or anastamosis). These growth processes lead to the development of a mycelium, an interconnected network of hyphae. Hyphae can be either septate or coenocytic. Septate hyphae are divided into compartments separated by cross walls (internal cell walls, called septa, that are formed at right angles to the cell wall giving the hypha its shape), with each compartment containing one or more nuclei; coenocytic hyphae are not compartmentalized. Septa have pores that allow cytoplasm, organelles, and sometimes nuclei to pass through; an example is the dolipore septum in fungi of the phylum Basidiomycota. Coenocytic hyphae are in essence multinucleate supercells.

Many species have developed specialized hyphal structures for nutrient uptake from living hosts; examples include haustoria in plant-parasitic species of most fungal phyla, and arbuscules of several mycorrhizal fungi, which penetrate into the host cells to consume nutrients.

Although fungi are opisthokonts—a grouping of evolutionarily related organisms broadly characterized by a single posterior flagellum—all phyla except for the chytrids have lost their posterior flagella. Fungi are unusual among the eukaryotes in having a cell wall that, in addition to glucans (e.g., β-1,3-glucan) and other typical components, also contains the biopolymer chitin.

Macroscopic Structures

Armillaria solidipes

Fungal mycelia can become visible to the naked eye, for example, on various surfaces and substrates, such as damp walls and spoiled food, where they are commonly called molds. Mycelia grown on solid agar media in laboratory petri dishes are usually referred to as colonies. These colonies can exhibit growth shapes and colors (due to spores or pigmentation) that can be used as diagnostic features in the identification of species or groups. Some individual fungal colonies can reach extraordinary dimensions and ages as in the case of a clonal colony of *Armillaria solidipes*, which extends over an area of more than 900 ha (3.5 square miles), with an estimated age of nearly 9,000 years.

The apothecium—a specialized structure important in sexual reproduction in the ascomycetes—is a cup-shaped fruit body that is often macroscopic and holds the hymenium, a layer of tissue containing the spore-bearing cells. The fruit bodies of the basidiomycetes (basidiocarps) and some ascomycetes can sometimes grow very large, and many are well known as mushrooms.

Growth and Physiology

Mold growth covering a decaying peach. The frames were taken approximately 12 hours apart over a period of six days.

The growth of fungi as hyphae on or in solid substrates or as single cells in aquatic environments is adapted for the efficient extraction of nutrients, because these growth forms have high surface area to volume ratios. Hyphae are specifically adapted for growth on solid surfaces, and to invade substrates and tissues. They can exert large penetrative mechanical forces; for example, many plant pathogens, including *Magnaporthe grisea*, form a structure called an appressorium that evolved to puncture plant tissues. The pressure generated by the appressorium, directed against the plant epidermis, can exceed 8 megapascals (1,200 psi). The filamentous fungus *Paecilomyces lilacinus* uses a similar structure to penetrate the eggs of nematodes.

The mechanical pressure exerted by the appressorium is generated from physiological processes that increase intracellular turgor by producing osmolytes such as glycerol. Adaptations such as these are complemented by hydrolytic enzymes secreted into the environment to digest large organic molecules—such as polysaccharides, proteins, and lipids—into smaller molecules that may then be absorbed as nutrients. The vast majority of filamentous fungi grow in a polar fashion (extending in one direction) by elongation at the tip (apex) of the hypha. Other forms of fungal growth include intercalary extension (longitudinal expansion of hyphal compartments that are below the apex) as in the case of some endophytic fungi, or growth by volume expansion during the development of mushroom stipes and other large organs. Growth of fungi as multicellular structures consisting of somatic and reproductive cells—a feature independently evolved in animals and plants—has several functions, including the development of fruit bodies for dissemination of sexual spores and biofilms for substrate colonization and intercellular communication.

The fungi are traditionally considered heterotrophs, organisms that rely solely on carbon fixed by other organisms for metabolism. Fungi have evolved a high degree of metabolic versatility that allows them to use a diverse range of organic substrates for growth, including simple compounds such as nitrate, ammonia, acetate, or ethanol. In some species the pigment melanin may play a role in extracting energy from ionizing radiation, such as gamma radiation. This form of "radiotrophic" growth has been described for only a few species, the effects on growth rates are small, and the underlying biophysical and biochemical processes are not well known. This process might bear similarity to CO_2 fixation via visible light, but instead uses ionizing radiation as a source of energy.

Reproduction

Polyporus squamosus

Fungal reproduction is complex, reflecting the differences in lifestyles and genetic makeup within this diverse kingdom of organisms. It is estimated that a third of all fungi reproduce using more than one method of propagation; for example, reproduction may occur in two well-differentiated stages within the life cycle of a species, the teleomorph and the anamorph. Environmental conditions trigger genetically determined developmental states that lead to the creation of specialized structures for sexual or asexual reproduction. These structures aid reproduction by efficiently dispersing spores or spore-containing propagules.

Asexual Reproduction

Asexual reproduction occurs via vegetative spores (conidia) or through mycelial fragmentation. Mycelial fragmentation occurs when a fungal mycelium separates into pieces, and each component grows into a separate mycelium. Mycelial fragmentation and vegetative spores maintain clonal populations adapted to a specific niche, and allow more rapid dispersal than sexual reproduction. The “Fungi imperfecti” (fungi lacking the perfect or sexual stage) or Deuteromycota comprise all the species that lack an observable sexual cycle. Deuteromycota is not an accepted taxonomic clade, and is now taken to mean simply fungi that lack a known sexual stage.

Sexual Reproduction

Sexual reproduction with meiosis has been directly observed in all fungal phyla except Glomeromycota (genetic analysis suggests meiosis in Glomeromycota as well). It differs in many aspects from sexual reproduction in animals or plants. Differences also exist between fungal groups and can be used to discriminate species by morphological differences in sexual structures and reproductive strategies. Mating experiments between fungal isolates may identify species on the basis of biological species concepts. The major fungal groupings have initially been delineated based on the morphology of their sexual structures and spores; for example, the spore-containing structures, asci and basidia, can be used in the identification of ascomycetes and basidiomycetes, respectively. Some species may allow mating only between individuals of opposite mating type, whereas others can mate and sexually reproduce with any other individual or itself. Species of the former mating system are called heterothallic, and of the latter homothallic.

Most fungi have both a haploid and a diploid stage in their life cycles. In sexually reproducing fungi, compatible individuals may combine by fusing their hyphae together into an interconnected network; this process, anastomosis, is required for the initiation of the sexual cycle. Many ascomycetes and basidiomycetes go through a dikaryotic stage, in which the nuclei inherited from the two parents do not combine immediately after cell fusion, but remain separate in the hyphal cells.

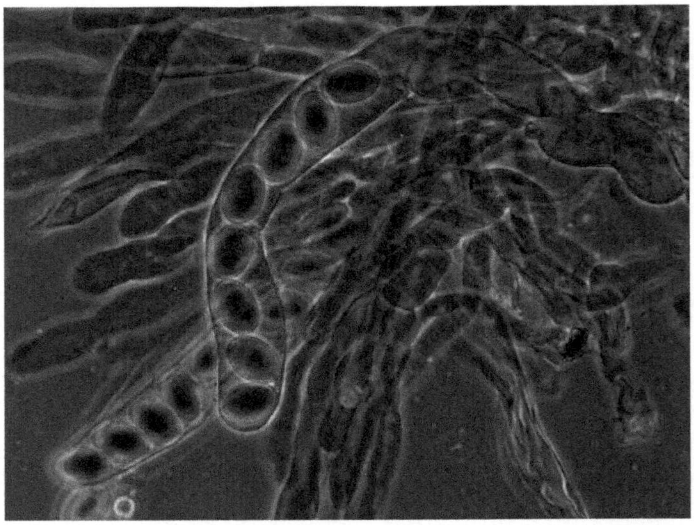

The 8-spore asci of *Morchella elata*, viewed with phase contrast microscopy

In ascomycetes, dikaryotic hyphae of the hymenium (the spore-bearing tissue layer) form a characteristic *hook* at the hyphal septum. During cell division, formation of the hook ensures proper distribution of the newly divided nuclei into the apical and basal hyphal compartments. An ascus (plural *asci*) is then formed, in which karyogamy (nuclear fusion) occurs. Asci are embedded in an ascocarp, or fruiting body. Karyogamy in the asci is followed immediately by meiosis and the production of ascospores. After dispersal, the ascospores may germinate and form a new haploid mycelium.

Sexual reproduction in basidiomycetes is similar to that of the ascomycetes. Compatible haploid hyphae fuse to produce a dikaryotic mycelium. However, the dikaryotic phase is more extensive in the basidiomycetes, often also present in the vegetatively growing mycelium. A specialized anatomical structure, called a clamp connection, is formed at each hyphal septum. As with the structurally similar hook in the ascomycetes, the clamp connection in the basidiomycetes is required for controlled transfer of nuclei during cell division, to maintain the dikaryotic stage with two genetically different nuclei in each hyphal compartment. A basidiocarp is formed in which club-like structures known as basidia generate haploid basidiospores after karyogamy and meiosis. The most commonly known basidiocarps are mushrooms, but they may also take other forms.

In glomeromycetes (formerly zygomycetes), haploid hyphae of two individuals fuse, forming a gametangium, a specialized cell structure that becomes a fertile gamete-producing cell. The gametangium develops into a zygospore, a thick-walled spore formed by the union of gametes. When the zygospore germinates, it undergoes meiosis, generating new haploid hyphae, which may then form asexual sporangiospores. These sporangiospores allow the fungus to rapidly disperse and germinate into new genetically identical haploid fungal mycelia.

Spore Dispersal

Both asexual and sexual spores or sporangiospores are often actively dispersed by forcible ejection from their reproductive structures. This ejection ensures exit of the spores from the reproductive structures as well as traveling through the air over long distances.

The bird's nest fungus *Cyathus stercoreus*

Specialized mechanical and physiological mechanisms, as well as spore surface structures (such as hydrophobins), enable efficient spore ejection. For example, the structure of the spore-bearing cells in some ascomycete species is such that the buildup of substances affecting cell volume and fluid balance enables the explosive discharge of spores into the air. The forcible discharge of single spores termed *ballistospores* involves formation of a small drop of water (Buller's drop), which upon contact with the spore leads to its projectile release with an initial acceleration of more than 10,000 g; the net result is that the spore is ejected 0.01–0.02 cm, sufficient distance for it to fall through the gills or pores into the air below. Other fungi, like the puffballs, rely on alternative mechanisms for spore release, such as external mechanical forces. The bird's nest fungi use the force of falling water drops to liberate the spores from cup-shaped fruiting bodies. Another strategy is seen in the stinkhorns, a group of fungi with lively colors and putrid odor that attract insects to disperse their spores.

Other Sexual Processes

Besides regular sexual reproduction with meiosis, certain fungi, such as those in the genera *Penicillium* and *Aspergillus*, may exchange genetic material via parasexual processes, initiated by anastomosis between hyphae and plasmogamy of fungal cells. The frequency and relative importance of parasexual events is unclear and may be lower than other sexual processes. It is known to play a role in intraspecific hybridization and is likely required for hybridization between species, which has been associated with major events in fungal evolution.

Evolution

In contrast to plants and animals, the early fossil record of the fungi is meager. Factors that likely contribute to the under-representation of fungal species among fossils include the nature of fungal fruiting bodies, which are soft, fleshy, and easily degradable tissues and the microscopic dimensions of most fungal structures, which therefore are not readily evident. Fungal fossils are difficult to distinguish from those of other microbes, and are most easily identified when they resemble extant fungi. Often recovered from a permineralized plant or animal host, these samples are typically studied by making thin-section preparations that can be examined with light microscopy or transmission electron microscopy. Researchers study compression fossils by dissolving the surrounding matrix with acid and then using light or scanning electron microscopy to examine surface details.

The earliest fossils possessing features typical of fungi date to the Proterozoic eon, some 1,430 million years ago (Ma); these multicellular benthic organisms had filamentous structures with septa, and were capable of anastomosis. More recent studies (2009) estimate the arrival of fungal organisms at about 760–1060 Ma on the basis of comparisons of the rate of evolution in closely related groups. For much of the Paleozoic Era (542–251 Ma), the fungi appear to have been aquatic and consisted of organisms similar to the extant chytrids in having flagellum-bearing spores. The evolutionary adaptation from an aquatic to a terrestrial lifestyle necessitated a diversification of ecological strategies for obtaining nutrients, including parasitism, saprobism, and the development of mutualistic relationships such as mycorrhiza and lichenization. Recent (2009) studies suggest that the ancestral ecological state of the Ascomycota was saprobism, and that independent lichenization events have occurred multiple times.

It is presumed that the fungi colonized the land during the Cambrian (542–488.3 Ma), long before land plants. Fossilized hyphae and spores recovered from the Ordovician of Wisconsin (460 Ma) resemble modern-day Glomerales, and existed at a time when the land flora likely consisted of only non-vascular bryophyte-like plants. Prototaxites, which was probably a fungus or lichen, would have been the tallest organism of the late Silurian. Fungal fossils do not become common and uncontroversial until the early Devonian (416–359.2 Ma), when they occur abundantly in the Rhynie chert, mostly as Zygomycota and Chytridiomycota. At about this same time, approximately 400 Ma, the Ascomycota and Basidiomycota diverged, and all modern classes of fungi were present by the Late Carboniferous (Pennsylvanian, 318.1–299 Ma).

Lichen-like fossils have been found in the Doushantuo Formation in southern China dating back to 635–551 Ma. Lichens formed a component of the early terrestrial ecosystems, and the estimated age of the oldest terrestrial lichen fossil is 400 Ma; this date corresponds to the age of the oldest known sporocarp fossil, a *Paleopyrenomycites* species found in the Rhynie Chert. The oldest fossil with microscopic features resembling modern-day basidiomycetes is *Palaeoancistrus*, found permineralized with a fern from the Pennsylvanian. Rare in the fossil record are the Homobasidiomycetes (a taxon roughly equivalent to the mushroom-producing species of the Agaricomycetes). Two amber-preserved specimens provide evidence that the earliest known mushroom-forming fungi (the extinct species *Archaeomarasmius leggetti*) appeared during the late Cretaceous, 90 Ma.

Some time after the Permian–Triassic extinction event (251.4 Ma), a fungal spike (originally thought to be an extraordinary abundance of fungal spores in sediments) formed, suggesting that fungi were the dominant life form at this time, representing nearly 100% of the available fossil record for this period. However, the relative proportion of fungal spores relative to spores formed by algal species is difficult to assess, the spike did not appear worldwide, and in many places it did not fall on the Permian–Triassic boundary.

Taxonomy

Although commonly included in botany curricula and textbooks, fungi are more closely related to animals than to plants and are placed with the animals in the monophyletic group of opisthokonts. Analyses using molecular phylogenetics support a monophyletic origin of the Fungi. The taxonomy of the Fungi is in a state of constant flux, especially due to recent research based on DNA comparisons. These current phylogenetic analyses often overturn classifications based on older and sometimes less discriminative methods based on morphological features and biological species concepts obtained from experimental matings.

There is no unique generally accepted system at the higher taxonomic levels and there are frequent name changes at every level, from species upwards. Efforts among researchers are now underway to establish and encourage usage of a unified and more consistent nomenclature. Fungal species can also have multiple scientific names depending on their life cycle and mode (sexual or asexual) of reproduction. Web sites such as Index Fungorum and ITIS list current names of fungal species (with cross-references to older synonyms).

The 2007 classification of Kingdom Fungi is the result of a large-scale collaborative research effort involving dozens of mycologists and other scientists working on fungal taxonomy. It recognizes seven phyla, two of which—the Ascomycota and the Basidiomycota—are contained within a branch representing subkingdom Dikarya, the most species rich and familiar group, including all the mushrooms, most food-spoilage molds, most plant pathogenic fungi, and the beer, wine, and bread yeasts. The accompanying cladogram depicts the major fungal taxa and their relationship to opisthokont and unikont organisms, based on the work of Philippe Silar and "The Mycota: A Comprehensive Treatise on Fungi as Experimental Systems for Basic and Applied Research". The lengths of the branches are not proportional to evolutionary distances.

Taxonomic Groups

The major phyla (sometimes called divisions) of fungi have been classified mainly on the basis of characteristics of their sexual reproductive structures. Currently, seven phyla are proposed: Microsporidia, Chytridiomycota, Blastocladiomycota, Neocallimastigomycota, Glomeromycota, Ascomycota, and Basidiomycota.

Phylogenetic analysis has demonstrated that the Microsporidia, unicellular parasites of animals and protists, are fairly recent and highly derived endobiotic fungi (living within the tissue of another species). One 2006 study concludes that the Microsporidia are a sister group to the true fungi; that is, they are each other's closest evolutionary relative. Hibbett and colleagues suggest that this analysis does not clash with their classification of the Fungi, and although the Microsporidia are elevated to phylum status, it is acknowledged that further analysis is required to clarify evolutionary relationships within this group.

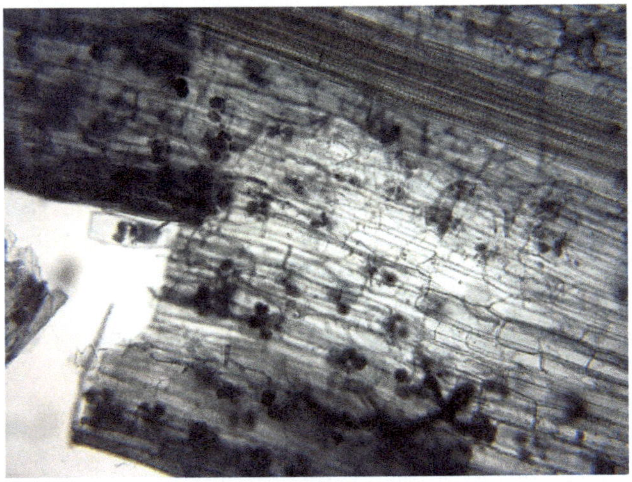

Arbuscular mycorrhiza seen under microscope. Flax root cortical cells containing paired arbuscules.

The Chytridiomycota are commonly known as chytrids. These fungi are distributed worldwide. Chytrids and their close relatives Neocallimastigomycota and Blastocladiomycota (below) are the only fungi with active motility, producing zoospores that are capable of active movement through aqueous phases with a single flagellum, leading early taxonomists to classify them as protists. Molecular phylogenies, inferred from rRNA sequences in ribosomes, suggest that the Chytrids are a basal group divergent from the other fungal phyla, consisting of four major clades with suggestive evidence for paraphyly or possibly polyphyly.

The Blastocladiomycota were previously considered a taxonomic clade within the Chytridiomycota. Recent molecular data and ultrastructural characteristics, however, place the Blastocladiomycota as a sister clade to the Zygomycota, Glomeromycota, and Dikarya (Ascomycota and Basidiomycota). The blastocladiomycetes are saprotrophs, feeding on decomposing organic matter, and they are parasites of all eukaryotic groups. Unlike their close relatives, the chytrids, most of which exhibit zygotic meiosis, the blastocladiomycetes undergo sporic meiosis.

The Neocallimastigomycota were earlier placed in the phylum Chytridomycota. Members of this small phylum are anaerobic organisms, living in the digestive system of larger herbivorous mammals and in other terrestrial and aquatic environments enriched in cellulose (e.g., domestic waste landfill sites). They lack mitochondria but contain hydrogenosomes of mitochondrial origin. As in the related chrytrids, neocallimastigomycetes form zoospores that are posteriorly uniflagellate or polyflagellate.

Members of the Glomeromycota form arbuscular mycorrhizae, a form of mutualist symbiosis wherein fungal hyphae invade plant root cells and both species benefit from the resulting increased supply of nutrients. All known Glomeromycota species reproduce asexually. The symbiotic association between the Glomeromycota and plants is ancient, with evidence dating to 400 million years ago. Formerly part of the Zygomycota (commonly known as 'sugar' and 'pin' molds), the Glomeromycota were elevated to phylum status in 2001 and now replace the older phylum Zygomycota. Fungi that were placed in the Zygomycota are now being reassigned to the Glomeromycota, or the subphyla incertae sedis Mucoromycotina, Kickxellomycotina, the Zoopagomycotina and the Entomophthoromycotina. Some well-known examples of fungi formerly in the Zygomycota include black bread mold (*Rhizopus stolonifer*), and *Pilobolus* species, capable of ejecting spores several meters through the air. Medically relevant genera include *Mucor*, *Rhizomucor*, and *Rhizopus*.

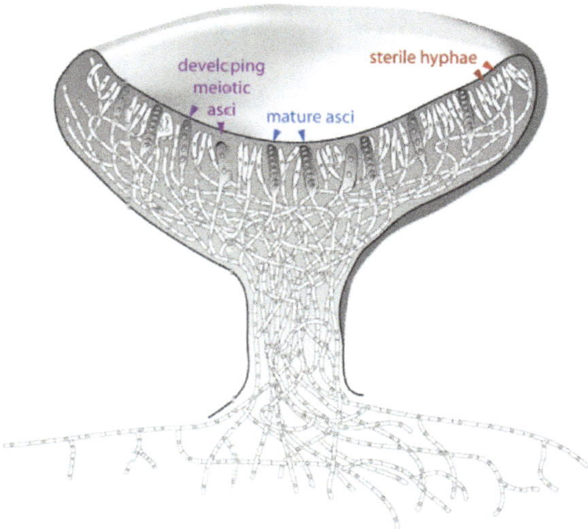

Diagram of an apothecium (the typical cup-like reproductive structure of Ascomycetes) showing sterile tissues as well as developing and mature asci.

The Ascomycota, commonly known as sac fungi or ascomycetes, constitute the largest taxonomic group within the Eumycota. These fungi form meiotic spores called ascospores, which are enclosed in a special sac-like structure called an ascus. This phylum includes morels, a few mushrooms and truffles, unicellular yeasts (e.g., of the genera *Saccharomyces*, *Kluyveromyces*, *Pichia*, and *Candida*), and many filamentous fungi living as saprotrophs, parasites, and mutualistic symbionts (e.g. lichens). Prominent and important genera of filamentous ascomycetes include *Aspergillus*, *Penicillium*, *Fusarium*, and *Claviceps*. Many ascomycete species have only been observed undergoing asexual reproduction (called anamorphic species), but analysis of molecular data has often been able to identify their closest teleomorphs in the Ascomycota. Because the products of meiosis are retained within the sac-like ascus, ascomycetes have been used for elucidating principles of genetics and heredity (e.g., *Neurospora crassa*).

Members of the Basidiomycota, commonly known as the club fungi or basidiomycetes, produce meiospores called basidiospores on club-like stalks called basidia. Most common mushrooms belong to this group, as well as rust and smut fungi, which are major pathogens of grains. Other important basidiomycetes include the maize pathogen *Ustilago maydis*, human commensal species of the genus *Malassezia*, and the opportunistic human pathogen, *Cryptococcus neoformans*.

Fungus-like Organisms

Because of similarities in morphology and lifestyle, the slime molds (mycetozoans, plasmodiophorids, acrasids, *Fonticula* and labyrinthulids, now in Amoebozoa, Rhizaria, Excavata, Opisthokonta and Stramenopiles, respectively), water molds (oomycetes) and hyphochytrids (both Stramenopiles) were formerly classified in the kingdom Fungi, in groups like Mastigomycotina, Gymnomycota and Phycomycetes. The slime molds were studied also as protozoans, leading to a ambiregnal, duplicated taxonomy.

Unlike true fungi, the cell walls of oomycetes contain cellulose and lack chitin. Hyphochytrids have both chitin and cellulose. Slime molds lack a cell wall during the assimilative phase (except labyrinthulids, which have a wall of scales), and ingest nutrients by ingestion (phagocytosis, except labyrinthulids) rather than absorption (osmotrophy, as fungi, labyrinthulids, oomycetes and hyphochytrids). Neither water molds nor slime molds are closely related to the true fungi, and, therefore, taxonomists no longer group them in the kingdom Fungi. Nonetheless, studies of the oomycetes and myxomycetes are still often included in mycology textbooks and primary research literature.

The Eccrinales and Amoebidiales are opisthokont protists, previously thought to be zygomycete fungi. Other groups now in Opisthokonta (e.g., *Corallochytrium*, Ichthyosporea) were also at given time classified as fungi. The genus *Blastocystis*, now in Stramenopiles, was originally classified as a yeast. *Ellobiopsis*, now in Alveolata, was considered a chytrid. The bacteria were also included in fungi in some classifications, as the group Schizomycetes.

The Rozellida clade, including the "ex-chytrid" *Rozella*, is a genetically disparate group known mostly from environmental DNA sequences that is a sister group to fungi. Members of the group that have been isolated lack the chitinous cell wall that is characteristic of fungi.

The nucleariids, protists currently grouped in the Choanozoa (Opisthokonta), may be the next sister group to the eumycete clade, and as such could be included in an expanded fungal kingdom.

Many Actinomycetales (Actinobacteria), a group with many filamentous bacteria, were also long believed to be fungi.

Ecology

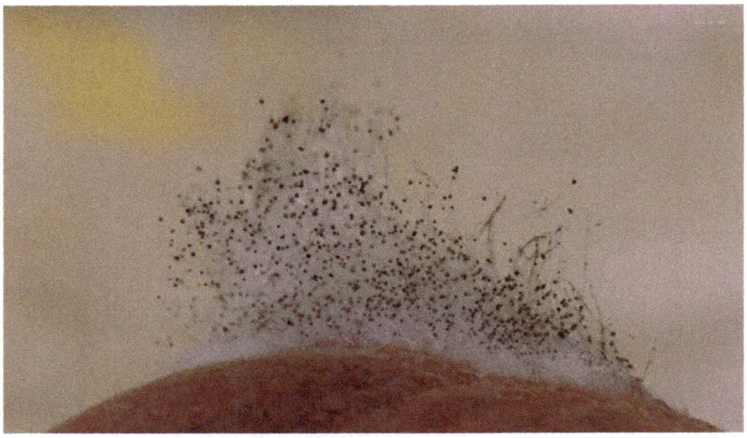

A pin mold decomposing a peach

Although often inconspicuous, fungi occur in every environment on Earth and play very important roles in most ecosystems. Along with bacteria, fungi are the major decomposers in most terrestrial (and some aquatic) ecosystems, and therefore play a critical role in biogeochemical cycles and in many food webs. As decomposers, they play an essential role in nutrient cycling, especially as saprotrophs and symbionts, degrading organic matter to inorganic molecules, which can then re-enter anabolic metabolic pathways in plants or other organisms.

Symbiosis

Many fungi have important symbiotic relationships with organisms from most if not all Kingdoms. These interactions can be mutualistic or antagonistic in nature, or in the case of commensal fungi are of no apparent benefit or detriment to the host.

With plants

Mycorrhizal symbiosis between plants and fungi is one of the most well-known plant–fungus associations and is of significant importance for plant growth and persistence in many ecosystems; over 90% of all plant species engage in mycorrhizal relationships with fungi and are dependent upon this relationship for survival.

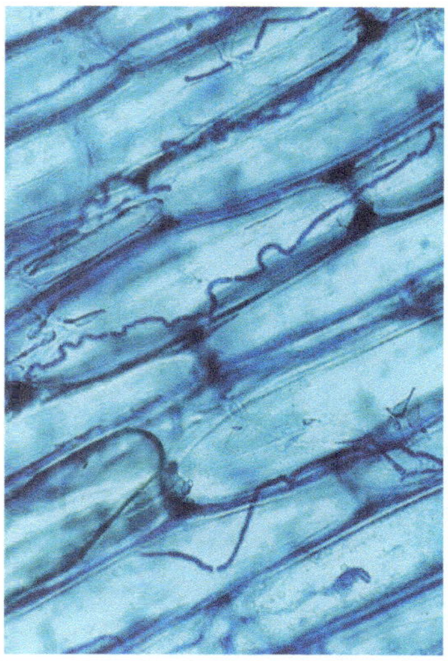

The dark filaments are hyphae of the endophytic fungus *Neotyphodium coenophialum* in the intercellular spaces of tall fescue leaf sheath tissue

The mycorrhizal symbiosis is ancient, dating to at least 400 million years ago. It often increases the plant's uptake of inorganic compounds, such as nitrate and phosphate from soils having low concentrations of these key plant nutrients. The fungal partners may also mediate plant-to-plant transfer of carbohydrates and other nutrients. Such mycorrhizal communities are called "common mycorrhizal networks". A special case of mycorrhiza is myco-heterotrophy, whereby the plant parasitizes the fungus, obtaining all of its nutrients from its fungal symbiont. Some fungal species inhabit the tissues inside roots, stems, and leaves, in which case they are called endophytes. Similar to mycorrhiza, endophytic colonization by fungi may benefit both symbionts; for example, endophytes of grasses impart to their host increased resistance to herbivores and other environmental stresses and receive food and shelter from the plant in return.

With Algae and Cyanobacteria

Lichens are a symbiotic relationship between fungi and photosynthetic algae or cyanobacteria. The photosynthetic partner in the relationship is referred to in lichen terminology as a "photobiont". The fungal part of the relationship is composed mostly of various species of ascomycetes and a few basidiomycetes. Lichens occur in every ecosystem on all continents, play a key role in soil formation and the initiation of biological succession, and are prominent in some extreme environments, including polar, alpine, and semiarid desert regions. They are able to grow on inhospitable surfaces, including bare soil, rocks, tree bark, wood, shells, barnacles and leaves. As in mycorrhizas, the photobiont provides sugars and other carbohydrates via photosynthesis to the fungus, while the fungus provides minerals and water to the photobiont. The functions of both symbiotic organisms are so closely intertwined that they function almost as a single organism; in most cases the resulting organism differs greatly from the individual components. Lichenization is a common mode of nutrition for fungi; around 20% of fungi—between 17,500 and 20,000 described species—are lichenized. Characteristics common to most lichens include obtaining organic carbon by photosynthesis, slow growth, small size, long life, long-lasting (seasonal) vegetative reproductive structures, mineral nutrition obtained largely from airborne sources, and greater tolerance of desiccation than most other photosynthetic organisms in the same habitat.

The lichen *Lobaria pulmonaria*, a symbiosis of fungal, algal, and cyanobacterial species

With Insects

Many insects also engage in mutualistic relationships with fungi. Several groups of ants cultivate fungi in the order Agaricales as their primary food source, while ambrosia beetles cultivate various species of fungi in the bark of trees that they infest. Likewise, females of several wood wasp species (genus *Sirex*) inject their eggs together with spores of the wood-rotting fungus *Amylostereum areolatum* into the sapwood of pine trees; the growth of the fungus provides ideal nutritional conditions for the development of the wasp larvae. At least one species of stingless bee has a relationship with a fungus in the genus *Monascus*, where the larvae consume and depend on fungus transferred from old to new nests. Termites on the African savannah are also known to cultivate fungi, and yeasts of the genera *Candida* and *Lachancea* inhabit the gut of a wide range of insects, including neuropterans, beetles, and cockroaches; it is not known whether these fungi benefit their hosts. Fungi ingrowing dead wood are essential for xylophagous insects (e.g. woodboring beetles). They deliver nutrients needed by xylophages to nutritionally scarce dead wood. Thanks to this nutritional enrichment the larvae of woodboring insect is able to grow and develop to adulthood. The larvae of many families of fungicolous flies, particularly those within the superfamily Sciaroidea such as the Mycetophilidae and some Keroplatidae feed on fungal fruiting bodies and sterile mycorrhizae.

As Pathogens and Parasites

The plant pathogen *Aecidium magellanicum* causes calafate rust, seen here on a *Berberis* shrub in Chile.

Many fungi are parasites on plants, animals (including humans), and other fungi. Serious pathogens of many cultivated plants causing extensive damage and losses to agriculture and forestry include the rice blast fungus *Magnaporthe oryzae*, tree pathogens such as *Ophiostoma ulmi* and *Ophiostoma novo-ulmi* causing Dutch elm disease and *Cryphonectria parasitica* responsible for chestnut blight, and plant pathogens in the genera *Fusarium, Ustilago, Alternaria*, and *Cochliobolus*. Some carnivorous fungi, like *Paecilomyces lilacinus*, are predators of nematodes, which they capture using an array of specialized structures such as constricting rings or adhesive nets.

Some fungi can cause serious diseases in humans, several of which may be fatal if untreated. These include aspergillosis, candidiasis, coccidioidomycosis, cryptococcosis, histoplasmosis, mycetomas, and paracoccidioidomycosis. Furthermore, persons with immuno-deficiencies are particularly susceptible to disease by genera such as *Aspergillus, Candida, Cryptoccocus, Histoplasma*, and *Pneumocystis*. Other fungi can attack eyes, nails, hair, and especially skin, the so-called dermatophytic and keratinophilic fungi, and cause local infections such as ringworm and athlete's foot. Fungal spores are also a cause of allergies, and fungi from different taxonomic groups can evoke allergic reactions.

As Targets of Mycoparasites

The organisms which parasitize fungi are known as mycoparasitic organisms. Certain species of the *Pythium* genus, which are oomycetes, have potential as biocontrol agents against certain fungi. Fungi can also act as mycoparasites or antagonists of other fungi, such as *Hypomyces chrysospermus*, which grows on bolete mushrooms.

Fungi can become the target of infection by Mycoviruses.

Mycotoxins

Many fungi produce biologically active compounds, several of which are toxic to animals or plants and are therefore called mycotoxins. Of particular relevance to humans are mycotoxins produced by molds causing food spoilage, and poisonous mushrooms. Particularly infamous are the lethal amatoxins in some *Amanita* mushrooms, and ergot alkaloids, which have a long history of causing serious epidemics of ergotism (St Anthony's Fire) in people consuming rye or related cereals contaminated with sclerotia of the ergot fungus, *Claviceps purpurea*. Other notable mycotoxins include the aflatoxins, which are insidious liver toxins and highly carcinogenic metabolites produced by certain *Aspergillus* species often growing in or on grains and nuts consumed by humans, ochratoxins, patulin, and trichothecenes (e.g., T-2 mycotoxin) and fumonisins, which have significant impact on human food supplies or animal livestock.

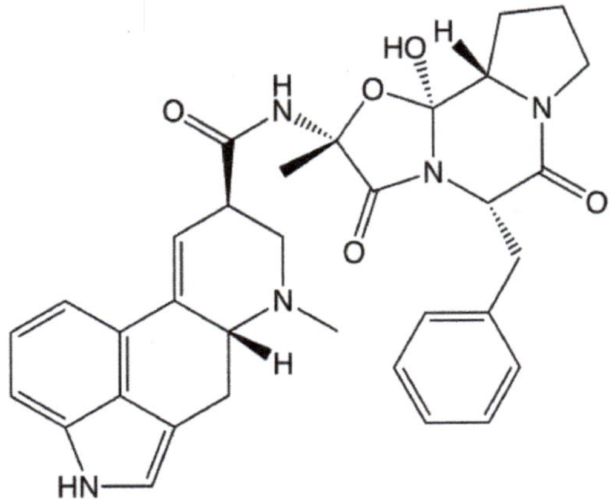

Ergotamine, a major mycotoxin produced by *Claviceps* species, which if ingested can cause gangrene, convulsions, and hallucinations

Mycotoxins are secondary metabolites (or natural products), and research has established the existence of biochemical pathways solely for the purpose of producing mycotoxins and other natural products in fungi. Mycotoxins may provide fitness benefits in terms of physiological adaptation, competition with other microbes and fungi, and protection from consumption (fungivory). Many fungal secondary metabolites (or derivatives) are used medically, as described under Human Use below.

Pathogenic Mechanisms

Ustilago maydis is a pathogenic plant fungus that causes smut disease in maize and teosinte. Plants have evolved efficient defense systems against pathogenic microbes such as *U. maydis*. A rapid defense reaction after pathogen attack is the oxidative burst where the plant produces reactive oxygen species at the site of the attempted invasion. *U. maydis* can respond to the oxidative burst with an oxidative stress response, regulated by the gene *YAP1*. The response protects *U. maydis* from the host defense, and is necessary for the pathogen's virulence. Furthermore, *U. maydis* has a well-established recombinational DNA repair system which acts during mitosis and meiosis. The system may assist the pathogen in surviving DNA damage arising from the host plant's oxidative defensive response to infection.

Cryptococcus neoformans is an encapsulated yeast that can live in both plants and animals. *C. neoformans* usually infects the lungs, where it is phagocytosed by alveolar macrophages. Some *C. neoformans* can survive inside macrophages, which appears to be the basis for latency, disseminated disease, and resistance to antifungal agents. One mechanism by which *C. neoformans* survives the hostile macrophage environment is by up-regulating the expression of genes involved in the oxidative stress response. Another mechanism involves meiosis. The majority of *C. neoformans* are mating "type a". Filaments of mating "type a" ordinarily have haploid nuclei, but they can become diploid (perhaps by endoduplication or by stimulated nuclear fusion) to form blastospores. The diploid nuclei of blastospores can undergo meiosis, including recombination, to form haploid basidiospores that can be dispersed. This process is referred to as monokaryotic fruiting. this process requires a gene called *DMC1*, which is a conserved homologue of genes *recA* in bacteria and *RAD51* in eukaryotes, that mediates homologous chromosome pairing during meiosis and repair of DNA double-strand breaks. Thus, *C. neoformans* can undergo a meiosis, monokaryotic fruiting, that promotes recombinational repair in the oxidative, DNA damaging environment of the host macrophage, and the repair capability may contribute to its virulence.

Human Use

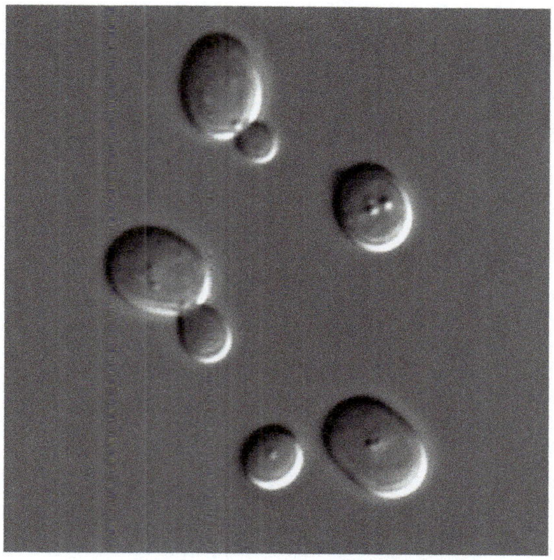

Saccharomyces cerevisiae cells shown with DIC microscopy

The human use of fungi for food preparation or preservation and other purposes is extensive and has a long history. Mushroom farming and mushroom gathering are large industries in many countries. The study of the historical uses and sociological impact of fungi is known as ethnomycology. Because of the capacity of this group to produce an enormous range of natural products with antimicrobial or other biological activities, many species have long been used or are being developed for industrial production of antibiotics, vitamins, and anti-cancer and cholesterol-lowering drugs. More recently, methods have been developed for genetic engineering of fungi, enabling metabolic engineering of fungal species. For example, genetic modification of yeast species—which are easy to grow at fast rates in large fermentation vessels—has opened up ways of pharmaceutical production that are potentially more efficient than production by the original source organisms.

Therapeutic Uses

Modern Chemotherapeutics

Many species produce metabolites that are major sources of pharmacologically active drugs. Particularly important are the antibiotics, including the penicillins, a structurally related group of β-lactam antibiotics that are synthesized from small peptides. Although naturally occurring penicillins such as penicillin G (produced by *Penicillium chrysogenum*) have a relatively narrow spectrum of biological activity, a wide range of other penicillins can be produced by chemical modification of the natural penicillins. Modern penicillins are semisynthetic compounds, obtained initially from fermentation cultures, but then structurally altered for specific desirable properties. Other antibiotics produced by fungi include: ciclosporin, commonly used as an immunosuppressant during transplant surgery; and fusidic acid, used to help control infection from methicillin-resistant *Staphylococcus aureus* bacteria. Widespread use of antibiotics for the treatment of bacterial diseases, such as tuberculosis, syphilis, leprosy, and others began in the early 20th century and continues to date. In nature, antibiotics of fungal or bacterial origin appear to play a dual role: at high concentrations they act as chemical defense against competition with other microorganisms in species-rich environments, such as the rhizosphere, and at low concentrations as quorum-sensing molecules for intra- or interspecies signaling. Other drugs produced by fungi include griseofulvin isolated from *Penicillium griseofulvum*, used to treat fungal infections, and statins (HMG-CoA reductase inhibitors), used to inhibit cholesterol synthesis. Examples of statins found in fungi include mevastatin from *Penicillium citrinum* and lovastatin from *Aspergillus terreus* and the oyster mushroom.

Traditional and Folk Medicine

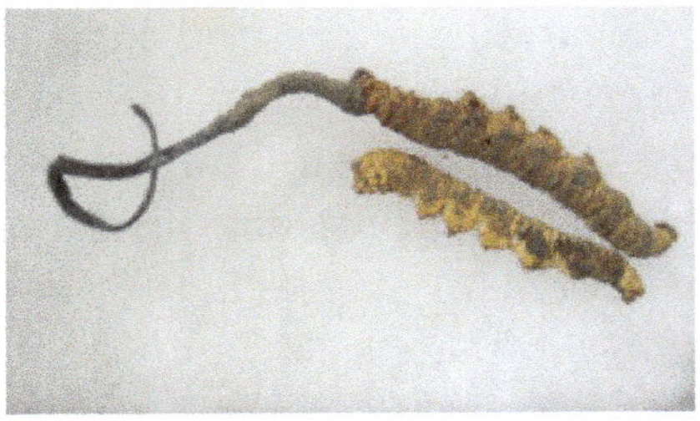

The medicinal fungi *Ganoderma lucidum* (left) and *Ophiocordyceps sinensis* (right)

Certain mushrooms enjoy usage as therapeutics in folk medicines, such as Traditional Chinese medicine. Notable medicinal mushrooms with a well-documented history of use include *Agaricus subrufescens*, *Ganoderma lucidum*, and *Ophiocordyceps sinensis*. Research has identified compounds produced by these and other fungi that have inhibitory biological effects against viruses and cancer cells. Specific metabolites, such as polysaccharide-K, ergotamine, and β-lactam antibiotics, are routinely used in clinical medicine. The shiitake mushroom is a source of lentinan, a clinical drug approved for use in cancer treatments in several countries, including Japan. In Europe and Japan, polysaccharide-K (brand name Krestin), a chemical derived from *Trametes versicolor*, is an approved adjuvant for cancer therapy.

Cultured Foods

Baker's yeast or *Saccharomyces cerevisiae*, a unicellular fungus, is used to make bread and other wheat-based products, such as pizza dough and dumplings. Yeast species of the genus *Saccharomyces* are also used to produce alcoholic beverages through fermentation. Shoyu koji mold (*Aspergillus oryzae*) is an essential ingredient in brewing Shoyu (soy sauce) and sake, and the preparation of miso, while *Rhizopus* species are used for making tempeh. Several of these fungi are domesticated species that were bred or selected according to their capacity to ferment food without producing harmful mycotoxins, which are produced by very closely related *Aspergilli*. Quorn, a meat substitute, is made from *Fusarium venenatum*.

Edible and Poisonous Species

Amanita phalloides accounts for the majority of fatal mushroom poisonings worldwide.

Edible mushrooms are well-known examples of fungi. Many are commercially raised, but others must be harvested from the wild. *Agaricus bisporus*, sold as button mushrooms when small or Portobello mushrooms when larger, is a commonly eaten species, used in salads, soups, and many other dishes. Many Asian fungi are commercially grown and have increased in popularity in the West. They are often available fresh in grocery stores and markets, including straw mushrooms (*Volvariella volvacea*), oyster mushrooms (*Pleurotus ostreatus*), shiitakes (*Lentinula edodes*), and enokitake (*Flammulina* spp.).

There are many more mushroom species that are harvested from the wild for personal consumption or commercial sale. Milk mushrooms, morels, chanterelles, truffles, black trumpets, and *porcini* mushrooms (*Boletus edulis*) (also known as king boletes) demand a high price on the market. They are often used in gourmet dishes.

Certain types of cheeses require inoculation of milk curds with fungal species that impart a unique flavor and texture to the cheese. Examples include the blue color in cheeses such as Stilton or Roquefort, which are made by inoculation with *Penicillium roqueforti*. Molds used in cheese production are non-toxic and are thus safe for human consumption; however, mycotoxins (e.g., aflatoxins, roquefortine C, patulin, or others) may accumulate because of growth of other fungi during cheese ripening or storage.

Stilton cheese veined with *Penicillium roqueforti*

Many mushroom species are poisonous to humans, with toxicities ranging from slight digestive problems or allergic reactions as well as hallucinations to severe organ failures and death. Genera with mushrooms containing deadly toxins include *Conocybe*, *Galerina*, *Lepiota*, and, the most infamous, *Amanita*. The latter genus includes the destroying angel *(A. virosa)* and the death cap *(A. phalloides)*, the most common cause of deadly mushroom poisoning. The false morel (*Gyromitra esculenta*) is occasionally considered a delicacy when cooked, yet can be highly toxic when eaten raw. *Tricholoma equestre* was considered edible until it was implicated in serious poisonings causing rhabdomyolysis. Fly agaric mushrooms (*Amanita muscaria*) also cause occasional non-fatal poisonings, mostly as a result of ingestion for its hallucinogenic properties. Historically, fly agaric was used by different peoples in Europe and Asia and its present usage for religious or shamanic purposes is reported from some ethnic groups such as the Koryak people of north-eastern Siberia.

As it is difficult to accurately identify a safe mushroom without proper training and knowledge, it is often advised to assume that a wild mushroom is poisonous and not to consume it.

Pest Control

In agriculture, fungi may be useful if they actively compete for nutrients and space with pathogenic microorganisms such as bacteria or other fungi via the competitive exclusion principle, or if they are parasites of these pathogens. For example, certain species may be used to eliminate or suppress the growth of harmful plant pathogens, such as insects, mites, weeds, nematodes, and other fungi that cause diseases of important crop plants. This has generated strong interest in practical applications that use these fungi in the biological control of these agricultural pests. Entomopathogenic fungi can be used as biopesticides, as they actively kill insects. Examples that have been used as biological insecticides are *Beauveria bassiana*, *Metarhizium* spp, *Hirsutella* spp, *Paecilomyces* (*Isaria*) spp, and *Lecanicillium lecanii*. Endophytic fungi of grasses of the genus *Neotyphodium*, such as *N. coenophialum*, produce alkaloids that are toxic to a range of invertebrate and vertebrate herbivores. These alkaloids protect grass plants from herbivory, but several endophyte alkaloids can poison grazing animals, such as cattle and sheep. Infecting cultivars of pasture or forage grasses with *Neotyphodium* endophytes is one approach being used in grass breeding programs; the fungal strains are selected for producing only alkaloids that increase resistance to herbivores such as insects, while being non-toxic to livestock.

Grasshoppers killed by *Beauveria bassiana*

Bioremediation

Certain fungi, in particular "white rot" fungi, can degrade insecticides, herbicides, pentachlorophenol, creosote, coal tars, and heavy fuels and turn them into carbon dioxide, water, and basic elements. Fungi have been shown to biomineralize uranium oxides, suggesting they may have application in the bioremediation of radioactively polluted sites.

Model Organisms

Several pivotal discoveries in biology were made by researchers using fungi as model organisms, that is, fungi that grow and sexually reproduce rapidly in the laboratory. For example, the one gene-one enzyme hypothesis was formulated by scientists using the bread mold *Neurospora crassa* to test their biochemical theories. Other important model fungi are *Aspergillus nidulans* and the yeasts *Saccaromyces cerevisiae* and *Schizosaccharomyces pombe*, each of which with a long history of use to investigate issues in eukaryotic cell biology and genetics, such as cell cycle regulation, chromatin structure, and gene regulation. Other fungal models have more recently emerged that address specific biological questions relevant to medicine, plant pathology, and industrial uses; examples include *Candida albicans*, a dimorphic, opportunistic human pathogen, *Magnaporthe grisea*, a plant pathogen,] and *Pichia pastoris*, a yeast widely used for eukaryotic protein production.

Others

Fungi are used extensively to produce industrial chemicals like citric, gluconic, lactic, and malic acids, and industrial enzymes, such as lipases used in biological detergents, cellulases used in making cellulosic ethanol and stonewashed jeans, and amylases, invertases, proteases and xylanases. Several species, most notably *Psilocybin mushrooms* (colloquially known as *magic mushrooms*), are ingested for their psychedelic properties, both recreationally and religiously.

Protozoa

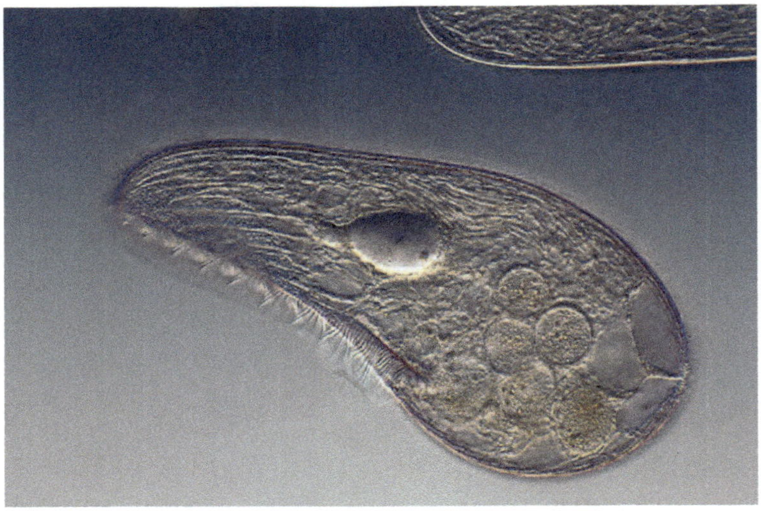

Blepharisma japonicum, a free-living ciliated protozoan.

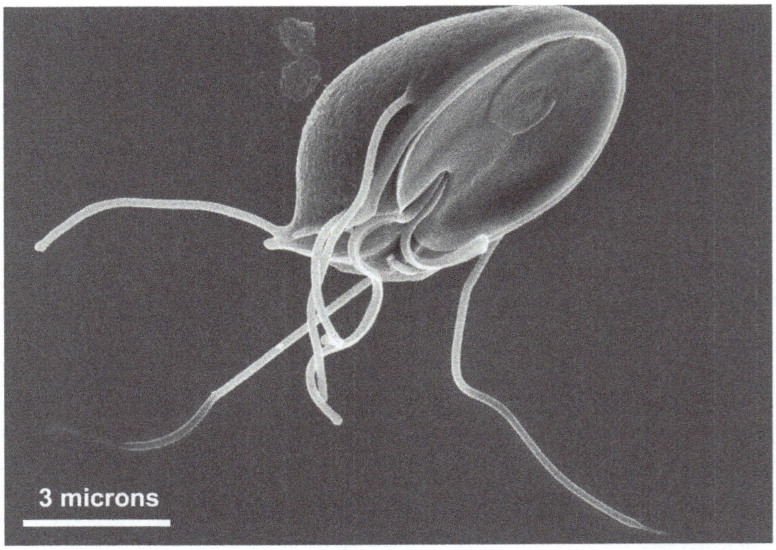

Giardia muris, a flagellate protozoan, is an intestinal parasite found in rodents, birds and reptiles.

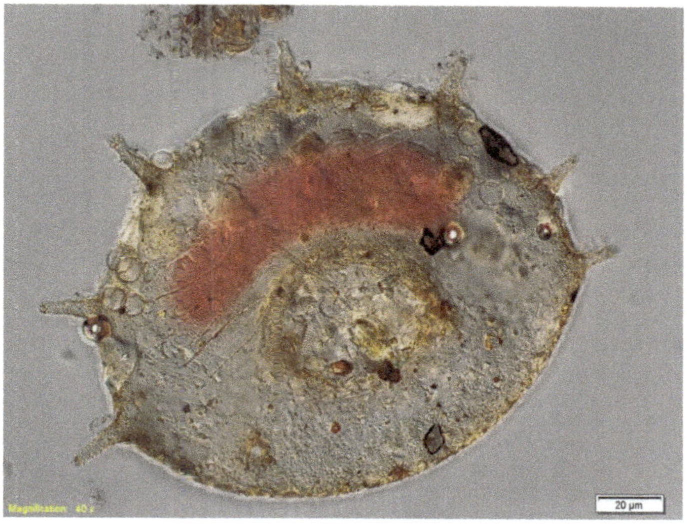

Centropyxis aculeata, a testate (shelled) amoeba

In 21st-century systems of biological classification, the Protozoa are defined as a diverse group of unicellular eukaryotic organisms. Historically, protozoa were defined as single-celled animals or organisms with animal-like behaviors, such as motility and predation. The group was regarded as the zoological counterpart to the "protophyta", which were considered to be plant-like, as they are capable of photosynthesis.

The terms *protozoa* and *protozoans* are now mostly used informally to designate single-celled, non-photosynthetic protists, such as the ciliates, amoebae and flagellates.

The term Protozoa was introduced in 1818 for a taxonomic class, but in later classification schemes the group was elevated to higher ranks, including phylum, subkingdom and kingdom. In several classification systems proposed by Thomas Cavalier-Smith and his collaborators since 1981, Protozoa is ranked as a kingdom. The seven-kingdom scheme proposed by Ruggiero et al. in 2015, places eight phyla under Protozoa: Euglenozoa, Amoebozoa, Metamonada, Choanozoa, Loukozoa, Percolozoa, Microsporidia and Sulcozoa. This kingdom does not form a clade, but an evolutionary grade or paraphyletic group, from which the fungi and animals are specifically excluded.

The use of Protozoa as a formal taxon has been discouraged by some recent researchers, mainly because the term, which is formed from the Greek *protos* "first" + *zoia*, plural of *zoion*, "animal", implies kinship with animals (metazoa) and promotes a separation of "animal-like" from "plant-like" organisms, neither of which they encourage. Modern ultrastructural, biochemical, and genetic techniques have shown that protozoa, as traditionally defined, belong to widely divergent lineages, and can no longer be regarded as "primitive animals." For this reason, the terms "protists," "Protista" or "Protoctista" are sometimes preferred for the high-level classification of eukaryotic microbes. In 2005, members of the Society of Protozoologists voted to change the name of that organization to the International Society of Protistologists.

History and Terminology

The word "protozoa" (singular *protozoon* or *protozoan*) was coined in 1818 by zoologist Georg August Goldfuss, as the Greek equivalent of the German *Urthiere*, meaning "primitive, or original animals" (*ur-* 'proto-' + *Thier* 'animal'). Goldfuss erected Protozoa as a class containing what he believed to be the simplest animals. Originally, the group included not only microbes, but also some "lower" multicellular animals, such as rotifers, corals, sponges, jellyfish, bryozoa and polychaete worms.

In 1848, in light of advancements in cell theory pioneered by Theodore Schwann and Matthias Schleiden, the anatomist and zoologist C.T. von Siebold proposed that the bodies of microbial organisms such as ciliates and amoebae were made up of single cells, similar to those from which the multicellular tissues of plants and animals were constructed. Von Siebold redefined Protozoa to include only such unicellular forms, to the exclusion of all metazoa. At the same time, he raised the group to the level of a phylum containing two broad classes of microbes: Infusoria (mostly ciliates and flagellated algae), and Rhizopoda (amoeboid organisms). The definition of Protozoa as a phylum or sub-kingdom made up of "unicellular animals" was adopted by the zoologist Otto Bütschli—celebrated at his centenary as the "architect of protozoology"—and the term came into wide use.

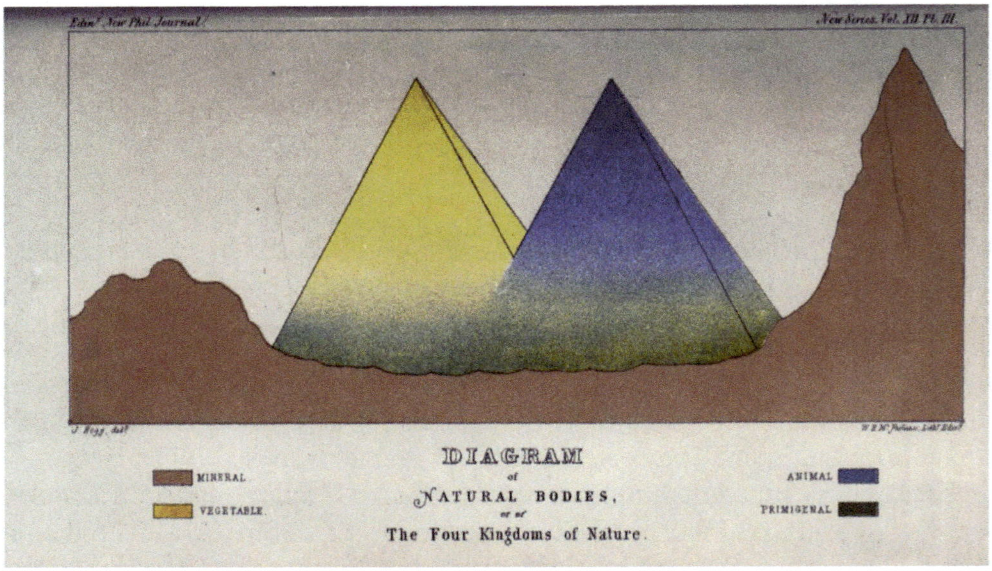

John Hogg's illustration of the Four Kingdoms of Nature, showing "Primigenal" as a greenish haze at the base of the Animals and Plants, 1860

As a phylum under Animalia, the Protozoa were firmly rooted in the old "two-kingdom" classification of life, according to which all living beings were classified as either animals or plants. As long as this scheme remained dominant, the protozoa were understood to be animals and studied in departments of Zoology, while photosynthetic microbes and microscopic fungi—the so-called Protophyta—were assigned to the Plants, and studied in departments of Botany.

Criticism of this system began in the latter half of the 19th century, with the realization that many organisms met the criteria for inclusion among both plants and animals. For example, the algae *Euglena* and *Dinobryon* have chloroplasts for photosynthesis, but can also feed on organic matter and are motile. In 1860, John Hogg argued against the use of "protozoa", on the grounds that "naturalists are divided in opinion—and probably some will ever continue so—whether many of these organisms, or living beings, are animals or plants." As an alternative, he proposed a new kingdom called Primigenum, consisting of both the protozoa and unicellular algae (protophyta), which he combined together under the name "Protoctista". In Hoggs's conception, the animal and plant kingdoms were likened to two great "pyramids" blending at their bases in the Kingdom Primigenum.

Six years later, Ernst Haeckel also proposed a third kingdom of life, which he named Protista. At first, Haeckel included a few multicellular organisms in this kingdom, but in later work he restricted the Protista to single-celled organisms, or simple colonies whose individual cells are not differentiated into different kinds of tissues.

Despite these proposals, Protozoa emerged as the preferred taxonomic placement for heterotrophic microbes such as amoebae and ciliates, and remained so for more than a century. In the course of the 20th century, however, the old "two kingdom" system began to weaken, with the growing awareness that fungi did not belong among the plants, and that most of the unicellular protozoa were no more closely related to the animals than they were to the plants. By mid-century, some biologists, such as Herbert Copeland, Robert H. Whittaker and Lynn Margulis, advocated the revival of Haeckel's Protista or Hogg's Protoctista as a kingdom-level eukaryotic group, alongside Plants, Animals and Fungi. A variety of multi-kingdom systems were proposed, and Kingdoms Protista and Protoctista became well established in biology texts and curricula.

While many taxonomists have abandoned Protozoa as a high-level group, Thomas Cavalier-Smith has retained it as a kingdom in the various classifications he has proposed. As of 2015, Cavalier-Smith's Protozoa excludes several major groups of organisms traditionally placed among the protozoa, including the ciliates, dinoflagellates and foraminifera (all members of the SAR supergroup). In its current form, his kingdom Protozoa is a paraphyletic group which includes a common ancestor and most of its descendents, but excludes two important clades that branch within it: the animals and fungi.

Characteristics

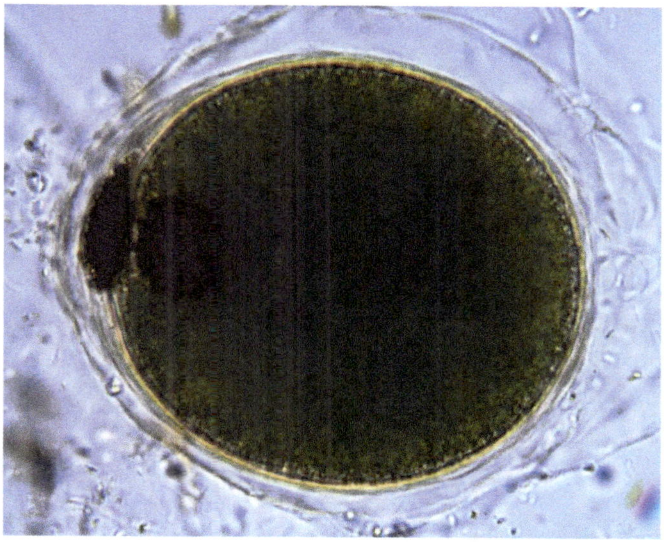

Resting cyst of ciliated protozoan *Dileptus viridis*.

Protozoa, as traditionally defined, are mainly microscopic organisms, ranging in size from 10 to 52 micrometers. Some, however, are significantly larger. Among the largest are the deep-sea–

dwelling xenophyophores, single-celled foraminifera whose shells can reach 20 cm in diameter. Free-living forms are restricted to moist environments, such as soils, mosses and aquatic habitats, although many form resting cysts which enable them to survive drying. Many protozoan species are symbionts, some are parasites, and some are predators of bacteria, algae and other protists.

Motility and Feeding

Organisms traditionally classified as protozoa are abundant in aqueous environments and soil, occupying a range of trophic levels. The group includes flagellates (which move with the help of whip-like structures called flagella), ciliates (which move by using hair-like structures called cilia) and amoebae (which move by the use of foot-like structures called pseudopodia). Some protozoa are sessile, and do not move at all.

Protozoa may take in food by osmotrophy, absorbing nutrients through their cell membranes; or they may feed by phagocytosis, either by engulfing particles of food with pseudopodia (as amoebae do), or taking in food through a mouth-like aperture called a cytostome. All protozoa digest their food in stomach-like compartments called vacuoles.

Pellicle

The pellicle is a thin layer supporting the cell membrane in various protozoa, such as ciliates, protecting them and allowing them to retain their shape, especially during locomotion, allowing the organism to be more hydrodynamic. The pellicle varies from flexible and elastic to rigid. Although somewhat stiff, the pellicle is also flexible and allows the protist to fit into tighter spaces. In ciliates and Apicomplexa, it is formed from closely packed vesicles called alveoli. In euglenids, it is formed from protein strips arranged spirally along the length of the body. Familiar examples of protists with a pellicle are the euglenoids and the ciliate *Paramecium*. In some protozoa, the pellicle hosts epibiotic bacteria that adhere to the surface by their fimbriae (attachment pili).

Life Cycle

Some protozoa have life phases alternating between proliferative stages (e.g., trophozoites) and dormant cysts. As cysts, protozoa can survive harsh conditions, such as exposure to extreme temperatures or harmful chemicals, or long periods without access to nutrients, water, or oxygen for periods of time. Being a cyst enables parasitic species to survive outside of a host, and allows their transmission from one host to another. When protozoa are in the form of trophozoites (Greek *tropho* = to nourish), they actively feed. The conversion of a trophozoite to cyst form is known as encystation, while the process of transforming back into a trophozoite is known as excystation. Protozoa reproduce asexually by binary fission or multiple fission. Many protozoan species exchange genetic material by sexual means (typically, through conjugation); however, sexuality is generally decoupled from the process of reproduction, and does not immediately result in increased population.

Although meiotic sex is widespread among present day eukaryotes, it has, until recently, been unclear whether or not eukaryotes were sexual early in their evolution. Due to recent advances in gene detection and other techniques, evidence has been found for some form of meiotic sex in an increasing number of protozoans of ancient lineage that diverged early in eukaryotic evolution.

Thus, such findings suggest that meiotic sex arose early in eukaryotic evolution. Examples of protozoan meiotic sexuality are described in the articles *Amoebozoa, Giardia lamblia, Leishmania, Plasmodium falciparum biology, Paramecium, Toxoplasma gondii, Trichomonas vaginalis* and *Trypanosoma brucei.*

Classification

The classification of protozoa has been and remains a problematic area of taxonomy. Where they are available, DNA sequences are used as the basis for classification; however, for the majority of described protozoa, such material is not available. Protozoa have been and still are organized mostly on the basis of their morphology, means of locomotion, and for the parasitic species their hosts.

- Flagellates (e.g., *Giardia lamblia*)
- Amoeboids (e.g., *Entamoeba histolytica*)
- Sporozoans (e.g., *Plasmodium knowlesi*)
    - Apicomplexa (now in Alveolata)
    - Microsporidia (now in Fungi)
    - Ascetosporea (now in Rhizaria)
    - Myxosporidia (now in Cnidaria)
- Ciliates (e.g., *Balantidium coli*)

As a phylum the Protozoa were, historically, divided into four subphyla reflecting the means of locomotion:

- Subphylum Sarcomastigophora
    - Superclass Mastigophora (includes flagellates)
    - Superclass Sarcodina
    - Superclass Opalinata
- Subphylum Sporozoa (includes apicomplexans)
- Subphylum Cnidospora
    - Class Microsporidea
- Subphylum Ciliophora (includes ciliates)

These systems are no longer considered to be valid.

Ecological Role

As components of the micro- and meiofauna, protozoa are an important food source for microinvertebrates. Thus, the ecological role of protozoa in the transfer of bacterial and algal production to successive trophic levels is important. As predators, they prey upon unicellular or filamentous algae, bacteria, and microfungi. Protozoan species include both herbivores and consumers in the decomposer link of the food chain. They also control bacteria populations and biomass to some extent. On average, protozoa eat ~ 100 to 1,000 bacteria per hour. Protozoa can stimulate decomposition of organic matter, digest cellulose in the rumen of cows and termite guts, and can play a role in nutrient mobilization.

Diseases

In Humans

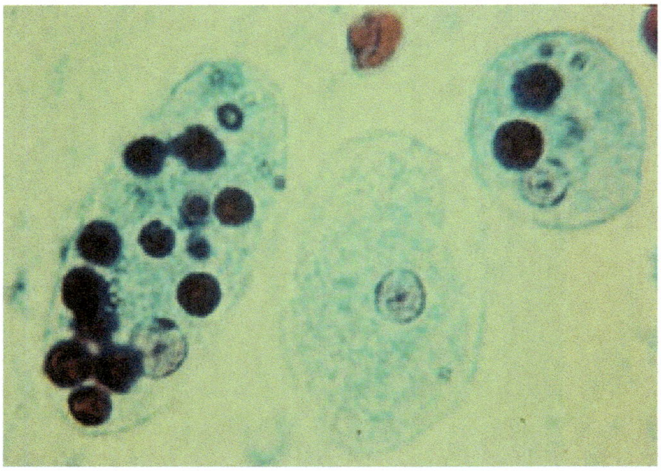

Trophozoites of the amoebic dysentery pathogen *Entamoeba histolytica* with ingested human red blood cells (dark circles)

A number of protozoan pathogens are human parasites, causing diseases such as malaria (by *Plasmodium*), amoebiasis, giardiasis, toxoplasmosis, cryptosporidiosis, trichomoniasis, Chagas disease, leishmaniasis, African trypanosomiasis (sleeping sickness), amoebic dysentery, acanthamoeba keratitis, and primary amoebic meningoencephalitis (naegleriasis).

In other Animals

The protozoan *Ophryocystis elektroscirrha* is a parasite of butterfly larvae, passed from female to caterpillar. Severely infected individuals are weak, unable to expand their wings, or unable to eclose, and have shortened lifespans, but parasite levels vary in populations. Infection creates a culling effect, whereby infected migrating animals are less likely to complete the migration. This results in populations with lower parasite loads at the end of the migration. This is not the case in laboratory or commercial rearing, where after a few generations, all individuals can be infected.

Importance of Microbes

Microbes influence human society in countless ways. Sometimes, the influence of microorganisms on human life is beneficial, whereas at other times, it is detrimental. For example, microorganisms are required for the production of bread, cheese, yogurt, alcohol, wine, beer, antibiotics (e.g., penicillin, streptomycin, and chloramphenicol), vaccines, vitamins and enzymes. Many products of microbes contribute to public health as aids to nutrition, other products are used to interrupt the spread of disease, and still others hold promise for improving the quality of life in the years ahead. Microbes are also an important and essential component of an ecosystem. Molds and bacteria play key roles in the cycling of important nutrients in plant nutrition particularly those of carbon, nitrogen and sulphur. Bacteria referred to as nitrogen fixers live in the soil where they convert vast quantities of nitrogen in air into a form that plants can use. Microorganisms also play major roles in energy production. Natural gas (methane) is a product of bacterial activity, arising from the metabolism of methanogenic bacteria. Microorganisms are also being used to clean up pollution caused by human activities, a process called bioremediation (the introduction of microbes to restore stability to disturbed or polluted environments). Bacteria and fungi have been used to consume spilled oil, solvents, pesticides and other environmentally toxic substances.

Agricultural microbiology – try to combat plant diseases that attack important food crops, work on methods to increase soil fertility and crop yields etc. Currently there is a great interest in using bacterial or viral insect pathogens as substitute for chemical pesticides.

Microbial ecology – biogeochemical cycles – bioremediation to reduce pollution effects

Food and dairy microbiology – try to prevent microbial spoilage of food and transmission of food borne diseases such as botulism and salmonellolis. Use microorganisms to make foods such as cheese, yogurt, pickles and beers.

Industrial microbiology – used to make products such as antibiotics, vaccines, steroids, alcohols and other solvents, vitamins, amino acids and enzymes.

Microbial physiology and Biochemistry – study the synthesis of antibiotics and toxins, microbial energy production, microbial nitrogen fixation, effects of chemical and physical agents on microbial growth and survival etc.

Microbial genetics and Molecular biology – nature of genetic information and how it regulated the development and function of cells and organisms. Development of new microbial strains that are more efficient in synthesizing useful products.

Genetic engineering – arisen from work of microbial genetics and molecular biology. Engineered microorganisms are used to make hormones, antibiotics, vaccines and other products. New genes can be inserted into plants and animals.

Areas impacted by microbes include:

Medicine : Microbes produce valuable chemicals such as antibiotics. Many antibiotics are produced by common soil bacteria called Streptomyces and actinomycetes. The ability of Streptomyces cultures to inhibit the growth of other bacteria leads to discovery of many antibiotics. Streptomycin is an antibiotic and was the first used for tuberculosis. It is produced by actinobacterium *Streptomyces griseus*. Many antibiotics are produced by microbes such as Rifampicin produced by *Amycolatopsis rifamycinica,* Chloramphenicol by bacterium *Streptomyces venezuelae* and Actinomycin D produced by genus *Streptomyces etc.*

Industry: - The microorganisms are used for the production of food, either human or animal. Yogurt, cheese, chocolate, and silage (animal food) are all produced by industrial microbiology processes. Lactic acid bacteria and *Bifidobacteria* are amongst the most important groups of microorganisms used in the food industry. The microorganisms used in industrial processes may be natural isolates; laboratory selected mutants or genetically engineered organisms.

Ecology: - Microbial life plays a primary role in regulating biogeochemical systems in all environment such as frozen environments and acidic lakes, to hydrothermal vents at the bottom of deepest oceans, and human small intestine. Microbes, often engage in symbiotic relationships (either positive or negative) with other organisms, and these relationships affect the ecosystem. They are the backbone of all ecosystems. Other microbes are decomposers, with the ability to recycle nutrients from other organisms' waste products. These microbes play a vital role in biogeochemical cycles. The nitrogen cycle , the phosphorus cycle and the carbon cycle all depend on microorganisms in one way or another. Presently, microbiologists facing many challenges to solve many of society's problems including combating disease, reducing environmental pollution, and maintaining improving the world's food supply.

Future of microbiology:

- Future challenges such as finding new ways to combat disease, reduce pollution and feed the world's population.
- AIDS, hemorrhagic fevers and other infectious diseases
- Create new drugs, vaccines. Use the techniques in molecular biology and rDNA to solve the problems
- Host-pathogen relationships
- Study the role of microorganisms as
- Sources of high-quality food and other practical products such as enzymes for industrial application
- Degrade pollutants and toxic wastes
- Used as vectors to treat diseases and enhance agricultural productivity

Microbial diversity – less than 1% of the earth's microbial population has been cultured. Develop isolation techniques and work needs to be done on microorganisms living in extreme environments. Discovery of new organisms may lead to further advances in industrial processes and enhanced environmental control

- Microbe – microbe interactions.
- Analysis of genome – advances in the field of bioinformatics
- Symbiotic relationships – knowledge can help improve our appreciation of the living world, and improvements in the health of plant, livestock's and humans.

References

- Howland, John L. (2000). The Surprising Archaea: Discovering Another Domain of Life. Oxford: Oxford University Press. pp. 25–30. ISBN 0-19-511183-4
- Paerl HW, Fulton RS, Moisander PH, Dyble J (2001). "Harmful freshwater algal blooms, with an emphasis on cyanobacteria". ScientificWorldJournal. 1: 76–113. doi:10.1100/tsw.2001.16. PMID 12805693
- Borenstein, Seth (19 October 2015). "Hints of life on what was thought to be desolate early Earth". Excite. Yonkers, NY: Mindspark Interactive Network. Associated Press. Retrieved 2015-10-20
- Bernstein, H; Bernstein, C (2013). Bernstein, Carol, ed. Evolutionary Origin and Adaptive Function of Meiosis, Meiosis. InTech. ISBN 978-953-51-1197-9
- Jurtshuk, Peter (1996). Medical Microbiology (4th ed.). Galveston (TX): University of Texas Medical Branch at Galveston. Retrieved 5 November 2014
- Martin MO (2002). "Predatory prokaryotes an emerging research opportunity". Journal of Microbiology and Biotechnology. 4 (5): 467–77. PMID 12432957
- Koch A (1998). "How did bacteria come to be?". Adv Microb Physiol. Advances in Microbial Physiology. 40: 353–99. doi:10.1016/S0065-2911(08)60135-6. ISBN 978-0-12-027740-7. PMID 9889982
- Patrick J. Keeling (2004). "Diversity and evolutionary history of plastids and their hosts". American Journal of Botany. 91 (10): 1481–1493. doi:10.3732/ajb.91.10.1481. PMID 21652304
- Shand RF; Leyva KJ (2008). "Archaeal Antimicrobials: An Undiscovered Country". In Blum P. Archaea: New Models for Prokaryotic Biology. Caister Academic Press. ISBN 978-1-904455-27-1
- Zimmer, Carl (1 March 2017). "Scientists Say Canadian Bacteria Fossils May Be Earth's Oldest". New York Times. Retrieved 2 March 2017
- Guiry, M. D. (2012). "How Many Species of Algae Are There?". Journal of Phycology. 48 (5): 1057–1063. doi:10.1111/j.1529-8817.2012.01222.x. PMID 27011267
- Silva PC, Basson PW and Moe RL (1996) Catalogue of the Benthic Marine Algae of the Indian Ocean page 2, University of California Press. ISBN 978-0-520-91581-7
- The University of Waikato (March 25, 2014). "Bacterial DNA – the role of plasmids". Themes — Bacteria in biotech. Biotechnology Learning Hub. Retrieved 2014-09-03
- Williams, B. A.; Keeling, P. J. (2003). "Cryptic organelles in parasitic protists and fungi". In Littlewood, D. T. J. The Evolution of Parasitism. London: Elsevier Academic Press. p. 46. ISBN 0-12-031754-0
- Euzéby JP (8 December 2011). "Number of published names". List of Prokaryotic names with Standing in Nomenclature. Archived from the original on 19 January 2012. Retrieved 10 December 2011
- Nabors, Murray W. (2004). Introduction to Botany. San Francisco, CA: Pearson Education, Inc. ISBN 978-0-8053-4416-5
- "The Nobel Prize in Physiology or Medicine 1905". Nobelprize.org. Archived from the original on 10 December 2006. Retrieved 22 November 2006
- Simoons, Frederick J (1991). "6, Seaweeds and Other Algae". Food in China: A Cultural and Historical Inquiry. CRC Press. pp. 179–190. ISBN 978-0-936923-29-1
- "Pasteur's Papers on the Germ Theory". LSU Law Center's Medical and Public Health Law Site, Historic Public Health Articles. Archived from the original on 18 December 2006. Retrieved 23 November 2006
- Mondragón, Jennifer; Mondragón, Jeff (2003). Seaweeds of the Pacific Coast. Monterey, California: Sea Challengers Publications. ISBN 978-0-930118-29-7

3

Prokaryotic Cell Structure

Unicellular microorganisms lacking a membrane-bound nucleus are known as prokaryotes, in which bacteria are considered the most prominent. These microbes lack membrane-bound organelles and usually divide by a process known as binary fusion. Prokaryotes are the smallest microorganisms and are found in spherical, rod-like and spiral shapes. The chapter on prokaryotic cell structure offers an insightful focus, keeping in mind the complex subject matter.

Prokaryote

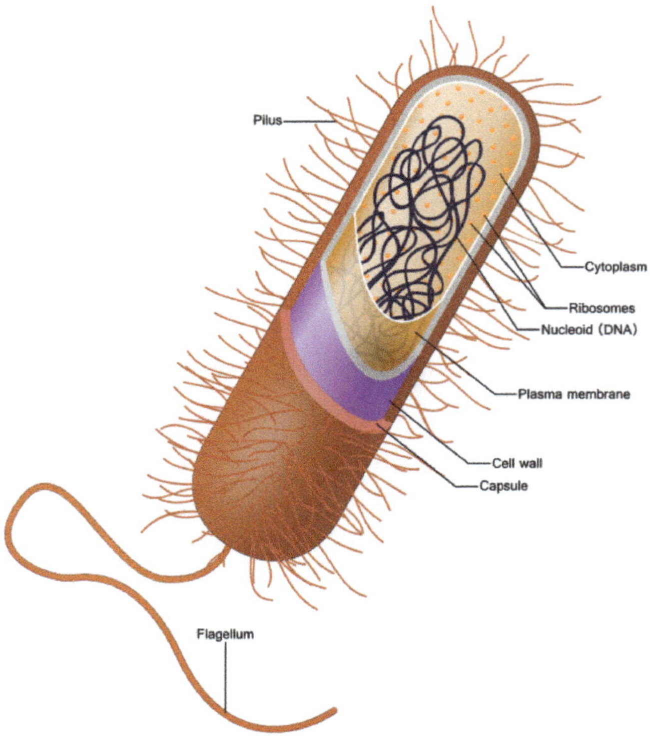

A diagram of a typical prokaryotic bacteria cell.

A prokaryote is a unicellular organism that lacks a membrane-bound nucleus (karyon), mitochondria, or any other membrane-bound organelle. The word *prokaryote* comes from the Greek (*pro*) "before" and (*karyon*) "nut or kernel". Prokaryotes can be divided into two domains, Archaea and bacteria. In contrast, species with nuclei and organelles are placed in the domain Eukaryota.

In the prokaryotes, all the intracellular water-soluble components (proteins, DNA and metabolites) are located together in the cytoplasm enclosed by the cell membrane, rather than in separate cellular compartments. Bacteria, however, do possess protein-based bacterial microcompartments, which are thought to act as primitive organelles enclosed in protein shells. Some prokaryotes, such as cyanobacteria may form large colonies. Others, such as myxobacteria, have multicellular stages in their life cycles.

Molecular studies have provided insight into the evolution and interrelationships of the three domains of biological species. Eukaryotes are organisms, including humans, whose cells have a well defined membrane-bound nucleus (containing chromosomal DNA) and organelles. The division between prokaryotes and eukaryotes reflects the existence of two very different levels of cellular organization. Distinctive types of prokaryotes include extremophiles and methanogens; these are common in some extreme environments.

Structure

Prokaryotes have a prokaryotic cytoskeleton, albeit more primitive than that of the eukaryotes. Besides homologues of actin and tubulin (MreB and FtsZ), the helically arranged building-block of the flagellum, flagellin, is one of the most significant cytoskeletal proteins of bacteria, as it provides structural backgrounds of chemotaxis, the basic cell physiological response of bacteria. At least some prokaryotes also contain intracellular structures that can be seen as primitive organelles. Membranous organelles (or intracellular membranes) are known in some groups of prokaryotes, such as vacuoles or membrane systems devoted to special metabolic properties, such as photosynthesis or chemolithotrophy. In addition, some species also contain carbohydrate-enclosed microcompartments, which have distinct physiological roles (e.g. carboxysomes or gas vacuoles).

Most prokaryotes are between 1 µm and 10 µm, but they can vary in size from 0.2 µm (*Mycoplasma genitalium*) to 750 µm (*Thiomargarita namibiensis*).

Prokaryotic Cell Structure
Flagellum (only in some types of prokaryotes) Long, whip-like protrusion that aids cellular locomotion.
Cell membrane Surrounds the cell's cytoplasm and regulates the flow of substances in and out of the cell.
Cell wall (except genera Mycoplasma and Thermoplasma) Outer covering of most cells that protects the bacterial cell and gives it shape.
Cytoplasm A gel-like substance composed mainly of water that also contains enzymes, salts, cell components, and various organic molecules.
Ribosome Cell structures responsible for protein production.
Nucleoid Area of the cytoplasm that contains the prokaryote's single DNA molecule.
Glycocalyx (only in some types of prokaryotes) A glycoprotein-polysaccharide covering that surrounds the cell membranes.
Inclusions

Morphology

Prokaryotic cells have various shapes; the four basic shapes of bacteria are:

- Cocci – spherical
- Bacilli – rod-shaped
- Spirochaete – spiral-shaped
- Vibrio – comma-shaped

The archaeon Haloquadratum has flat square-shaped cells.

Reproduction

Bacteria and archaea reproduce through asexual reproduction, usually by binary fission. Genetic exchange and recombination still occur, but this is a form of horizontal gene transfer and is not a replicative process, simply involving the transference of DNA between two cells, as in bacterial conjugation.

DNA Transfer

DNA transfer between prokaryotic cells occurs in bacteria and archaea, although it has been mainly studied in bacteria. In bacteria, gene transfer occurs by three processes. These are (1) bacterial virus (bacteriophage)-mediated transduction, (2) plasmid-mediated conjugation, and (3) natural transformation. Transduction of bacterial genes by bacteriophage appears to reflect an occasional error during intracellular assembly of virus particles, rather than an adaptation of the host bacteria. The transfer of bacterial DNA is under the control of the bacteriophage's genes rather than bacterial genes. Conjugation in the well-studied *E. coli* system is controlled by plasmid genes, and is an adaptation for distributing copies of a plasmid from one bacterial host to another. Infrequently during this process, a plasmid may integrate into the host bacterial chromosome, and subsequently transfer part of the host bacterial DNA to another bacterium. Plasmid mediated transfer of host bacterial DNA (conjugation) also appears to be an accidental process rather than a bacterial adaptation.

Natural bacterial transformation involves the transfer of DNA from one bacterium to another through the intervening medium. Unlike transduction and conjugation, transformation is clearly a bacterial adaptation for DNA transfer, because it depends on numerous bacterial gene products that specifically interact to perform this complex process. For a bacterium to bind, take up and recombine donor DNA into its own chromosome, it must first enter a special physiological state called competence. About 40 genes are required in *Bacillus subtilis* for the development of competence. The length of DNA transferred during *B. subtilis* transformation can be as much as a third to the whole chromosome. Transformation is a common mode of DNA transfer, and 67 prokaryotic species are thus far known to be naturally competent for transformation.

Among archaea, *Halobacterium volcanii* forms cytoplasmic bridges between cells that appear to be used for transfer of DNA from one cell to another. Another archaeon, *Sulfolobus solfataricus*, transfers DNA between cells by direct contact. Frols et al. found that exposure of *S. solfataricus* to

DNA damaging agents induces cellular aggregation, and suggested that cellular aggregation may enhance DNA transfer among cells to provide increased repair of damaged DNA via homologous recombination.

Sociality

While prokaryotes are considered strictly unicellular, most can form stable aggregate communities. When such communities are encased in a stabilizing polymer matrix ("slime"), they may be called "biofilms". Cells in biofilms often show distinct patterns of gene expression (phenotypic differentiation) in time and space. Also, as with multicellular eukaryotes, these changes in expression often appear to result from cell-to-cell signaling, a phenomenon known as quorum sensing.

Biofilms may be highly heterogeneous and structurally complex and may attach to solid surfaces, or exist at liquid-air interfaces, or potentially even liquid-liquid interfaces. Bacterial biofilms are often made up of microcolonies (approximately dome-shaped masses of bacteria and matrix) separated by "voids" through which the medium (e.g., water) may flow easily. The microcolonies may join together above the substratum to form a continuous layer, closing the network of channels separating microcolonies. This structural complexity—combined with observations that oxygen limitation (a ubiquitous challenge for anything growing in size beyond the scale of diffusion) is at least partially eased by movement of medium throughout the biofilm—has led some to speculate that this may constitute a circulatory system and many researchers have started calling prokaryotic communities multicellular (for example). Differential cell expression, collective behavior, signaling, programmed cell death, and (in some cases) discrete biological dispersal events all seem to point in this direction. However, these colonies are seldom if ever founded by a single founder (in the way that animals and plants are founded by single cells), which presents a number of theoretical issues. Most explanations of co-operation and the evolution of multicellularity have focused on high relatedness between members of a group (or colony, or whole organism). If a copy of a gene is present in all members of a group, behaviors that promote cooperation between members may permit those members to have (on average) greater fitness than a similar group of selfish individuals.

Should these instances of prokaryotic sociality prove to be the rule rather than the exception, it would have serious implications for the way we view prokaryotes in general, and the way we deal with them in medicine. Bacterial biofilms may be 100 times more resistant to antibiotics than free-living unicells and may be nearly impossible to remove from surfaces once they have colonized them. Other aspects of bacterial cooperation—such as bacterial conjugation and quorum-sensing-mediated pathogenicity, present additional challenges to researchers and medical professionals seeking to treat the associated diseases.

Environment

Prokaryotes have diversified greatly throughout their long existence. The metabolism of prokaryotes is far more varied than that of eukaryotes, leading to many highly distinct prokaryotic types. For example, in addition to using photosynthesis or organic compounds for energy, as eukaryotes do, prokaryotes may obtain energy from inorganic compounds such as hydrogen sulfide. This enables prokaryotes to thrive in harsh environments as cold as the snow surface of Antarctica, studied in cryobiology or as hot as undersea hydrothermal vents and land-based hot springs.

Prokaryotes live in nearly all environments on Earth. Some archaea and bacteria thrive in harsh conditions, such as high temperatures (thermophiles) or high salinity (halophiles). Organisms such as these are referred to as extremophiles. Many archaea grow as plankton in the oceans. Symbiotic prokaryotes live in or on the bodies of other organisms, including humans.

Classification

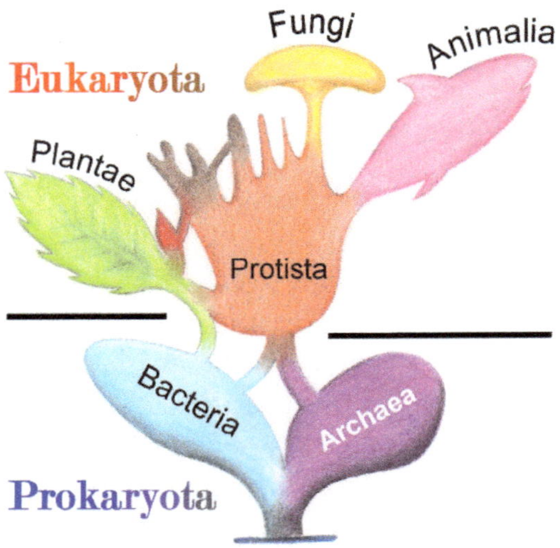

Phylogenetic and symbiogenetic tree of living organisms, showing the origins of eukaryotes and prokaryotes

In 1977, Carl Woese proposed dividing prokaryotes into the Bacteria and Archaea (originally Eubacteria and Archaebacteria) because of the major differences in the structure and genetics between the two groups of organisms. Archaea were originally thought to be extremophiles, living only in inhospitable conditions such as extremes of temperature, pH, and radiation but have since been found in all types of habitats. The resulting arrangement of Eukaryota (also called "Eukarya"), Bacteria, and Archaea is called the three-domain system, replacing the traditional two-empire system.

One criticism of the dichotomous prokaryote-eukaryote distinction points out that the word "prokaryote" is based on what these organisms are not (they are not eukaryotic), rather than what they are (either archaea or bacteria).

Evolution

The current model of the evolution of the first living organisms is that these were some form of prokaryotes, which may have evolved out of protocells. In general, the eukaryotes are thought to have evolved later in the history of life.

However, some authors have questioned this conclusion, arguing that the current set of prokaryotic species may have evolved from more complex eukaryotic ancestors through a process of simplification.

Others have argued that the three domains of life arose simultaneously, from a set of varied cells that formed a single gene pool.

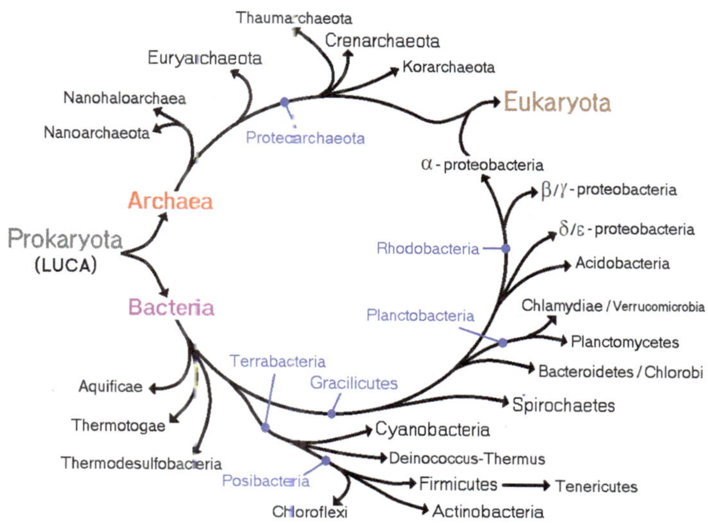

Phylogenetic ring showing the diversity of prokaryotes, and symbiogenetic origins of eukaryotes

This controversy was summarized in 2005:

There is no consensus among biologists concerning the position of the eukaryotes in the overall scheme of cell evolution. Current opinions on the origin and position of eukaryotes span a broad spectrum including the views that eukaryotes arose first in evolution and that prokaryotes descend from them, that eukaryotes arose contemporaneously with eubacteria and archeabacteria and hence represent a primary line of descent of equal age and rank as the prokaryotes, that eukaryotes arose through a symbiotic event entailing an endosymbiotic origin of the nucleus, that eukaryotes arose without endosymbiosis, and that eukaryotes arose through a symbiotic event entailing a simultaneous endosymbiotic origin of the flagellum and the nucleus, in addition to many other models, which have been reviewed and summarized elsewhere.

The oldest known fossilized prokaryotes were laid down approximately 3.5 billion years ago, only about 1 billion years after the formation of the Earth's crust. Eukaryotes only appear in the fossil record later, and may have formed from endosymbiosis of multiple prokaryote ancestors. The oldest known fossil eukaryotes are about 1.7 billion years old. However, some genetic evidence suggests eukaryotes appeared as early as 3 billion years ago.

While Earth is the only place in the universe where life is known to exist, some have suggested that there is evidence on Mars of fossil or living prokaryotes. However, this possibility remains the subject of considerable debate and skepticism.

Relationship to Eukaryotes

The division between prokaryotes and eukaryotes is usually considered the most important distinction or difference among organisms. The distinction is that eukaryotic cells have a "true" nucleus containing their DNA, whereas prokaryotic cells do not have a nucleus. Both eukaryotes and prokaryotes contain large RNA/protein structures called ribosomes, which produce protein.

Another difference is that ribosomes in prokaryotes are smaller than in eukaryotes. However, two organelles found in many eukaryotic cells, mitochondria and chloroplasts, contain ribosomes similar in size and makeup to those found in prokaryotes. This is one of many pieces of evidence that mitochondria and chloroplasts are themselves descended from free-living bacteria. This theory holds that early eukaryotic cells took in primitive prokaryotic cells by phagocytosis and adapted themselves to incorporate their structures, leading to the mitochondria we see today.

The genome in a prokaryote is held within a DNA/protein complex in the cytosol called the nucleoid, which lacks a nuclear envelope. The complex contains a single, cyclic, double-stranded molecule of stable chromosomal DNA, in contrast to the multiple linear, compact, highly organized chromosomes found in eukaryotic cells. In addition, many important genes of prokaryotes are stored in separate circular DNA structures called plasmids.

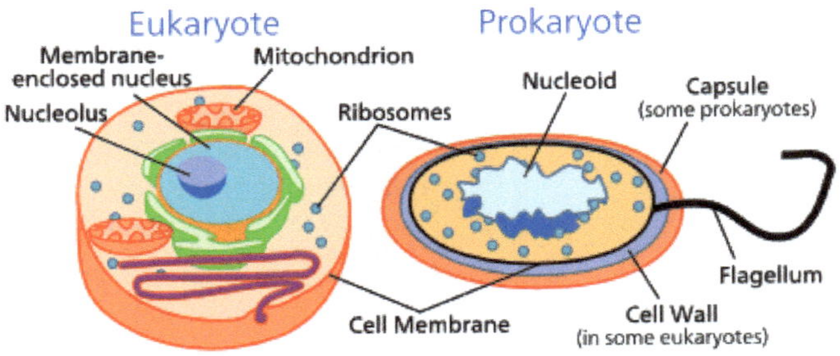

Comparison of eukaryotes vs. prokaryotes

Prokaryotes lack mitochondria and chloroplasts. Instead, processes such as oxidative phosphorylation and photosynthesis take place across the prokaryotic cell membrane. However, prokaryotes do possess some internal structures, such as prokaryotic cytoskeletons. It has been suggested that the bacterial order Planctomycetes have a membrane around their nucleoid and contain other membrane-bound cellular structures. However, further investigation revealed that Planctomycetes cells are not compartmentalized or nucleated and like the other bacterial membrane systems are all interconnected.

Prokaryotic cells are usually much smaller than eukaryotic cells. Therefore, prokaryotes have a larger surface-area-to-volume ratio, giving them a higher metabolic rate, a higher growth rate, and as a consequence, a shorter generation time than eukaryotes.

The chief distinguishing characteristics of prokaryotic cells are:

- Their genetic material (DNA) is not enclosed within a membrane.
- They lack other membrane-bounded organelles
- Their DNA is not associated with histone proteins (special chromosomal proteins found in eukaryotes).
- Their cell walls most of the time contains the complex polysaccharide peptidoglycan.
- They usually divide by binary fission. During this process, the DNA is copied and the cell splits into two cells.

Bacterial Cell Structure

Prokaryotes are among the smallest of all organisms (0.5 to 2.0 m m). Because of their small size, bacteria have a large surface-to-volume ratio. The smallest member of the genus is about 0.3μm in diameter. Even smaller cells have been reported like the nanobacteria or ultramicrobacteria appear to range from around 0.2μm to not less than 0.05μm. *E. coli* , a bacillusof about average size is 1.1 to 1.5 μm wide by 2.0 to 6.0 μm long. Spirochaetes occasionally reach 500 μm in length and the cyano-bacterium *Oscillatoria* is about 7 μm in diameter. The bacterium, *Epulosiscium fishelsoni* , can be seen with the naked eye (600 m m long by 80 m m in diameter). Thus a few bacteria are much larger than the average eukaryotic cell (typical plant and animal cells are around 10 to 50 μm in diameter).

Shape and arrangement:

Typically bacteria have three basic shapes – spherical, rod like and spiral.

Spherical bacteria :

Coccus (pluralcocci) in pairs– diploccccic; in chains– Streptococci (Streptococcus); in cube like groups of eight– Sarcinae (Sarcina); in grape like structures– Staphylococci (*Staphylococcus*).

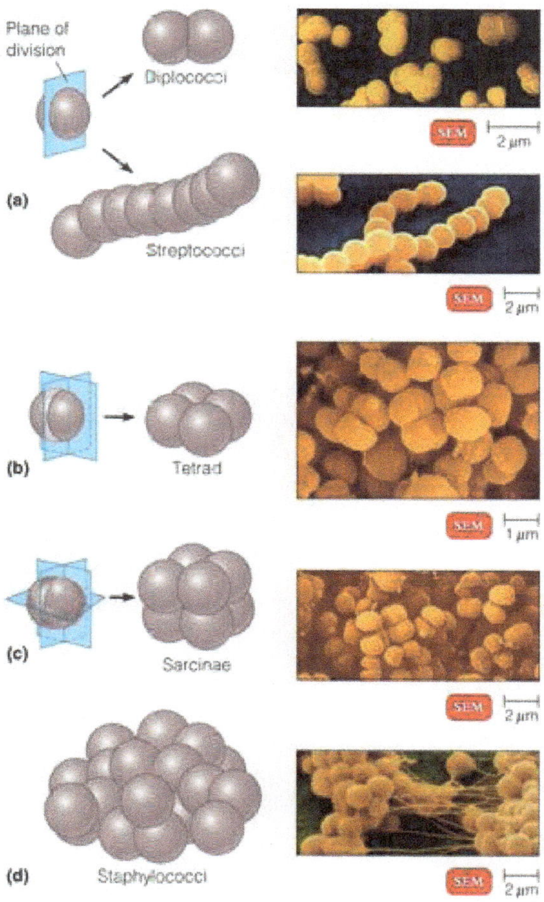

Arrangement of spherical cells

Rod like bacteria - Bacillus (plural bacilli)

In pairs – diplococci; In chains – streptococci. Still others look like cocci and are called coccoba-cilli.

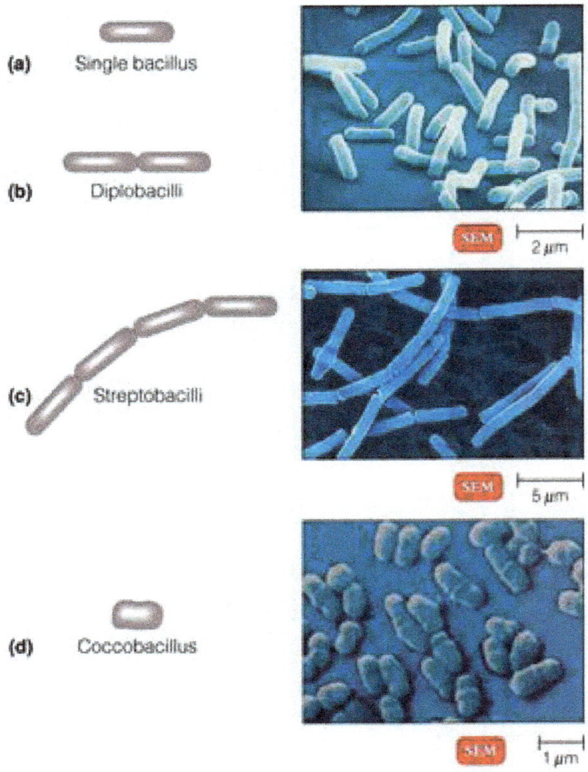

Arrangement of rod like cells

Spiral Bacteria

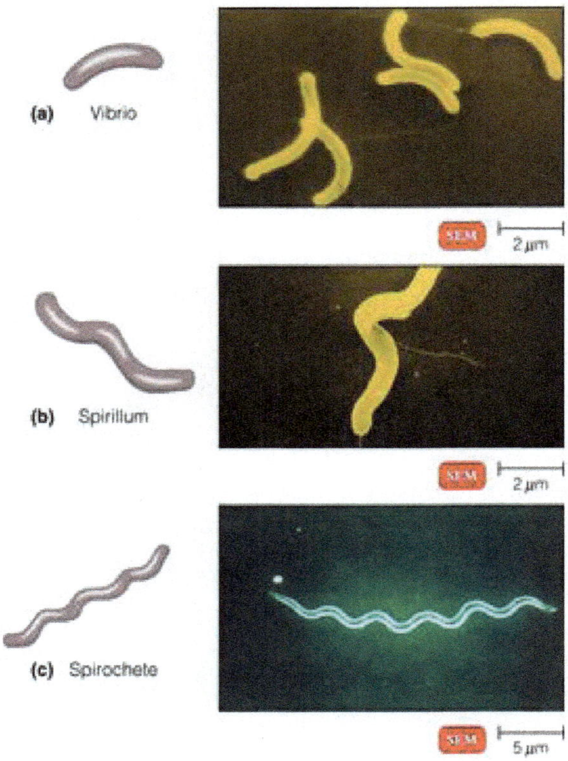

Sprial bacteria

Comma shaped – Vibrio (*Vibrio*); Helical, long and curved – Spirilla (*Rhodospirillium*); Helical and flexible – Sprirochetes (*Treponema* and *Borrelia*) movement by axial filaments.

In addition there are: Star shaped cells (genus *Stella*); Square flat cells (halophilic archaebacteria); Triangular cells (*Haloarcula*); Branching mycelium – like filaments (*Streptomyces*).

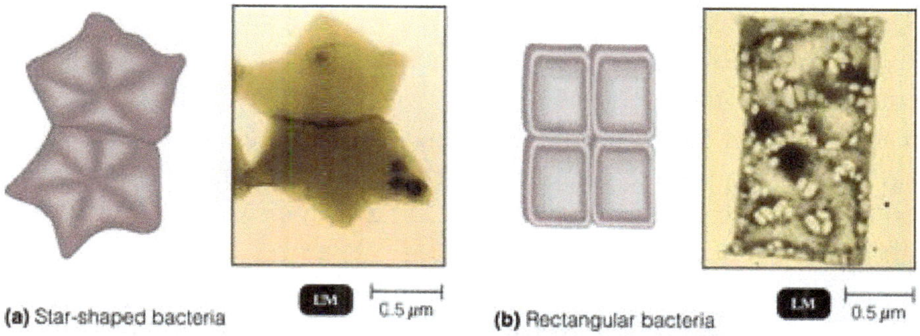

Star shaped and rectangular shaped bacteria

Genetically most bacteria are monomorphic (single shape), but some have many shapes and are called pleomorphic (*Rhizobium* and *Corynebacterium*).

Structure:

A variety of structures are found I prokaryotic cells. The cells are almost always are bounded by a chemically complex cell wall. In between the cell wall and the periplasmic space lies the plasma membrane. Internal membrane bound organelles are absent in prokaryotes and hence the internal is morphologically simple. The genetic material is localized in a discrete region called nucleoid and not separated from the surrounding cytoplasm by membrane. Ribososmes and inclusions are found scattered in the cytoplasmic matrix. Bacteria use flagella for locomotion, and in addition, many cells are surrounded by a capsule or slime layer external to the cell wall.

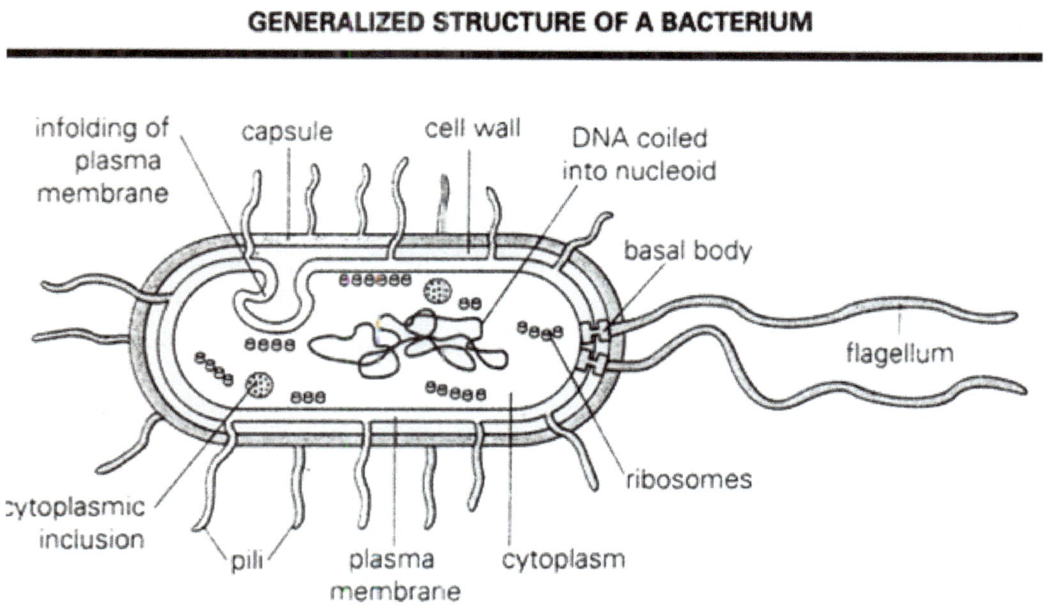

Generalized structure of a bacterium

The Bacterial Cell Wall

Is a complex, semi-rigid structure that surrounds the underlying plasma membrane and protects the cell and the internal parts from adverse surrounding environment. Except for Mycoplasmas and some Achaea most bacteria have strong walls that give them shape and protect them from osmotic lysis; wall shape and strength is primarily due to peptidoglycan. The cell walls of pathogenic bacteria have components that contribute to their pathogenicity and protect a cell from toxic substances and are the site of action of several antibiotics. Major functions – prevent cells from rupturing, maintains shape and serves as a point of anchorage for flagella. Christian Gram developed the Gram stain in 1884, and it became evident that bacteria could be divided into two major groups based on their response to staining. Gram positive bacteria stained purple and the gram negative ones colored pink or red by the technique. The gram positive cell wall consists of a single thick layer of 20 to 80 nm of peptidoglycan or murein layer lying outside the cell membrane. In gram negative cells, the peptidoglycan layer is thin of about 2 to 7 nm surrounded by a 7 to 8 nm thick outer membrane. The gram positive cells are stronger than gram negative ones because of the peptidoglycan layer.

A space is seen between the plasma membrane and the outer membrane in electron micrographs of gram negative bacteria and a similar one but smaller gap may be observed between the plasma membrane and wall in gram positive bacteria. This space is called periplasmic space. The substance that occupies the space is the periplasm. The periplasmic space in gram negative bacteria ranges from 1nm to as great as 71 nm. It contains many proteins and participates in nutrient acquisition, hydrolytic enzymes attacking nucleic acids and phosphorylated molecules. The periplasmic space contains enzymes involved in peptidoglycan synthesis and modification of toxic compounds that could harm the cell. Gram positive bacteria may not have a visible periplasmic space but appear to have as many periplasmic proteins and several enzymes that are secreted are often called exozymes.

The Archaea differs from other prokaryotes. Thewalls lack peptidoglycan and are composed of proteins, glycoprotein or polysaccharides.

Structure of Peptidoglycan

The cell wall of bacteria is composed of a macromolecular network of polymer called peptidoglycan (murein). The polymer contains two sugar derivatives, Monosaccharides- N-acetylglucosamine (NAG) and N-acetylmuramicacid (NAM). In addition, several different amino acids, three of which – D-Glutamic acid, D-alanine, and mesodiaminopimelic acid – not found in proteins are present. The pretense of D-amino acids protects against attack by most peptidases. The backbone of this polymer is composed of alternating N-acetylglucosamine and N-acetylmuramic acid residues. A peptide chain of four alternating D- and L- amino acids is connected to the carboxyl group of N-acetylmuramic acid. Many bacteria substitute another diaminoacid, usually L-Lysine, in the third position for meso-diaminopimelic acid. Chains of linked peptidoglycan subunits are joined by cross-links between the peptides. Often the carboxyl group of the terminal D-alanine is connected directly to the amino group of diaminopimelic acid, but a peptide interbridge may be used instead. Most gram-negative cell wall peptidoglycan lacks the peptide interbridge.

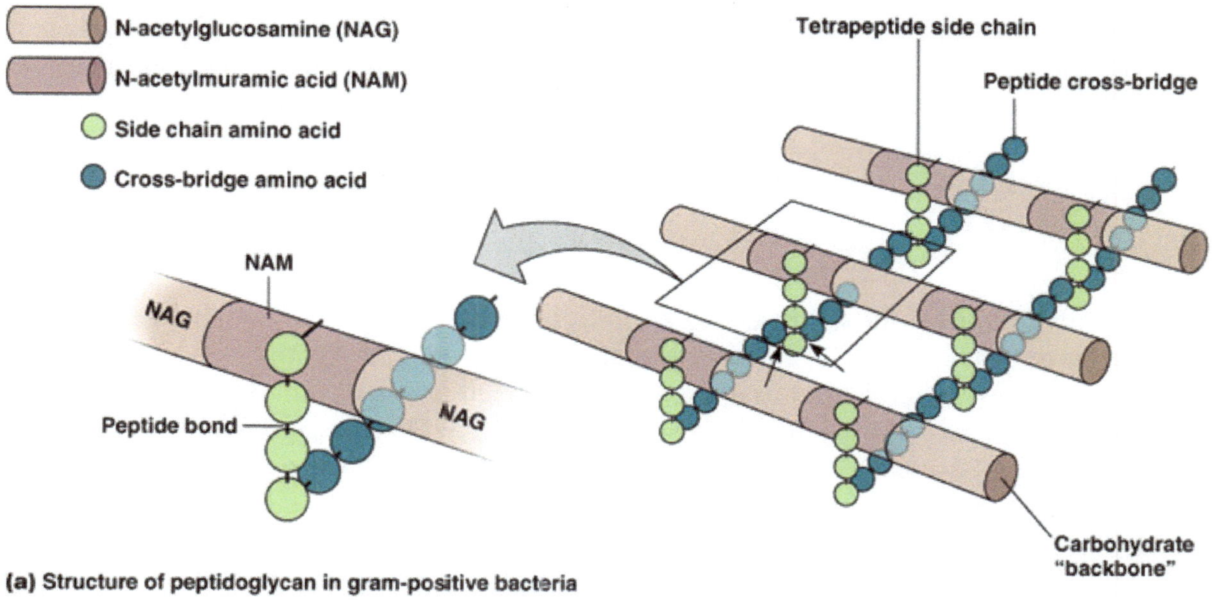

Peptidoglycan structure in gram-positive cell

Gram Positive Cell Walls

The cell wall consists of many layers of peptidoglycan, forming a thick, rigid structure, which contains a peptide interbridge. Gram positive cell walls also contain teichoic acids, polymers of glycerol and ribitol joined by phosphate groups. To the glycerol and ribotol groups, the amino acids such as D-alanine or sugars like glucose are attached. There are two classes to teichoic acids; lipoteichoic acid (if they are attached to the lipid of the plasma membrane) and wall teichoic acid (which extend to the surface of the peptidoglycan and are negatively charged). Teichoic acids are not found in gram-negative bacteria.

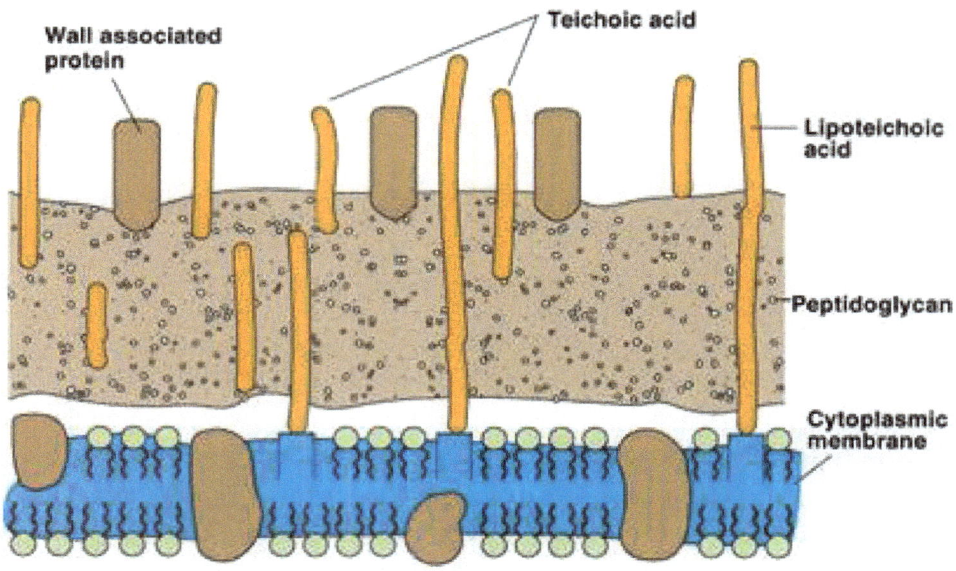

Cell wall of Gram positive bacteria

Some bacteria like *Staphylococci* and most other gram-positive bacteria have a layer of proteins on the surface of their cell wall peptidoglycan. Some are noncovalently attached by binding to the peptidoglycan, teichoic acids and other receptors and these proteins are involved in the interactions of the cell with its environment. Ex. S-layer proteins. Other surface proteins are covalently attached to the peptidoglycan and in gram positive pathogens these have roles such as aiding in adhesion to host tissues, preventing opsonization and blocking phagocytosis.

Gram Negative Cell Walls

The cell wall contains only a thin layer of peptidoglycan. In *E. coli,* it is about 2-3nmthick and contains only one or two layers of sheets of peptidoglycan. The peptidoglycan is bounded to lipoproteins in the outer membrane and is embedded in a soft material, the periplasmic gel. Gram negative cell walls do not contain teichoic acids. They are more susceptible to mechanical breakage because of small amount of peptidoglycan. Outer membrane - found primarily in gram-negative cell consists of lipoproteins, liposaccharides and phospholipids and lies outside the thin peptidoglycan layer. The most abundant membrane protein is Braun's lipoprotein, a small lipoprotein covalently joined to the underlying peptidoglycan and embedded in the outer membrane by its hydrophobic end. The outer membrane e and plasma membrane appear to be in direct contact at many locations in the cell wall. The adhesion sites may be regions of direct contact or possibly true membrane fusions. It is proposed that substances can move into the cell through these adhesions sites rather than travelling through the periplasm.

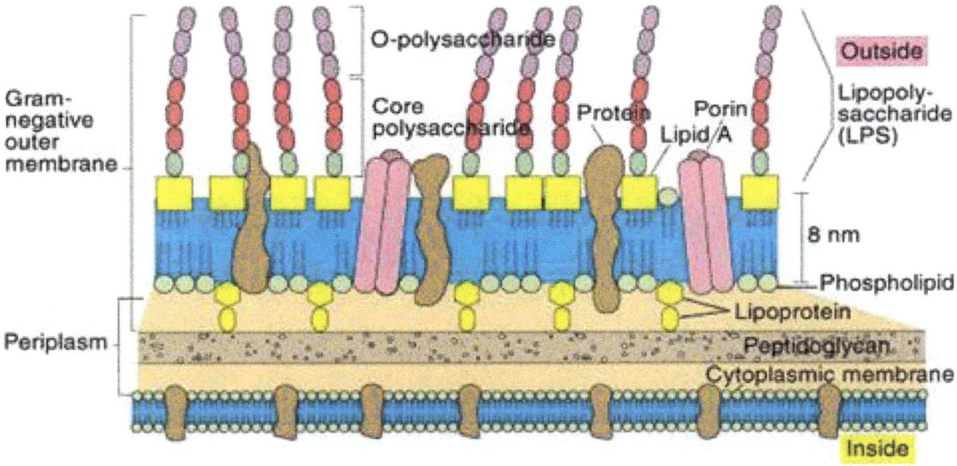

Cell wall of gram negative bacteria

The most unusual constituents of the outermembrane are its lipopolysaccharides. (LPSs). These contain both lipid and carbohydrate and consist of three parts: 1. Lipid A, 2. the core polysaccharide, and 3. the O side chain. The lipid A region contains two glucosamine sugar derivatives, each with three fatty acids and phosphate or pyrophosphate attached. This region is buried in the outer membrane and the remaining portion projects from the surface. The second region or the core polysaccharide is joined to lipid A and the third region or the O side chain which is a polysaccharide chain extends outward from the core. The O side chain is readily recognized by host antibodies, gram negative bacteria may thwart host defenses by readily changing the nature of their O side chains to avoid detection. The core polysaccharide contains charged sugars and phosphate, LPS contributes to the negative charge on the bacterial surface. Lipid A is a major constituent of the outer membrane, and the LPS helps stabilize the membrane structure. Lipid A is often toxic and as a result LPS can act as an endotoxin and cause some symptoms that arise in gram-negative bacterial infections. The function of the outer membrane is to serve as a protective barrier. It prevents or slows the entry of bile salts, antibiotics, and other toxic substances that might kill or injure the bacterium. Provides a barrier to certain antibiotics (like penicillin), lysozyme detergents, heavy metals, bile salts and digestive. Porins are proteins that permit small molecules to pass through the outer membrane (about 800 MW). S pecific channel proteins – allow specific substances like vitamin B12, iron, nucleotides and maltose.

Cell wall of gram negative bacteria Atypical cell walls:

- *Mycoplasma* – is a bacterial genus that naturally lacks cell walls. Their plasma membranes have lipids called sterols, which protect them from osmotic lysis.
- Archaeabacteria – have pseudomurein (N-acetylalosaminuronic acid) but no peptidoglycan.
- L forms - are mutant bacteria with defective cell walls .

Damage to the Cell Wall

The presence of cell wall is essential to protect bacteria against destruction by osmotic pressure. The bacterial cytoplasm is much more concentrated with solutes than in most microbial habitats which are hypotonic. During osmosis, water moves across selectively permeable membranes such as the plasma membrane from dilute solutions (higher water concentration) to more concentrated solutions (lower water concentration). Usually water generally enters the bacterial cells and the osmotic pressure may reach 20 atmospheres. Plasma membrane cannot resist such high pressures and the cell will swell and be physically disrupted and destroyed, a process called lysis. In hypertonic habitats, the water flows outward, and the cytoplasm shrivels up and pulls away from the cell wall. This phenomenon is called plamolysis and is useful in food preservation because many microorganisms cannot grow in dried foods and jellies as they cannot avoid plasmolysis. The importance of the cell wall in protecting bacteria can be demonstrated by treatment with lysozyme (naturally occurs in eukaryotic cells and is a constituent of tears, mucus and saliva), which attacks the peptidoglycan by hydrolyzing the bond that connects NAM and NAG units. Penicillin inhibits peptidoglycan synthesis. Gram positive cell walls are destroyed and the remaining cellular contents are referred to as protoplast .Gram negative cells are not completely destroyed and the remaining cellular contents are referred to as spheroplast. Protoplast and spheroplast are subject to osmotic lysis. Antibiotics such as penicillin destroy bacteria by interfering with the formation of the peptide cross bridges of peptidoglycan and ultimately cell wall synthesis.

Gram-positive Bacteria

Gram-positive bacteria are bacteria that give a positive result in the Gram stain test. Gram-positive bacteria take up the crystal violet stain used in the test, and then appear to be purple-coloured when seen through a microscope. This is because the thick peptidoglycan layer in the bacterial cell wall retains the stain after it is washed away from the rest of the sample, in the decolorization stage of the test.

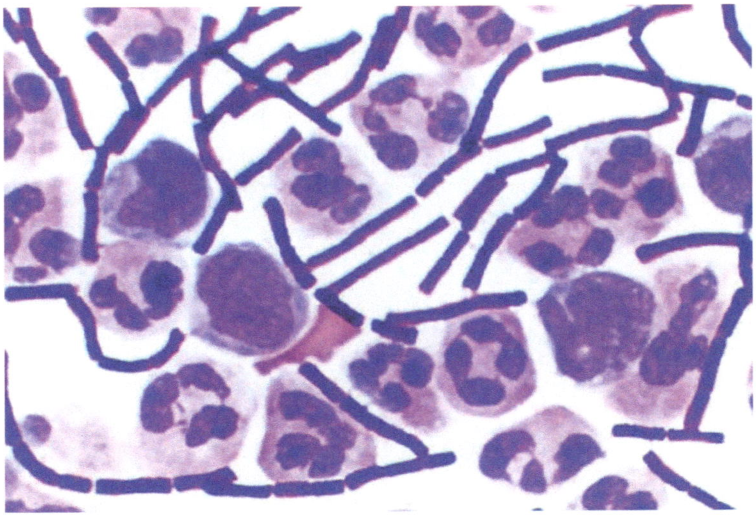

Rod-shaped gram-positive *Bacillus anthracis* bacteria in a cerebrospinal fluid sample stand out from round white blood cells.

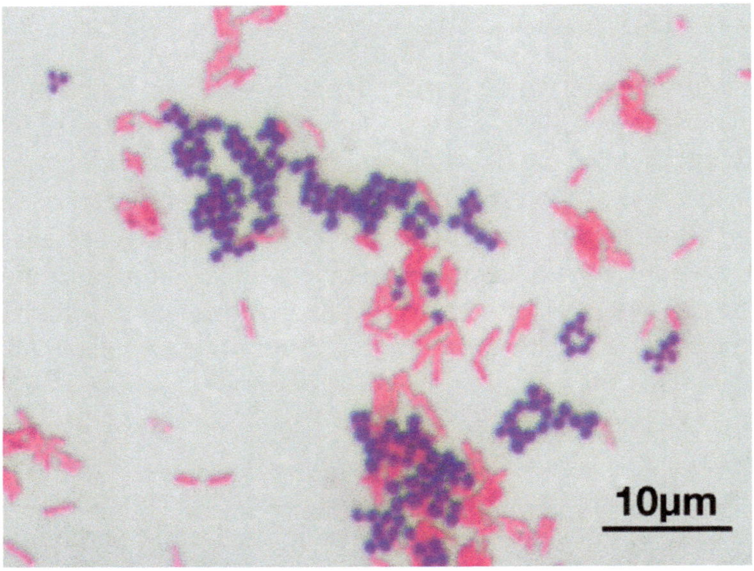

Violet stained gram-positive cocci and pink stained gram-negative rod-shaped bacteria

Gram-negative bacteria cannot retain the violet stain after the decolorization step; alcohol used in this stage degrades the outer membrane of gram-negative cells making the cell wall more porous and incapable of retaining the crystal violet stain. Their peptidoglycan layer is much thinner and sandwiched between an inner cell membrane and a bacterial outer membrane, causing them to take up the counterstain (safranin or fuchsine) and appear red or pink.

Despite their thicker peptidoglycan layer, gram-positive bacteria are more receptive to antibiotics than gram-negative, due to the absence of the outer membrane.

Characteristics

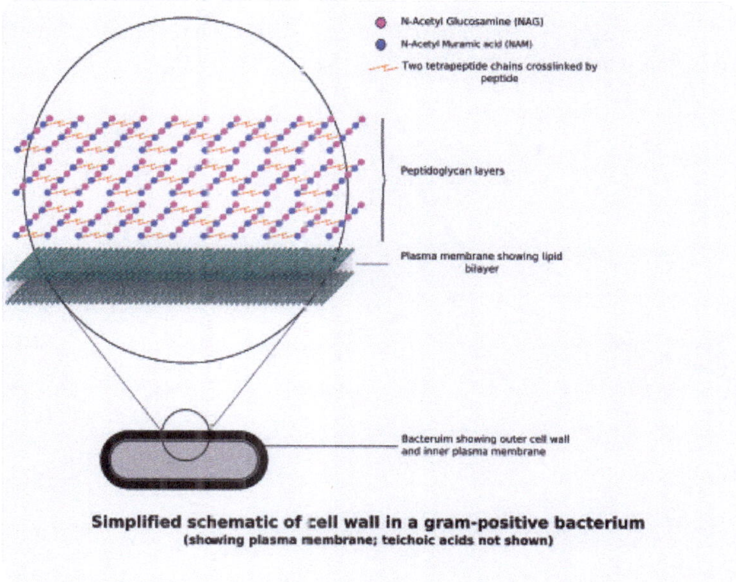

Structure of gram-positive cell wall

In general, the following characteristics are present in gram-positive bacteria:

1. Cytoplasmic lipid membrane
2. Thick peptidoglycan layer
3. Teichoic acids and lipoids are present, forming lipoteichoic acids, which serve as chelating agents, and also for certain types of adherence.
4. Peptidoglycan chains are cross-linked to form rigid cell walls by a bacterial enzyme DD-transpeptidase.
5. A much smaller volume of periplasm than that in gram-negative bacteria.

Only some species have a capsule usually consisting of polysaccharides. Also only some species are flagellates, and when they do have flagella they only have two basal body rings to support them (gram-negative have four). Both gram-positive and gram-negative bacteria commonly have a surface layer called an S-layer. In gram-positive bacteria, the S-layer is attached to the peptidoglycan layer (in gram-negative bacteria, the S-layer is attached directly to the outer membrane). Specific to gram-positive bacteria is the presence of teichoic acids in the cell wall. Some of these are lipoteichoic acids, which have a lipid component in the cell membrane that can assist in anchoring the peptidoglycan.

Classification

Along with cell shape, Gram staining is a rapid method used to differentiate bacterial species. Such staining, together with growth requirement and antibiotic susceptibility testing, and other macroscopic and physiologic tests, forms the full basis for classification and subdivision of the bacteria.

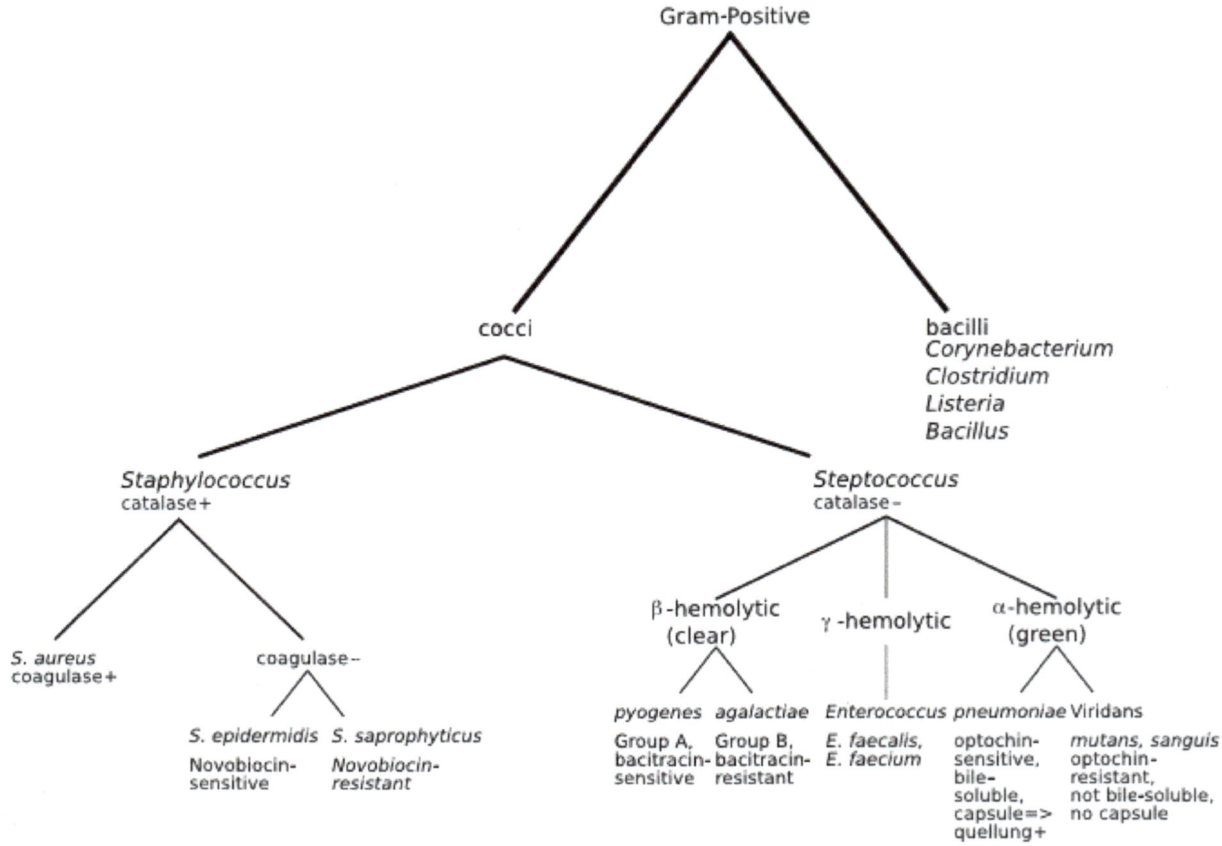

Species identification hierarchy in clinical settings

Historically, the kingdom Monera was divided into four divisions based primarily on Gram staining: Firmicutes (positive in staining), Gracilicutes (negative in staining), Mollicutes (neutral in staining) and Mendocutes (variable in staining). Based on 16S ribosomal RNA phylogenetic studies of the late microbiologist Carl Woese and collaborators and colleagues at the University of Illinois, the monophyly of the gram-positive bacteria was challenged, with major implications for the therapeutic and general study of these organisms. Based on molecular studies of the 16S sequences, Woese recognised twelve bacterial phyla. Two of these were both gram-positive and were divided on the proportion of the guanine and cytosine content in their DNA. The high G + C phylum was made up of the Actinobacteria and the low G + C phylum contained the Firmicutes. The Actinobacteria include the *Corynebacterium*, *Mycobacterium*, *Nocardia* and *Streptomyces* genera. The (low G + C) Firmicutes, have a 45–60% GC content, but this is lower than that of the Actinobacteria.

Importance of the Outer Cell Membrane in Bacterial Classification

Although bacteria are traditionally divided into two main groups, gram-positive and gram-negative, based on their Gram stain retention property, this classification system is ambiguous as it refers to three distinct aspects (staining result, envelope organization, taxonomic group), which do not necessarily coalesce for some bacterial species. The gram-positive and gram-negative staining response is also not a reliable characteristic as these two kinds of bacteria do not form phylogenetic coherent groups. However, although Gram staining response is an empirical criterion, its basis lies in the marked differences in the ultrastructure and chemical composition of the bacterial cell wall, marked by the absence or presence of an outer lipid membrane.

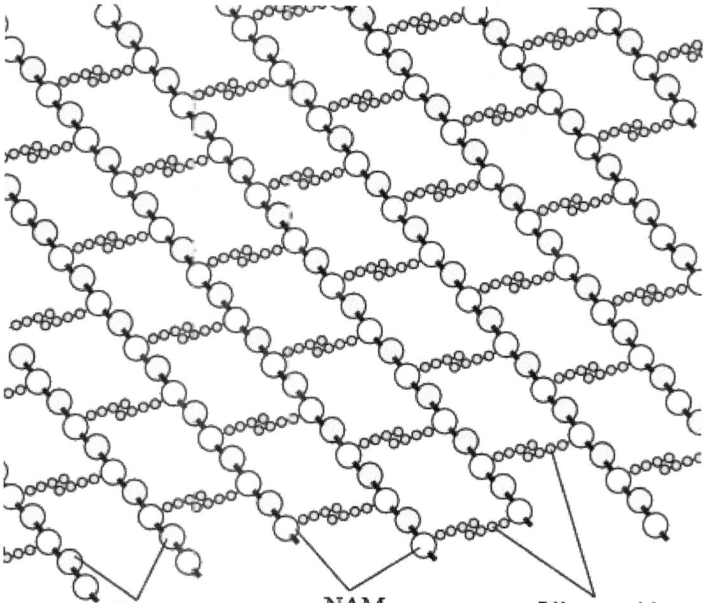

The structure of peptidoglycan, composed of N-acetylglucosamine and N-acetylmuramic acid

All gram-positive bacteria are bounded by a single-unit lipid membrane, and, in general, they contain a thick layer (20–80 nm) of peptidoglycan responsible for retaining the Gram stain. A number of other bacteria—that are bounded by a single membrane, but stain Gram-negative due to either lack of the peptidoglycan layer, as in the Mycoplasmas, or their inability to retain the Gram stain because of their cell wall composition—also show close relationship to the Gram-positive bacteria. For the bacterial cells bounded by a single cell membrane, the term "monoderm bacteria" or "monoderm prokaryotes" has been proposed.

In contrast to gram-positive bacteria, all archetypical gram-negative bacteria are bounded by a cytoplasmic membrane and an outer cell membrane; they contain only a thin layer of peptidoglycan (2–3 nm) between these membranes. The presence of inner and outer cell membranes defines a new compartment in these cells: the periplasmic space or the periplasmic compartment. These bacteria have been designated as "diderm bacteria." The distinction between the monoderm and diderm bacteria is supported by conserved signature indels in a number of important proteins (viz. DnaK, GroEL). Of these two structurally distinct groups of bacteria, monoderms are indicated to be ancestral. Based upon a number of observations including that the gram-positive bacteria are the major producers of antibiotics and that, in general, gram-negative bacteria are resistant to them, it has been proposed that the outer cell membrane in gram-negative bacteria (diderms) has evolved as a protective mechanism against antibiotic selection pressure. Some bacteria, such as *Deinococcus*, which stain gram-positive due to the presence of a thick peptidoglycan layer and also possess an outer cell membrane are suggested as intermediates in the transition between monoderm (gram-positive) and diderm (gram-negative) bacteria. The diderm bacteria can also be further differentiated between simple diderms lacking lipopolysaccharide, the archetypical diderm bacteria where the outer cell membrane contains lipopolysaccharide, and the diderm bacteria where outer cell membrane is made up of mycolic acid.

Exceptions

In general, gram-positive bacteria are monoderms and have a single lipid bilayer whereas gram-negative bacteria are diderms and have two bilayers. Some taxa lack peptidoglycan (such as the domain Archaea, the class Mollicutes, some members of the Rickettsiales, and the insect-endosymbionts of the Enterobacteriales) and are gram-variable. This, however, does not always hold true. The *Deinococcus-Thermus* bacteria have gram-positive stains, although they are structurally similar to gram-negative bacteria with two layers. The Chloroflexi have a single layer, yet (with some exceptions) stain negative. Two related phyla to the Chloroflexi, the TM7 clade and the Ktedonobacteria, are also monoderms.

Some Firmicute species are not gram-positive. These belong to the class Mollicutes (alternatively considered a class of the phylum Tenericutes), which lack peptidoglycan (gram-indeterminate), and the class Negativicutes, which includes Selenomonas and stain gram-negative. Additionally, a number of bacterial taxa (viz. Negativicutes, Fusobacteria, Synergistetes, and Elusimicrobia) that are either part of the phylum Firmicutes or branch in its proximity are found to possess a diderm cell structure. However, a conserved signature indel (CSI) in the HSP60 (GroEL) protein distinguishes all traditional phyla of gram-negative bacteria (e.g., Proteobacteria, Aquificae, Chlamydiae, Bacteroidetes, Chlorobi, Cyanobacteria, Fibrobacteres, Verrucomicrobia, Planctomycetes, Spirochetes, Acidobacteria, etc.) from these other atypical diderm bacteria, as well as other phyla of monoderm bacteria (e.g., Actinobacteria, Firmicutes, Thermotogae, Chloroflexi, etc.). The presence of this CSI in all sequenced species of conventional LPS (lipopolysaccharide)-containing gram-negative bacterial phyla provides evidence that these phyla of bacteria form a monophyletic clade and that no loss of the outer membrane from any species from this group has occurred.

Pathogenesis

Colonies of a gram-positive pathogen of the oral cavity, *Actinomyces* sp.

In the classical sense, six gram-positive genera are typically pathogenic in humans. Two of these, *Streptococcus* and *Staphylococcus*, are cocci (sphere-shaped). The remaining organisms are bacilli (rod-shaped) and can be subdivided based on their ability to form spores. The non-spore formers are *Corynebacterium* and *Listeria* (a coccobacillus), whereas *Bacillus* and *Clostridium* produce spores. The spore-forming bacteria can again be divided based on their respiration: *Bacillus* is a facultative anaerobe, while *Clostridium* is an obligate anaerobe. Also, *Rathybacter*, *Leifsonia*, and *Clavibacter* are three gram-positive genera that cause plant disease. Gram-positive bacteria are capable of causing serious and sometimes fatal infections in newborn infants.

Bacterial Transformation

Transformation is one of three processes for horizontal gene transfer, in which exogenous genetic material passes from bacterium to another, the other two being conjugation (transfer of genetic material between two bacterial cells in direct contact) and transduction (injection of foreign DNA by a bacteriophage virus into the host bacterium). In transformation, the genetic material passes through the intervening medium, and uptake is completely dependent on the recipient bacterium.

As of 2014 about 80 species of bacteria were known to be capable of transformation, about evenly divided between Gram-positive and Gram-negative bacteria; the number might be an overestimate since several of the reports are supported by single papers. Transformation among gram-positive bacteria has been studied in medically important species such as *Streptococcus pneumoniae*, *Streptococcus mutans*, *Staphylococcus aureus* and *Streptococcus sanguinis* and in gram-positive soil bacterium *Bacillus subtilis*.

Orthographic Note

The adjectives 'gram-positive' and 'gram-negative' derive from the surname of Hans Christian Gram; as eponymous adjectives, their initial letter can be either lowercase 'g' or capital 'G', depending on whether or not a style guide (for example that of the CDC) governs the document being written. This is further explained at *Gram staining Orthographic note*.

Gram-negative Bacteria

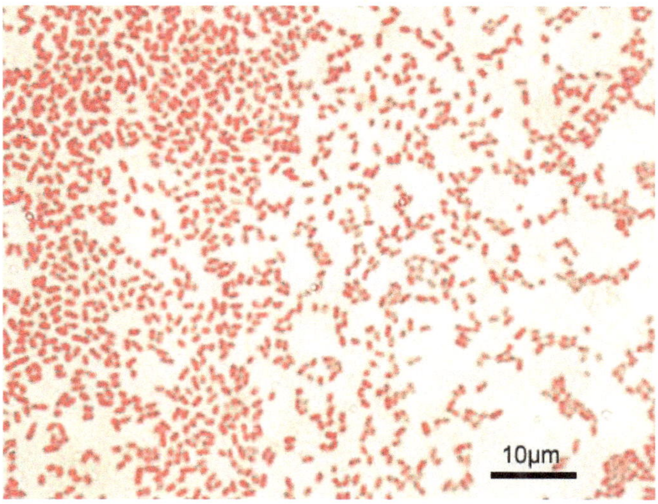

Microscopic image of gram-negative *Pseudomonas aeruginosa* bacteria (pink-red rods)

Gram-negative bacteria are a group of bacteria that do not retain the crystal violet stain used in the Gram staining method of bacterial differentiation. They are characterized by their cell envelopes, which are composed of a thin peptidoglycan cell wall sandwiched between an inner cytoplasmic cell membrane and a bacterial outer membrane.

Gram-negative bacteria are spread worldwide, in virtually all environments that support life. The gram-negative bacteria include the model organism *Escherichia coli*, as well as many pathogenic bacteria, such as *Pseudomonas aeruginosa*, *Neisseria gonorrhoeae*, Chlamydia trachomatis, and *Yersinia pestis*. Several classes of antibiotics target gram-negative bacteria specifically, including aminoglycosides and carbapenems.

Characteristics

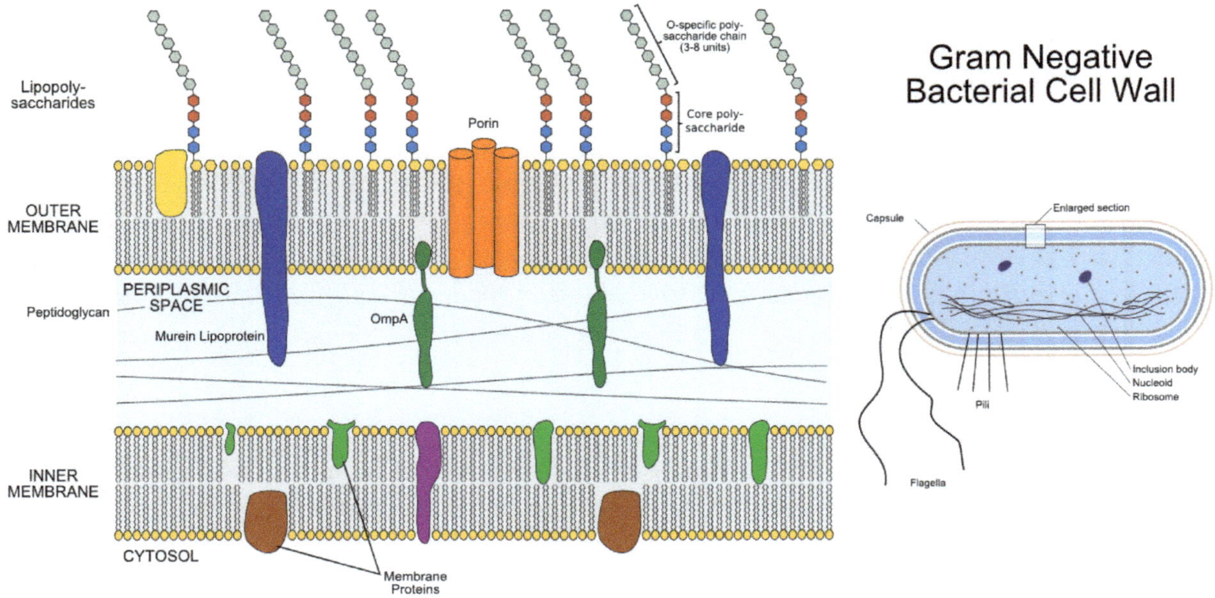

Gram-negative cell wall structure

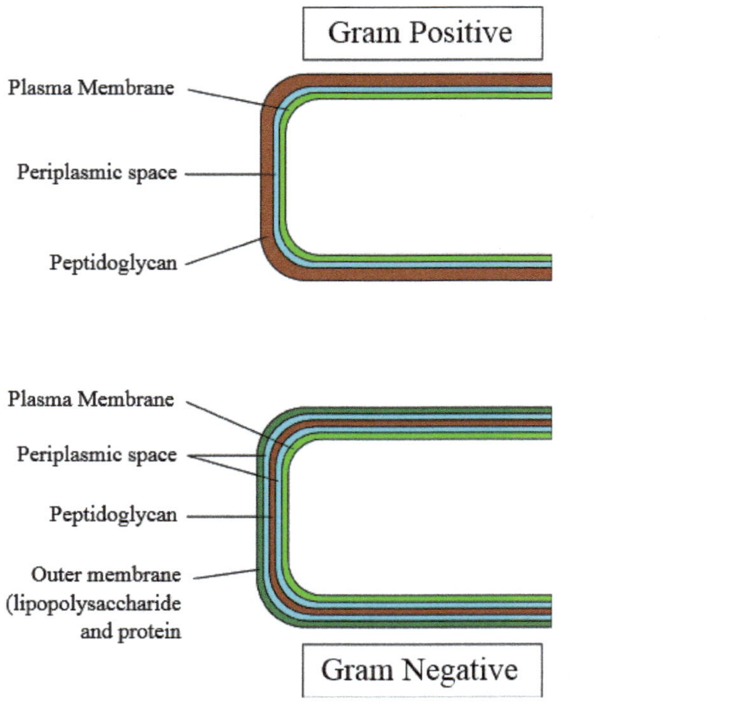

Gram-positive and -negative bacteria are differentiated chiefly by their cell wall structure

Gram-negative bacteria display these characteristics:

1. An inner cell membrane is present (cytoplasmic)
2. A thin peptidoglycan layer is present (This is much thicker in gram-positive bacteria)
3. Has outer membrane containing lipopolysaccharides (LPS, which consists of lipid A, core polysaccharide, and O antigen) in its outer leaflet and phospholipids in the inner leaflet
4. Porins exist in the outer membrane, which act like pores for particular molecules
5. Between the outer membrane and the cytoplasmic membrane there is a space filled with a concentrated gel-like substance called periplasm
6. The S-layer is directly attached to the outer membrane rather than to the peptidoglycan
7. If present, flagella have four supporting rings instead of two
8. Teichoic acids or lipoteichoic acids are absent
9. Lipoproteins are attached to the polysaccharide backbone
10. Some contain Braun's lipoprotein, which serves as a link between the outer membrane and the peptidoglycan chain by a covalent bond
11. Most, with very few exceptions, do not form spores

Classification

Along with cell shape, Gram staining is a rapid diagnostic tool and once was used to group species at the subdivision of Bacteria. Historically, the kingdom Monera was divided into four divisions based on Gram staining: Firmacutes (+), Gracillicutes (–), Mollicutes (0) and Mendocutes (var.). Since 1987, the monophyly of the gram-negative bacteria has been disproven with molecular studies. However some authors, such as Cavalier-Smith still treat them as a monophyletic taxon (though not a clade; his definition of monophyly requires a single common ancestor but does not require holophyly, the property that all descendents be encompassed by the taxon) and refer to the group as subkingdom "Negibacteria".

Taxonomy

Bacteria are traditionally divided into the two groups: gram-positive and gram-negative, based on their Gram stain retention. These groups are often thought of as lineages, with gram-negative bacteria more closely related to one another than to gram-positive bacteria. While this is often true, the classification system breaks down in some cases. A given bacteria's Gram stain result, bacterial membrane organization, and lineage groupings do not always match up. As such, the Gram stain cannot be reliably used to assess familial relationships of bacteria. That said, Gram staining does often give reliable information about the composition of the cell membrane, distinguishing between the presence or absence of an outer lipid membrane.

Of these two structurally distinct groups of prokaryotic organisms, monoderm prokaryotes are indicated to be ancestral. Based upon a number of different observations including that the gram-positive bacteria are the major reactors to antibiotics and that gram-negative bacteria are, in general, resistant to them, it has been proposed that the outer cell membrane in gram-negative bacteria (diderms) evolved as a protective mechanism against antibiotic selection pressure. Some bacteria such as Deinococcus, which stain gram-positive due to the presence of a thick peptidoglycan layer, but also possess an outer cell membrane are suggested as intermediates in the transition between monoderm (gram-positive) and diderm (gram-negative) bacteria. The diderm bacteria can also be further differentiated between simple diderms lacking lipopolysaccharide; the archetypical diderm bacteria, in which the outer cell membrane contains lipopolysaccharide; and the diderm bacteria, in which the outer cell membrane is made up of mycolic acid (e. g. Mycobacterium).

In addition, a number of bacterial taxa (viz. Negativicutes, Fusobacteria, Synergistetes, and Elusimicrobia) that are either part of the phylum Firmicutes or branches in its proximity are also found to possess a diderm cell structure. However, a conserved signature indel (CSI) in the HSP60 (GroEL) protein distinguishes all traditional phyla of gram-negative bacteria (e.g., Proteobacteria, Aquificae, Chlamydiae, Bacteroidetes, Chlorobi, Cyanobacteria, Fibrobacteres, Verrucomicrobia, Planctomycetes, Spirochetes, Acidobacteria) from these other atypical diderm bacteria as well as other phyla of monoderm bacteria (e.g., Actinobacteria, Firmicutes, Thermotogae, Chloroflexi). The presence of this CSI in all sequenced species of conventional lipopolysaccharide-containing gram-negative bacterial phyla provides evidence that these phyla of bacteria form a monophyletic clade and that no loss of the outer membrane from any species from this group has occurred.

Example Species

The proteobacteria are a major group of gram-negative bacteria, including *Escherichia coli* (*E. coli*), *Salmonella*, *Shigella*, and other Enterobacteriaceae, *Pseudomonas*, *Moraxella*, *Helicobacter*, *Stenotrophomonas*, *Bdellovibrio*, acetic acid bacteria, *Legionella* etc. Other notable groups of gram-negative bacteria include the cyanobacteria, spirochaetes, green sulfur, and green non-sulfur bacteria.

Medically relevant gram-negative cocci include the four types that cause a sexually transmitted disease (*Neisseria gonorrhoeae*), a meningitis (*Neisseria meningitidis*), and respiratory symptoms (*Moraxella catarrhalis*, *Haemophilus influenzae*).

Medically relevant gram-negative bacilli include a multitude of species. Some of them cause primarily respiratory problems (*Klebsiella pneumoniae*, *Legionella pneumophila*, *Pseudomonas aeruginosa*), primarily urinary problems (*Escherichia coli*, *Proteus mirabilis*, *Enterobacter cloacae*, *Serratia marcescens*), and primarily gastrointestinal problems (*Helicobacter pylori*, *Salmonella enteritidis*, *Salmonella typhi*).

Gram-negative bacteria associated with hospital-acquired infections include *Acinetobacter baumannii*, which cause bacteremia, secondary meningitis, and ventilator-associated pneumonia in hospital intensive-care units.

Bacterial Transformation

Transformation is one of three processes for horizontal gene transfer, in which exogenous genetic material passes from bacterium to another, the other two being conjugation (transfer of genetic material between two bacterial cells in direct contact) and transduction (injection of foreign DNA by a bacteriophage virus into the host bacterium). In transformation, the genetic material passes through the intervening medium, and uptake is completely dependent on the recipient bacterium.

As of 2014 about 80 species of bacteria were known to be capable of transformation, about evenly divided between Gram-positive and Gram-negative bacteria; the number might be an overestimate since several of the reports are supported by single papers. Transformation has been studied in medically important Gram-negative bacteria species such as *Helicobacter pylori*, *Legionella pneumophila*, *Neisseria meningitidis*, *Neisseria gonorrhoeae*, *Haemophilus influenzae* and *Vibrio cholerae*. It has also been studied in gram-negative species found in soil such as *Pseudomonas stutzeri*, *Acinetobacter baylyi*, and gram-negative plant pathogens such as *Ralstonia solanacearum* and *Xylella fastidiosa*.

Medical Treatment

One of the several unique characteristics of gram-negative bacteria is the structure of the bacterial outer membrane. The outer leaflet of this membrane comprises a complex lipopolysaccharide (LPS) whose lipid portion acts as an endotoxin. If gram-negative bacteria enter the circulatory system, the liposaccharide can cause a toxic reaction. This results in fever, an increased respiratory rate, and low blood pressure. This may lead to life-threatening condition of endotoxic shock.

The outer membrane protects the bacteria from several antibiotics, dyes, and detergents that would normally damage either the inner membrane or the cell wall (made of peptidoglycan). The outer membrane provides these bacteria with resistance to lysozyme and penicillin. However, alternative medicinal treatments such as lysozyme with ethylenediaminetetraacetic acid (EDTA) and the antibiotic ampicillin have been developed to combat the protective outer membrane of some pathogenic gram-negative organisms. Other drugs can also be used, significant ones being chloramphenicol, streptomycin, and nalidixic acid. Chloramphenicol is rarely used in the EU due to the association with drug induced pancytopenia.

The pathogenic capability of gram-negative bacteria is often associated with certain components of their membrane, in particular, the LPS. In humans, the presence of LPS triggers an innate immune response, activating the immune system and producing cytokines (hormonal regulators). Inflammation is a common reaction to cytokine production, which can also produce host toxicity. The innate immune response to LPS, however, is not synonymous with pathogenicity, or the ability to cause disease.

Orthographic note

The adjectives 'gram-positive' and 'gram-negative' derive from the surname of Hans Christian Gram; as eponymous adjectives, their initial letter can be either lowercase 'g' or capital 'G', depending on whose style guide (if any) governs the document being written. This is further explained at *Gram staining § Orthographic note*.

Vibrio

Vibrio is a genus of Gram-negative bacteria, possessing a curved-rod shape (comma shape), several species of which can cause foodborne infection, usually associated with eating undercooked seafood. Typically found in salt water, *Vibrio* species are facultative anaerobes that test positive for oxidase and do not form spores. All members of the genus are motile and have polar flagella with sheaths. *Vibrio* species typically possess two chromosomes, which is unusual for bacteria. Each chromosome has a distinct and independent origin of replication, and are conserved together over time in the genus. Recent phylogenies have been constructed based on a suite of genes (multilocus sequence analysis).

O. F. Müller (1773, 1786) described eight species of the genus *Vibrio* (included in Infusoria), three of which were spirilliforms. Some of the other species are today assigned to eukaryote taxa, e.g., to the euglenoid *Peranema* or to the diatom *Bacillaria*. However, *Vibrio* Müller, 1773 became regarded as the name of a zoological genus, and the name of the bacterial genus became *Vibrio* Pacini, 1854. Filippo Pacini isolated micro-organisms he called "vibrions" from cholera patients in 1854, because of their motility.

Pathogenic Strains

Several species of *Vibrio* are pathogens. Most disease-causing strains are associated with gastroenteritis, but can also infect open wounds and cause septicemia. They can be carried by numerous marine animals, such as crabs or prawns, and have been known to cause fatal infections in humans during exposure. Pathogenic *Vibrio* species include *V. cholerae* (the causative agent of cholera), *V. parahaemolyticus*, and *V. vulnificus*. *V. cholerae* is generally transmitted by contaminated water. Pathogenic *Vibrio* species can cause foodborne illness (infection), usually associated with eating undercooked seafood. The pathogenic features can be linked to quorum sensing where bacteria are able to express their virulence factor via their signalling molecules.

V. vulnificus outbreaks commonly occur in warm climates and small, generally lethal, outbreaks occur regularly. An outbreak occurred in New Orleans after Hurricane Katrina, and several lethal cases occur most years in Florida. As of 2013 in the United States, *Vibrio* infections as a whole were up 43% when compared with the rates observed in 2006-2008. *V. vulnificus*, the most severe strain, has not increased. Foodborne *Vibrio* infections are most often associated with eating raw shellfish.

V. parahaemolyticus is also associated with the Kanagawa phenomenon, in which strains isolated from human hosts (clinical isolates) are hemolytic on blood agar plates, while those isolated from nonhuman sources are not hemolytic.

Many *Vibrio* species are also zoonotic. They cause disease in fish and shellfish, and are common causes of mortality among domestic marine life.

Treatment

Medical care depends on the clinical presentation and the presence of underlying medical conditions.

Vibrio Gastroenteritis

Because *Vibrio* gastroenteritis is self-limited in most patients, no specific medical therapy is required. Patients who cannot tolerate oral fluid replacement may require intravenous fluid therapy.

Although most *Vibrio* species are sensitive to antibiotics such as doxycycline or quinolones, antibiotic therapy does not shorten the course of the illness or the duration of pathogen excretion. However, if the patient is ill and has a high fever or an underlying medical condition, oral antibiotic therapy with doxycycline or quinolone can be initiated.

Noncholera Vibrio Infections

Patients with noncholera *Vibrio* wound infection or septicemia are much more ill and frequently have other medical conditions. Medical therapy consists of:

- Prompt initiation of effective antibiotic therapy (doxycycline or a quinolone)
- Intensive medical therapy with aggressive fluid replacement and vasopressors for hypotension and septic shock to correct acid-base and electrolytes abnormalities that may be associated with severe sepsis
- Early fasciotomy within 24 hours after development of clinical symptoms can be life-saving in patients with necrotizing fasciitis.
- Early debridement of the infected wound has an important role in successful therapy and is especially indicated to avoid amputation of fingers, toes, or limbs.
- Expeditious and serial surgical evaluation and intervention are required because patients may deteriorate rapidly, especially those with necrotizing fasciitis or compartment syndrome.
- Reconstructive surgery, such as skin grafts, are used in the recovery phase.

Other Strains

V. harveyi is a pathogen of several aquatic animals, and is notable as a cause of luminous vibriosis in shrimp (prawns)

Flagella

The "typical", early-discovered *Vibrio* species, such as *V. cholerae*, have a single polar flagellum (monotrichous) with sheath. Some species, such as *V. parahaemolyticus* and *V. alginolyticus*, have both a single polar flagellum with sheath and thin flagella projecting in all directions (peritrichous), and the other species, such as *V. fischeri*, have tufts of polar flagella with sheath (lophotrichous).

Plasma (Cytoplasmic) Membrane

Membranes are absolute requirement of all living organisms. It is the chief point of contact with the cell's environment and thus is responsible for much of its relationship with the outside world.

Plasma membrane – encloses the cytoplasm and consists of phospholipids and proteins (fluid mosaic model).Most membrane-associated lipids are structurally asymmetric with polar and non-polar ends. The polar ends interact with water and are hydrophilic and the nonpolar hydrophobic ends are insoluble in water. The lipid composition of bacterial membranes varies with environmental temperature in such a way that the membrane remains fluid during growth. Bacterial membranes usually differ from eukaryotic membranes in lacking sterols such as cholesterol and they contain pentacyclic sterol-like molecules called hopanoids and these are said to stabilize the bacterial membranes. Cell membranes are very thin structures about 5 to 10 nm thick and can be seen only with electron microscope. Plasma membranes have a complex internal structure; the small globular particles seen in these membranes are thought to be membrane proteins that lie within the membrane lipid bilayer.

The most widely accepted current model for membrane structure is the fluid mosaic model of S. Jonathan Singer and Garth Nicholson. Two types of membrane proteins are seen, Peripheralproteins - which are loosely connected to the membrane and can be easily removed and are soluble in aqueous solutions and make up about 20 to 30% of total membrane protein. About 70 to 80% of membrane proteins are integral proteins. These cannot be easily extracted from membranes and are insoluble in aqueous solutions when freed of lipids. Integral proteins, like membrane lipids are amphipathic; their hydrophobic regions are buried in the lipid while the hydrophilic portions project from the membrane surface. The plasma membrane retains the cytoplasm, particularly in cells without cell walls, and separates it from the surroundings. Plasma membranes serve as a selectively permeable barrier; it allows particular ions and molecules to pass, either into or out of the cell, while preventing the movement of others. Transport systems can be used for such tasks as nutrient uptake, waste excretion, and protein secretion. The plasma membrane also is the location of a variety of crucial metabolic processes; respiration, photosynthesis, the synthesis of lipids and cell wall constituents, and probably chromosome segregation.

The bacterial plasma membrane can be destroyed by alcohols and polymixins which cause leakage of intracellular contents and subsequent cell death of the organism.

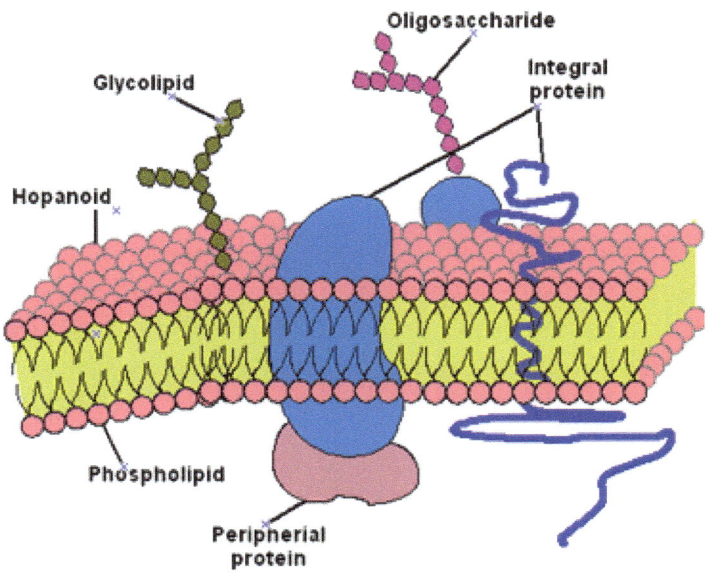

Plasma membrane strcuture

Internal membrane systems:

Prokaryotes do not contain complex membrane systems as present in eukaryotes like chloroplast and mitochondria. They contain membranous structures like the one observed most is mesosome. Mesosomes – irregular infoldings or invaginations of the plasma membrane in the shape of vesicles, tubules, or lamellae. They can be seen in both gram positive and gram-negative bacteria. These are often found next to the septa or cross-walls in dividing bacteria and sometimes seems attached to the bacterial chromosome. Thus they seem to be involved in cell wall formation during division or play a role in chromosome replication and distribution to daughter cells.

Some bacteria have internal membrane systems quite different from the mesosomes. The infoldings of the plasma membrane can become extensive and complex in photosynthetic bacteria such as the cyanobacteria and purple bacteria or in bacteria with very high respiratory activity like the nitrifying bacteria. They may be aggregates of spherical vesicles, flattened vesicles, or tubular membranes. Their function may be to provide a larger membrane surface for greater metabolic activity.

Cytoplasm

In cell biology, the cytoplasm is the material within a living cell, excluding the cell nucleus. It comprises cytosol (the gel-like substance enclosed within the cell membrane) and the organelles – the cell's internal sub-structures. All of the contents of the cells of prokaryote organisms (such as bacteria, which lack a cell nucleus) are contained within the cytoplasm. Within the cells of eukaryote organisms the contents of the cell nucleus are separated from the cytoplasm, and are then called the nucleoplasm. The cytoplasm is about 80% water and usually colorless.

The submicroscopic ground cell substance or cytoplasmatic matrix which remains after exclusion the cell organelles and particles is groundplasm. It is the hyaloplasm of light microscopy, and high complex, polyphasic system in which all of resolvable cytoplasmic elements of are suspended, including the larger organelles such the ribosomes, mitochondria, the plant plastids, lipid droplets, and vacuoles.

It is within the cytoplasm that most cellular activities occur, such as many metabolic pathways including glycolysis, and processes such as cell division. The concentrated inner area is called the endoplasm and the outer layer is called the cell cortex or the ectoplasm.

Movement of calcium ions in and out of the cytoplasm is a signaling activity for metabolic processes.

In plants, movement of the cytoplasm around vacuoles is known as cytoplasmic streaming.

History

The term was introduced by Rudolf von Kölliker in 1863, originally as a synonym for protoplasm, but later it has come to mean the cell substance and organelles outside the nucleus.

There has been certain disagreement on the definition of cytoplasm, as some authors prefer to exclude from it some organelles, especially the vacuoles and sometimes the plastids.

Physical Nature

The physical properties of the cytoplasm have been contested in recent years. It remains uncertain how the varied components of the cytoplasm interact to allow movement of particles and organelles while maintaining the cell's structure. The flow of cytoplasmic components plays an important role in many cellular functions which are dependent on the permeability of the cytoplasm. An example of such function is cell signalling, a process which is dependent on the manner in which signaling molecules are allowed to diffuse across the cell. While small signaling molecules like calcium ions are able to diffuse with ease, larger molecules and subcellular structures often require aid in moving through the cytoplasm. The irregular dynamics of such particles have given rise to various theories on the nature of the cytoplasm.

As a Sol-gel

There has long been evidence that the cytoplasm behaves like a sol-gel. It is thought that the component molecules and structures of the cytoplasm behave at times like a disordered colloidal solution (sol) and at other times like an integrated network, forming a solid mass (gel). This theory thus proposes that the cytoplasm exists in distinct fluid and solid phases depending on the level of interaction between cytoplasmic components, which may explain the differential dynamics of different particles observed moving through the cytoplasm.

As a Glass

Recently it has been proposed that the cytoplasm behaves like a glass-forming liquid approaching the glass transition. In this theory, the greater the concentration of cytoplasmic components, the less the cytoplasm behaves like a liquid and the more it behaves as a solid glass, freezing larger cytoplasmic components in place (it is thought that the cell's metabolic activity is able to fluidize the cytoplasm to allow the movement of such larger cytoplasmic components). A cell's ability to vitrify in the absence of metabolic activity, as in dormant periods, may be beneficial as a defence strategy. A solid glass cytoplasm would freeze subcellular structures in place, preventing damage, while allowing the transmission of very small proteins and metabolites, helping to kickstart growth upon the cell's revival from dormancy.

Other Perspectives

There has been research examining the motion of cytoplasmic particles independent of the nature of the cytoplasm. In such an alternative approach, the aggregate random forces within the cell caused by motor proteins explain the non-Brownian motion of cytoplasmic constituents.

Constituents

The three major elements of the cytoplasm are the cytosol, organelles and inclusions.

Cytosol

The cytosol is the portion of the cytoplasm not contained within membrane-bound organelles. Cytosol makes up about 70% of the cell volume and is a complex mixture of cytoskeleton filaments, dissolved molecules, and water. The cytosol's filaments include the protein filaments such as actin filaments and microtubules that make up the cytoskeleton, as well as soluble proteins and small structures such as ribosomes, proteasomes, and the mysterious vault complexes. The inner, granular and more fluid portion of the cytoplasm is referred to as endoplasm.

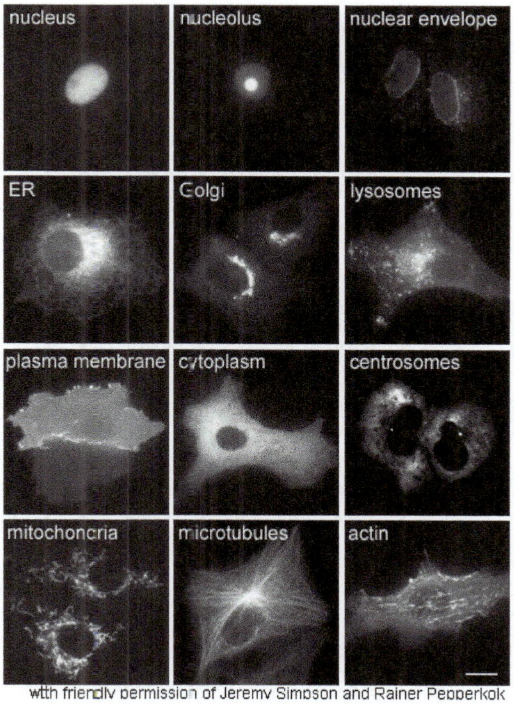

with friendly permission of Jeremy Simpson and Rainer Pepperkok

Proteins in different cellular compartments and structures tagged with green fluorescent protein

Due to this network of fibres and high concentrations of dissolved macromolecules, such as proteins, an effect called macromolecular crowding occurs and the cytosol does not act as an ideal solution. This crowding effect alters how the components of the cytosol interact with each other.

Organelles

Organelles (literally "little organs"), are usually membrane-bound structures inside the cell that have specific functions. Some major organelles that are suspended in the cytosol are the mitochondria, the endoplasmic reticulum, the Golgi apparatus, vacuoles, lysosomes, and in plant cells, chloroplasts.

Cytoplasmic inclusions

The inclusions are small particles of insoluble substances suspended in the cytosol. A huge range of inclusions exist in different cell types, and range from crystals of calcium oxalate or silicon dioxide in plants, to granules of energy-storage materials such as starch, glycogen, or polyhydroxybutyrate. A particularly widespread example are lipid droplets, which are spherical droplets composed of lipids and proteins that are used in both prokaryotes and eukaryotes as a way of storing lipids such as fatty acids and sterols. Lipid droplets make up much of the volume of adipocytes, which are specialized lipid-storage cells, but they are also found in a range of other cell types.

Controversy and Research

The cytoplasm, mitochondria and most organelles are contributions to the cell from the maternal gamete. Contrary to the older information that disregards any notion of the cytoplasm being active, new research has shown it to be in control of movement and flow of nutrients in and out of the cell by viscoplastic behavior and a measure of the reciprocal rate of bond breakage within the cytoplasmic network.

The material properties of the cytoplasm remain an ongoing investigation. Recent measurements using force spectrum microscopy reveal that the cytoplasm can be likened to an elastic solid, rather than a viscoelastic fluid.

Cytosol

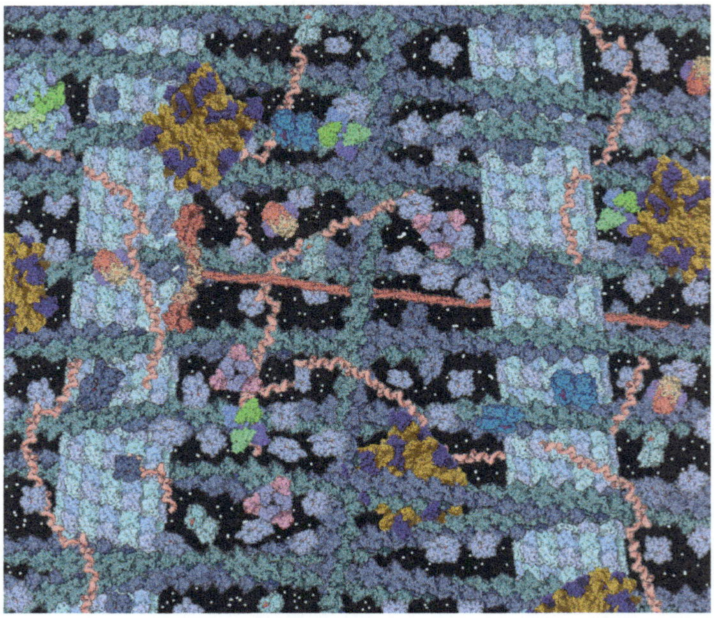

The cytosol is a crowded solution of many different types of molecules that fills much of the volume of cells.

The cytosol or cytoplasmic matrix is the liquid found inside cells. It constitutes most of the intracellular fluid (ICF). It is separated into compartments by membranes. For example, the mitochondrial matrix separates the mitochondrion into many compartments.

In the eukaryotic cell, the cytosol is within the cell membrane and is part of the cytoplasm, which also comprises the mitochondria, plastids, and other organelles (but not their internal fluids and structures); the cell nucleus is separate. The cytosol is thus a liquid matrix around the organelles. In prokaryotes, most of the chemical reactions of metabolism take place in the cytosol, while a few take place in membranes or in the periplasmic space. In eukaryotes, while many metabolic pathways still occur in the cytosol, others are contained within organelles.

The cytosol is a complex mixture of substances dissolved in water. Although water forms the large majority of the cytosol, its structure and properties within cells is not well understood. The concentrations of ions such as sodium and potassium are different in the cytosol than in the extracellular fluid; these differences in ion levels are important in processes such as osmoregulation, cell signaling, and the generation of action potentials in excitable cells such as endocrine, nerve and muscle cells. The cytosol also contains large amounts of macromolecules, which can alter how molecules behave, through macromolecular crowding.

Although it was once thought to be a simple solution of molecules, the cytosol has multiple levels of organization. These include concentration gradients of small molecules such as calcium, large complexes of enzymes that act together to carry out metabolic pathways, and protein complexes such as proteasomes and carboxysomes that enclose and separate parts of the cytosol.

Definition

The term "cytosol" was first introduced in 1965 by H. A. Lardy, and initially referred to the liquid that was produced by breaking cells apart and pelleting all the insoluble components by ultracentrifugation. Such a soluble cell extract is not identical to the soluble part of the cell cytoplasm and is usually called a cytoplasmic fraction.

The term *cytosol* is now used to refer to the liquid phase of the cytoplasm in an intact cell. This excludes any part of the cytoplasm that is contained within organelles. Due to the possibility of confusion between the use of the word "cytosol" to refer to both extracts of cells and the soluble part of the cytoplasm in intact cells, the phrase "aqueous cytoplasm" has been used to describe the liquid contents of the cytoplasm of living cells.

Prior to this, other terms were used for the cell fluid, not always synonymously, as its nature was not very clear.

Properties and Composition

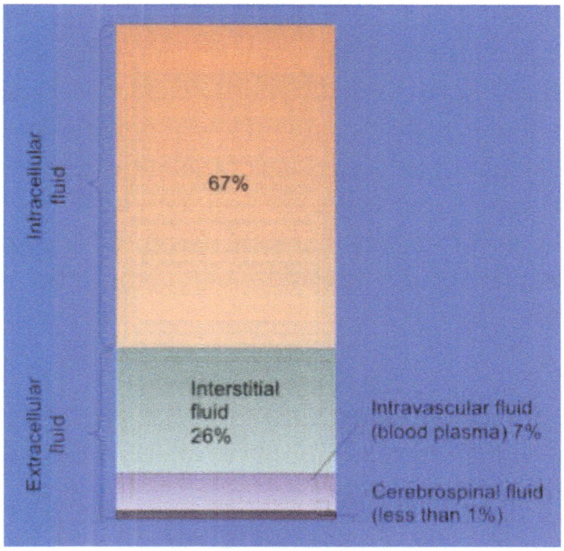

Intracellular fluid content in humans

The proportion of cell volume that is cytosol varies: for example while this compartment forms the bulk of cell structure in bacteria, in plant cells the main compartment is the large central vacuole. The cytosol consists mostly of water, dissolved ions, small molecules, and large water-soluble molecules (such as proteins). The majority of these non-protein molecules have a molecular mass of less than 300 Da. This mixture of small molecules is extraordinarily complex, as the variety of molecules that are involved in metabolism (the metabolites) is immense. For example, up to 200,000 different small molecules might be made in plants, although not all these will be present in the same species, or in a single cell. Estimates of the number of metabolites in single cells such as *E. coli* and baker's yeast predict that under 1,000 are made.

Water

Most of the cytosol is water, which makes up about 70% of the total volume of a typical cell. The pH of the intracellular fluid is 7.4. while human cytosolic pH ranges between 7.0 - 7.4, and is usually higher if a cell is growing. The viscosity of cytoplasm is roughly the same as pure water, although diffusion of small molecules through this liquid is about fourfold slower than in pure water, due mostly to collisions with the large numbers of macromolecules in the cytosol. Studies in the brine shrimp have examined how water affects cell functions; these saw that a 20% reduction in the amount of water in a cell inhibits metabolism, with metabolism decreasing progressively as the cell dries out and all metabolic activity halting when the water level reaches 70% below normal.

Although water is vital for life, the structure of this water in the cytosol is not well understood, mostly because methods such as nuclear magnetic resonance spectroscopy only give information on the average structure of water, and cannot measure local variations at the microscopic scale. Even the structure of pure water is poorly understood, due to the ability of water to form structures such as water clusters through hydrogen bonds.

The classic view of water in cells is that about 5% of this water is strongly bound in by solutes or macromolecules as water of solvation, while the majority has the same structure as pure water. This water of solvation is not active in osmosis and may have different solvent properties, so that some dissolved molecules are excluded, while others become concentrated. However, others argue that the effects of the high concentrations of macromolecules in cells extend throughout the cytosol and that water in cells behaves very differently from the water in dilute solutions. These ideas include the proposal that cells contain zones of low and high-density water, which could have widespread effects on the structures and functions of the other parts of the cell. However, the use of advanced nuclear magnetic resonance methods to directly measure the mobility of water in living cells contradicts this idea, as it suggests that 85% of cell water acts like that pure water, while the remainder is less mobile and probably bound to macromolecules.

Ions

The concentrations of the other ions in cytosol are quite different from those in extracellular fluid and the cytosol also contains much higher amounts of charged macromolecules such as proteins and nucleic acids than the outside of the cell structure.

Typical ion concentrations in mammalian cytosol and blood.		
Ion	Concentration in cytosol (**millimolar**)	Concentration in blood (**millimolar**)
Potassium	139	4
Sodium	12	145

Chloride	4	116
Bicarbonate	12	29
Amino acids in proteins	138	9
Magnesium	0.8	1.5
Calcium	<0.0002	1.8

In contrast to extracellular fluid, cytosol has a high concentration of potassium ions and a low concentration of sodium ions. This difference in ion concentrations is critical for osmoregulation, since if the ion levels were the same inside a cell as outside, water would enter constantly by osmosis - since the levels of macromolecules inside cells are higher than their levels outside. Instead, sodium ions are expelled and potassium ions taken up by the Na^+/K^+-ATPase, potassium ions then flow down their concentration gradient through potassium-selection ion channels, this loss of positive charge creates a negative membrane potential. To balance this potential difference, negative chloride ions also exit the cell, through selective chloride channels. The loss of sodium and chloride ions compensates for the osmotic effect of the higher concentration of organic molecules inside the cell.

Cells can deal with even larger osmotic changes by accumulating osmoprotectants such as betaines or trehalose in their cytosol. Some of these molecules can allow cells to survive being completely dried out and allow an organism to enter a state of suspended animation called cryptobiosis. In this state the cytosol and osmoprotectants become a glass-like solid that helps stabilize proteins and cell membranes from the damaging effects of desiccation.

The low concentration of calcium in the cytosol allows calcium ions to function as a second messenger in calcium signaling. Here, a signal such as a hormone or an action potential opens calcium channels so that calcium floods into the cytosol. This sudden increase in cytosolic calcium activates other signalling molecules, such as calmodulin and protein kinase C. Other ions such as chloride and potassium may also have signaling functions in the cytosol, but these are not well understood.

Macromolecules

Protein molecules that do not bind to cell membranes or the cytoskeleton are dissolved in the cytosol. The amount of protein in cells is extremely high, and approaches 200 mg/ml, occupying about 20-30% of the volume of the cytosol. However, measuring precisely how much protein is dissolved in cytosol in intact cells is difficult, since some proteins appear to be weakly associated with membranes or organelles in whole cells and are released into solution upon cell lysis. Indeed, in experiments where the plasma membrane of cells were carefully disrupted using saponin, without damaging the other cell membranes, only about one quarter of cell protein was released. These cells were also able to synthesize proteins if given ATP and amino acids, implying that many of the enzymes in cytosol are bound to the cytoskeleton. However, the idea that the majority of the proteins in cells are tightly bound in a network called the microtrabecular lattice is now seen as unlikely.

In prokaryotes the cytosol contains the cell's genome, within a structure known as a nucleoid. This is an irregular mass of DNA and associated proteins that control the transcription and replication of the bacterial chromosome and plasmids. In eukaryotes the genome is held within the cell nucleus, which is separated from the cytosol by nuclear pores that block the free diffusion of any molecule larger than about 10 nanometres in diameter.

This high concentration of macromolecules in cytosol causes an effect called macromolecular crowding, which is when the effective concentration of other macromolecules is increased, since they have less volume to move in. This crowding effect can produce large changes in both the rates and the position of chemical equilibrium of reactions in the cytosol. It is particularly important in its ability to alter dissociation constants by favoring the association of macromolecules, such as when multiple proteins come together to form protein complexes, or when DNA-binding proteins bind to their targets in the genome.

Organization

Although the components of the cytosol are not separated into regions by cell membranes, these components do not always mix randomly and several levels of organization can localize specific molecules to defined sites within the cytosol.

Concentration Gradients

Although small molecules diffuse rapidly in the cytosol, concentration gradients can still be produced within this compartment. A well-studied example of these are the "calcium sparks" that are produced for a short period in the region around an open calcium channel. These are about 2 micrometres in diameter and last for only a few milliseconds, although several sparks can merge to form larger gradients, called "calcium waves". Concentration gradients of other small molecules, such as oxygen and adenosine triphosphate may be produced in cells around clusters of mitochondria, although these are less well understood.

Protein Complexes

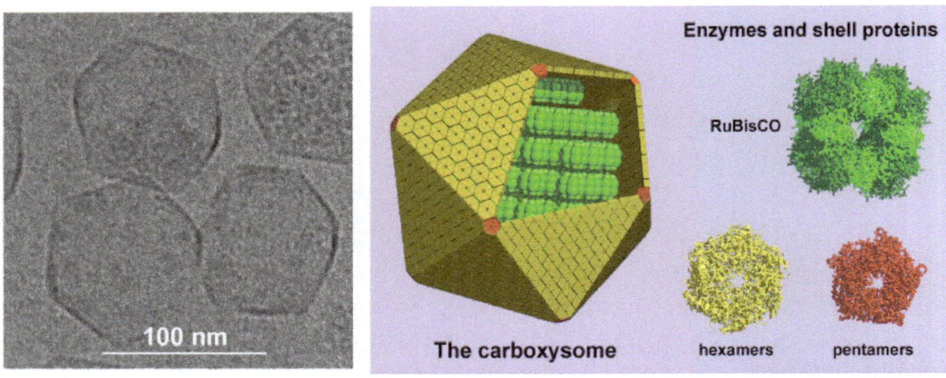

Carboxysomes are protein-enclosed bacterial microcompartments within the cytosol. On the left is an electron microscope image of carboxysomes, and on the right a model of their structure.

Proteins can associate to form protein complexes, these often contain a set of proteins with similar functions, such as enzymes that carry out several steps in the same metabolic pathway. This organization can allow substrate channeling, which is when the product of one enzyme is passed directly to the next enzyme in a pathway without being released into solution. Channeling can make a pathway more rapid and efficient than it would be if the enzymes were randomly distributed in the cytosol, and can also prevent the release of unstable reaction intermediates. Although a wide variety of metabolic pathways involve enzymes that are tightly bound to each other, others may involve more loosely associated complexes that are very difficult to study outside the cell. Consequently, the importance of these complexes for metabolism in general remains unclear.

Protein Compartments

Some protein complexes contain a large central cavity that is isolated from the remainder of the cytosol. One example of such an enclosed compartment is the proteasome. Here, a set of subunits form a hollow barrel containing proteases that degrade cytosolic proteins. Since these would be damaging if they mixed freely with the remainder of the cytosol, the barrel is capped by a set of regulatory proteins that recognize proteins with a signal directing them for degradation (a ubiquitin tag) and feed them into the proteolytic cavity.

Another large class of protein compartments are bacterial microcompartments, which are made of a protein shell that encapsulates various enzymes. These compartments are typically about 100-200 nanometres across and made of interlocking proteins. A well-understood example is the carboxysome, which contains enzymes involved in carbon fixation such as RuBisCO.

Cytoskeletal Sieving

Although the cytoskeleton is not part of the cytosol, the presence of this network of filaments restricts the diffusion of large particles in the cell. For example, in several studies tracer particles larger than about 25 nanometres (about the size of a ribosome) were excluded from parts of the cytosol around the edges of the cell and next to the nucleus. These "excluding compartments" may contain a much denser meshwork of actin fibres than the remainder of the cytosol. These microdomains could influence the distribution of large structures such as ribosomes and organelles within the cytosol by excluding them from some areas and concentrating them in others.

Function

The cytosol has no single function and is instead the site of multiple cell processes. Examples of these processes include signal transduction from the cell membrane to sites within the cell, such as the cell nucleus, or organelles. This compartment is also the site of many of the processes of cytokinesis, after the breakdown of the nuclear membrane in mitosis. Another major function of cytosol is to transport metabolites from their site of production to where they are used. This is relatively simple for water-soluble molecules, such as amino acids, which can diffuse rapidly through the cytosol. However, hydrophobic molecules, such as fatty acids or sterols, can be transported through the cytosol by specific binding proteins, which shuttle these molecules between cell membranes. Molecules taken into the cell by endocytosis or on their way to be secreted can also be transported through the cytosol inside vesicles, which are small spheres of lipids that are moved along the cytoskeleton by motor proteins.

The cytosol is the site of most metabolism in prokaryotes, and a large proportion of the metabolism of eukaryotes. For instance, in mammals about half of the proteins in the cell are localized to the cytosol. The most complete data are available in yeast, where metabolic reconstructions indicate that the majority of both metabolic processes and metabolites occur in the cytosol. Major metabolic pathways that occur in the cytosol in animals are protein biosynthesis, the pentose phosphate pathway, glycolysis and gluconeogenesis. The localization of pathways can be different in other organisms, for instance fatty acid synthesis occurs in chloroplasts in plants and in apicoplasts in apicomplexa.

Nuclear Area

The striking difference between prokaryotic and eukaryotic systems is the way in which their genetic material is packaged. Prokaryotes lack a membrane-delimited nucleus. Contains a single long circular molecule of double-stranded DNA bacterial chromosomes do not include histones and are not surrounded by a nuclear envelope and located in an irregularly shaped region called nucleiod . The nuclear area can be spherical, elongated, or dumb-bell shaped. It is attached to the plasma membrane and proteins of plasma membrane are believed to be responsible for replication of the DNA. It has been discovered recently that *Vibrio cholera* has more than one chromosome. Electron microscope studies have shown the nucleiod in contact with either the mesosome or the plasma membrane and hence evidence that the membranes may be involved in the separation of DNA into daughter cells during division. Chemical analysis reveals that they nucleoids are composed to about 60% DNA, 30% RNA and 10% protein. *E. coli*, which is about 2 to 6um long, the closed DNA circle measures approximately 1400um. Hence, it is evident that the DNA is efficiently packaged to fit within the nucleoid and the DA is looped and coiled extensively.

Many bacteria possess extra chromosomal double stranded, circular DNA molecules called plasmids in addition to their chromosome. They replicate independently and are associated with plasma membrane proteins. Plasmids usually contain from five to 100 genes. Plasmids may carry genes for such activities as antibiotic resistance, tolerance to toxic metals, production of toxins, and synthesis of enzymes. Plasmid DNA is used for gene manipulation in biotechnology. Because plasmids move between different bacteria, drug resistance can spread throughout a population.

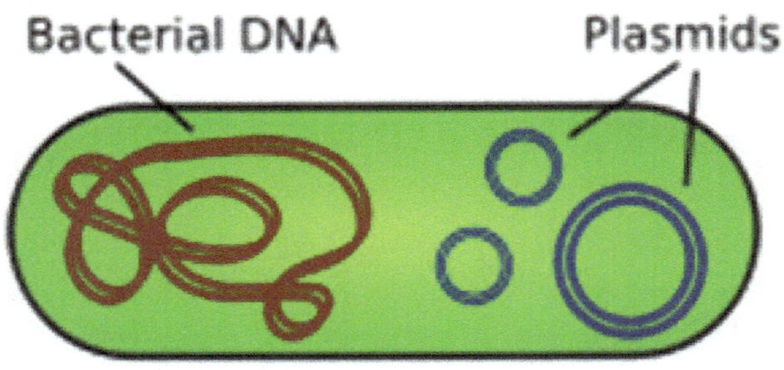

Bacterial DNA and plasmids

Ribosomes:

The cytoplasmic matrix is also packaged with ribosomes, they also may be loosely attached to the plasma membrane. They look like small, featureless particles at low magnification electron microscope. They are made up of both protein and ribonucleic acid (RNA). Ribosome's function as sites of protein synthesis; matrix ribosomes synthesize proteins destined to remain within the cell, whereas the plasma membrane ribosomes make proteins for transport to the outside.The shape of each protein is determined by its amino acids sequence and the special proteins called molecular chaperones or chaperones aid the polypeptide in folding to its proper shape. Prokaryotic ribosomes are smaller than eukaryotic ribosomes. Ribosomes are composed of two subunits, each subunit being composed of protein and a type of RNA called ribosomal RNA (rRNA).They are comm. Only 70S: 30S subunit (1 molecule of rRNA) and 50S subunit (2 molecules of rRNA) and have dimensions of about 14 to 15 nm, a molecular weight of approximately 2.7 million.The S in 70S stands for Svedberg value or sedimentation coefficient. It is the sedimentation velocity in a centrifuge; the faster a particle travels when centrifuged, the greater is its Svedberg value. The Sedimentation coefficient is a function of a particle's molecular weight, volume and shape. Several antibiotics, such as streptomycin, neomycin and tetracyclines, exert their antimicrobial effects by inhibiting protein synthesis on ribosomes.

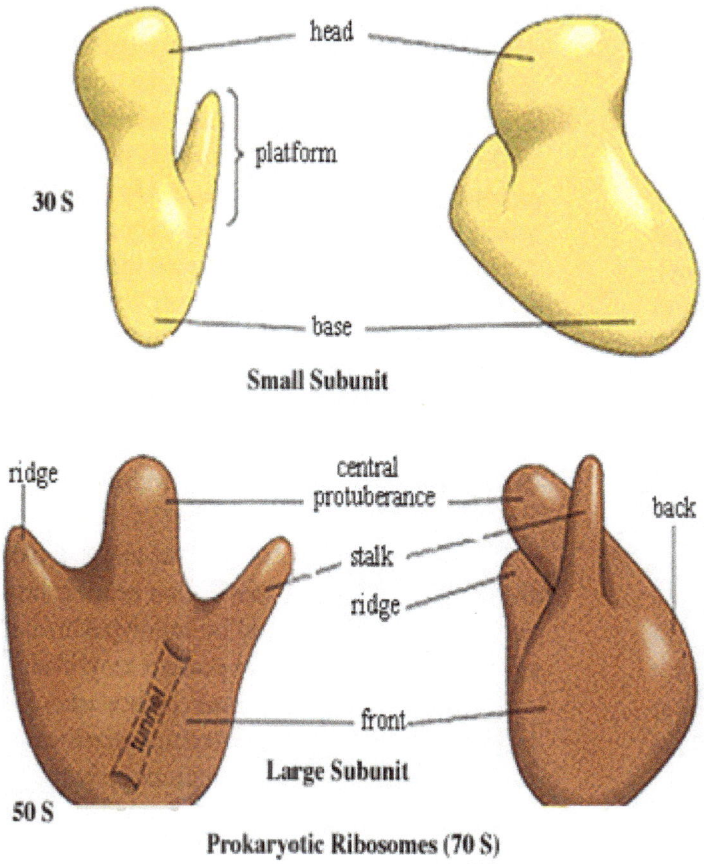

Ribosmes in bacteria

Inclusions:

Inclusion bodies can be divided into two types:

Inclusion bodies not bounded by a membrane and lie free in the cytoplasm. Ex. Polyphosphate granules, cyanophycingranules and some glycogen granules.

Inclusion bodies enclosed by a membrane about 2-4nm thick. Ex.PolyB-hydroxybutyrate granules, some glycogen and sulfur granules, carboxysomes and gas vacuoles.

Organic inclusion bodies:

Glycogen:

Polymer of glucose units composed of long chains formed by alpha (1-4) glycosidic bonds and branching chains connected to themby alpha (1-6)glycosidic bonds. Ex. glycogen and starch, and their presence can be demonstrated when iodine is applied to the cells (glycogen granules appear reddish brown and starch granules appear blue).

Poly B- hydroxybutyrate:

Contains beta-hydroxybutyrate molecules joined by ester bonds between the carboxyl and hydroxyl groups of adjacent molecules. Appear in various species of *Mycobacterium, Bacillus, Azotobacter* , *Spirillum* and other genera. Lipid inclusions are revealed by use of fat-soluble dyes, such as Sudan dyes.

Glycogen and PHB are carbon storage reservoirs providing material for energy and biosynthesis.

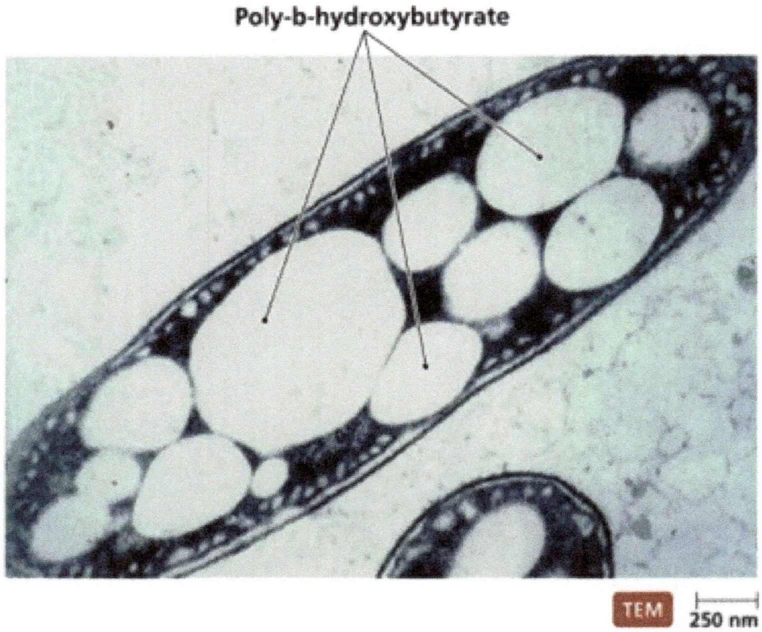

Poly B- hydroxybutyrate inclusions

Cyanophycin granules:

Cyanobacteria are composed of large amino acids containing approximately equal amounts of amino acids arginine and aspartic acid. These are used to store extra nitrogen for the bacteria.

Carboxysomes:

These are polyhedral and hexagonal inclusions that contain the enzyme ribulose 1,5-diphosphate carboxylase. ·Bacteria that use carbon dioxide as their sole source of carbon require this enzyme for carbon dioxide fixation during photosynthesis (Ex.nitrifying bacteria, cyanobacteria, and *Thiobacilli*).

Gas vacuoles:

These are hollow cavities found in many aquatic prokaryotes, including cyanobacteria, anoxygenic photosynthetic bacteria and halobacteria. Each vacuole consists of rows of several individual gas vesicles, which are hollow cylinders covered by protein. Their function is to maintain buoyancy so that the cells can remain at the depth in the water appropriate for them to receive sufficient amounts of oxygen, light and nutrients. They are impermeable to water and permeable to atmospheric gases.

Inorganic inclusion bodies:

Polyphosphate granules orMetachromatic granules:

Linear polymer of organo phosphates joined by ester bonds. Reservoirs for phosphate, an important component of cell nucleic acids and also energy reserves. Represents a reserve of inorganic phosphate (polyphosphate) that can be used in the synthesis of ATP. Stain red with certain blue dyes, such as methylene blue, and are collectively known as volutin. Found in algae, fungi and protozoans, as well as bacteria. These granules are quite large and are characteristic of *Corynebacterium diphtheriae*, the causative agent of diphtheria, thus they have diagnostic significance.

Sulphur granules:

Sulphur bacteria, which belong to the genus *Thiobacillus*, derive energy by oxidizing sulfur and sulfur containing compounds. These bacteria may deposit sulfur granules in the cell, where they serve as an energy reserve. Purple photosynthetic bacteria use H_2S as electron donor and accumulate resulting sulfurin either the periplasmic space or in special cytoplasmic globules.

Magnetosomes:

Not for storage, but these are used by some bacteria to orient in the earth's magnetic field. These inclusion bodies contain iron in the form of magnetite (greigite or pyrite). Ex. *Aquaspirillum magnetotacticum*. Also present in heads of birds, dolphins, and turtles etc which aid in navigation.

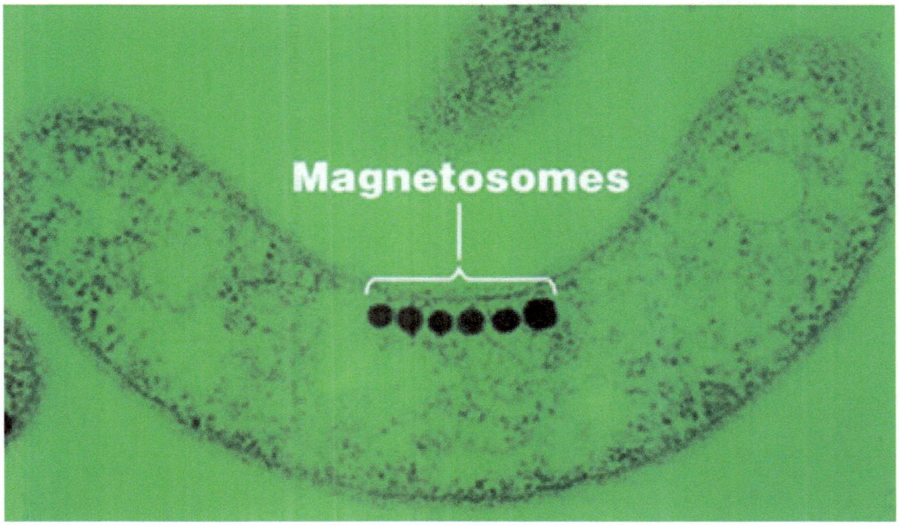

Magnetosomes

Endospores "An escape pod for DNA"

Endospores are a survival mechanism:they are triggered to form during adverse environmental conditions. They are NOT reproductive structures as only one cell gives rise to one spore and endospores can be identified with special stains and differentiated from the vegetative cell. Endospores are resistant to: heat: withstand boiling for over one hour; desiccation, UV radiation and chemical disinfectants. The resistance of these spores has serious consequence and some very pathogenic bacteria have the ability of produce such spores.

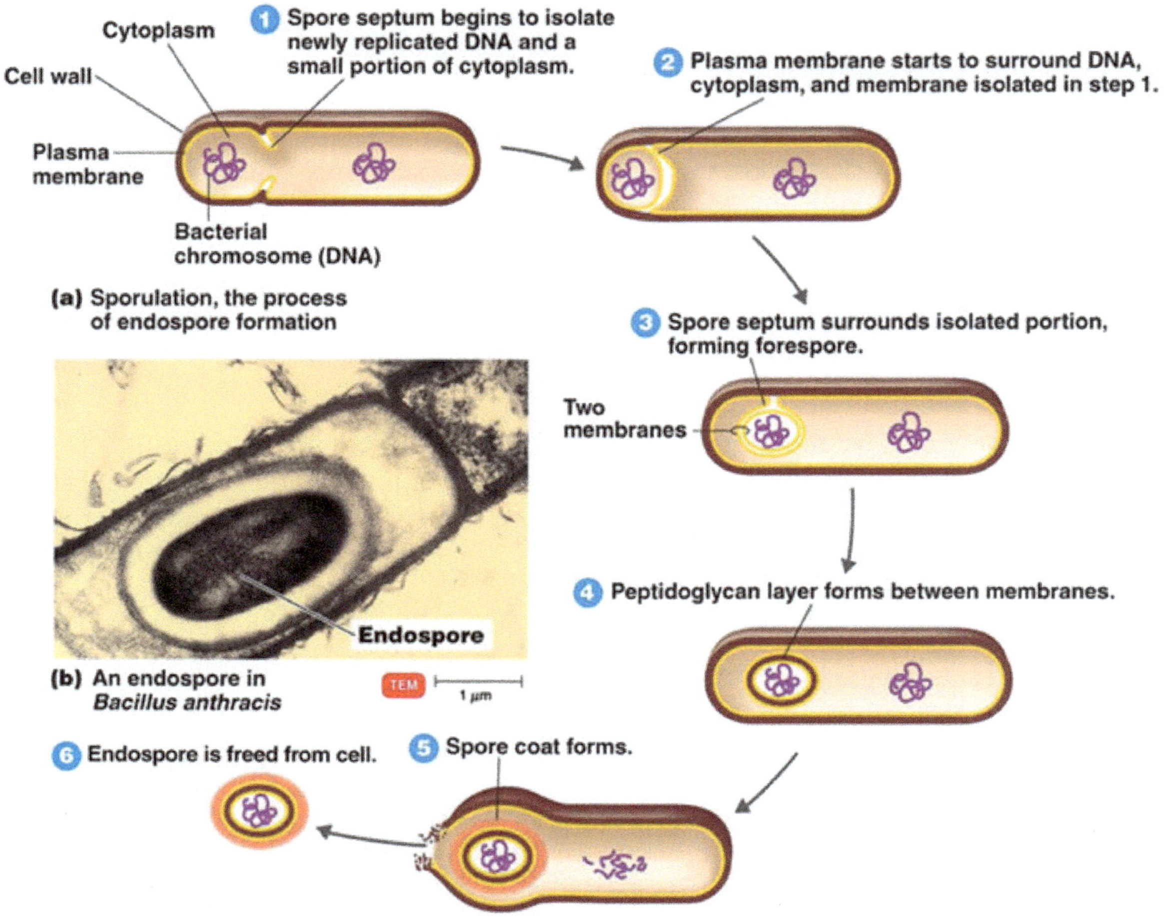

Endospore formation

References

- Kwang Hoon Sung & Hyun Kyu Song (July 22, 2014). "Insights into the Molecular Evolution of HslU ATPase through Biochemical and Mutational Analyses". PLOS. 9: e103027. doi:10.1371/journal.pone.0103027
- Bauman, Robert W.; Tizard, Ian R.; Machunis-Masouka, Elizabeth (2006). Microbiology. San Francisco: Pearson Benjamin Cummings. ISBN 0-8053-7693-3
- Kaiser D (October 2003). "Coupling cell movement to multicellular development in myxobacteria". Nat. Rev. Microbiol. 1 (1): 45–54. doi:10.1038/nrmicro733. PMID 15040179
- Taylor, C. V. (1923). "The contractile vacuole in Euplotes: An example of the sol-gel reversibility of cytoplasm". Journal of Experimental Zoology. 37 (3): 259–289. doi:10.1002/jez.1400370302

- Madigan, Michael T. (2012). Brock biology of microorganisms (13th ed.). San Francisco: Benjamin Cummings. ISBN 9780321649638
- "Infections from some foodborne germs increased, while others remained unchanged in 2012". Centers for Disease Control. April 18, 2013. Retrieved April 19, 2013
- Chen I; Dubnau D (March 2004). "DNA uptake during bacterial transformation". Nat. Rev. Microbiol. 2 (3): 241–9. doi:10.1038/nrmicro844. PMID 15083159
- Feneberg, Wolfgang; Sackmann, Erich; Westphal, Monika (2001). "Dictyostelium cells' cytoplasm as an active viscoplastic body". European Biophysics Journal. 30 (4): 284–94. doi:10.1007/s002490100135. PMID 11548131
- "Direct Observations". The Biofilm Primer. Springer Series on Biofilms. 1. 2007. pp. 3–4. doi:10.1007/978-3-540-68022-2_2. ISBN 978-3-540-68021-5
- Solomon JM; Grossman AD (April 1996). "Who's competent and when: regulation of natural genetic competence in bacteria". Trends Genet. 12 (4): 150–5. doi:10.1016/0168-9525(96)10014-7. PMID 8901420
- Hamilton WD (July 1964). "The genetical evolution of social behaviour. II". J. Theor. Biol. 7 (1): 17–52. doi:10.1016/0022-5193(64)90039-6. PMID 5875340
- The Molecular Biology of the Cell, fourth edition. Bruce Alberts, et al. Garland Science (2002) pg. 808 ISBN 0-8153-3218-1
- Zimmer C (August 2009). "Origins. On the origin of eukaryotes". Science. 325 (5941): 666–8. doi:10.1126/science.325_666. PMID 19661396
- Gladwin, Mark; Bill Trattler (2007). Clinical Microbiology made ridiculously simple. Miami, FL: MedMaster, Inc. pp. 4–5. ISBN 978-0-940780-81-1
- Brown, J.R. (February 2003). "Ancient Horizontal Gene Transfer". Nature Reviews Genetics. 4 (2): 121–132. doi:10.1038/nrg1000. PMID 12560809
- MacDonald, Mhairi (2015). Avery's neonatology : pathophysiology and management of the newborn. Philadelphia: Wolters Kluwer. ISBN 9781451192681; Access provided by the University of Pittsburgh
- Forterre P; Philippe H (October 1999). "Where is the root of the universal tree of life?". BioEssays. 21 (10): 871–9. doi:10.1002/(SICI)1521-1878(199910)21:10<871::AID-BIES10>3.0.CO;2-Q. PMID 10497338
- Faruque SM; Nair GB (editors). (2008). Vibrio cholerae: Genomics and Molecular Biology. Caister Academic Press. ISBN 978-1-904455-33-2
- Shih YL; Rothfield L (2006). "The bacterial cytoskeleton". Microbiol. Mol. Biol. Rev. 70 (3): 729–54. doi:10.1128/MMBR.00017-06. PMC 1594594. PMID 16959967
- Madigan, Michael; Martinko, John (editors) (2005). Brock Biology of Microorganisms (11th ed.). Prentice Hall. ISBN 0-13-144329-1
- Fuerst J (2005). "Intracellular compartmentation in planctomycetes". Annu Rev Microbiol. 59: 299–328. doi:10.1146/annurev.micro.59.030804.121258. PMID 15910279

Microscopy: An Overview

4

Since microorganisms cannot be seen with the naked eye, microscopy is used to study them using microscopes. Optical, electron, and scanning probe microscopy are the branches involved in the study. Optical or light microscope uses light and lenses to enlarge the samples. As light microscopes can achieve 200nm resolution, electron microscopes with a resolution higher than 50 pm resolution are used to study the structure of microbes. This chapter is a compilation of the various branches of microscopy that form an integral part of the broader subject matter.

Microscopy

Scanning electron microscope image of pollen

Microscopic examination in a biochemical laboratory

Microscopy is the technical field of using microscopes to view objects and areas of objects that cannot be seen with the naked eye (objects that are not within the resolution range of the normal eye). There are three well-known branches of microscopy: optical, electron, and scanning probe microscopy.

Optical & electron microscopy involve the diffraction, reflection, or refraction of electromagnetic radiation/electron beams interacting with the specimen, and the collection of the scattered radiation or another signal in order to create an image. This process may be carried out by wide-field irradiation of the sample (for example standard light microscopy and transmission electron microscopy) or by scanning of a fine beam over the sample (for example confocal laser scanning microscopy and scanning electron microscopy). Scanning probe microscopy involves the interaction of a scanning probe with the surface of the object of interest. The development of microscopy revolutionized biology, gave rise to the field of histology and so remains an essential technique in the life and physical sciences.

Optical Microscopy

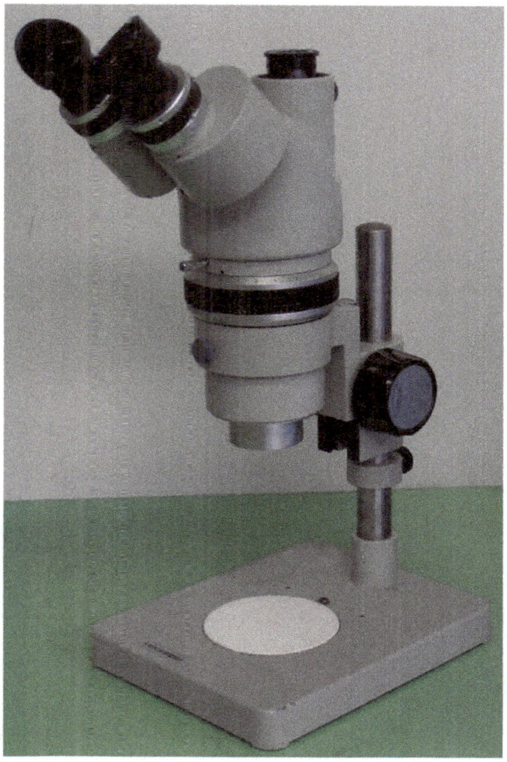

Stereo microscope

Optical or light microscopy involves passing visible light transmitted through or reflected from the sample through a single or multiple lenses to allow a magnified view of the sample. The resulting image can be detected directly by the eye, imaged on a photographic plate or captured digitally. The single lens with its attachments, or the system of lenses and imaging equipment, along with the appropriate lighting equipment, sample stage and support, makes up the basic light microscope. The most recent development is the digital microscope, which uses a CCD camera to focus on the exhibit of interest. The image is shown on a computer screen, so eye-pieces are unnecessary.

Limitations

Limitations of standard optical microscopy (bright field microscopy) lie in three areas:

- This technique can only image dark or strongly refracting objects effectively.
- Diffraction limits resolution to approximately 0.2 micrometres. This limits the practical magnification limit to ~1500x.
- Out of focus light from points outside the focal plane reduces image clarity.

Live cells in particular generally lack sufficient contrast to be studied successfully, since the internal structures of the cell are colourless and transparent. The most common way to increase contrast is to stain the different structures with selective dyes, but this often involves killing and fixing the sample. Staining may also introduce artifacts, apparent structural details that are caused by the processing of the specimen and are thus not legitimate features of the specimen. In general, these techniques make use of differences in the refractive index of cell structures. It is comparable to looking through a glass window: you (bright field microscopy) don't see the glass but merely the dirt on the glass. There is a difference, as glass is a denser material, and this creates a difference in phase of the light passing through. The human eye is not sensitive to this difference in phase, but clever optical solutions have been thought out to change this difference in phase into a difference in amplitude (light intensity).

Techniques

In order to improve specimen contrast or highlight certain structures in a sample special techniques must be used. A huge selection of microscopy techniques are available to increase contrast or label a sample.

- Four examples of transillumination techniques used to generate contrast in a sample of tissue paper. 1.559 μm/pixel.

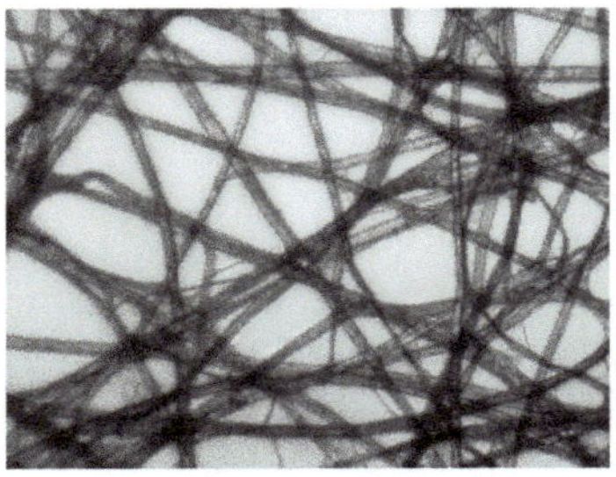

Bright field illumination, sample contrast comes from absorbance of light in the sample.

Cross-polarized light illumination, sample contrast comes from rotation of polarized light through the sample.

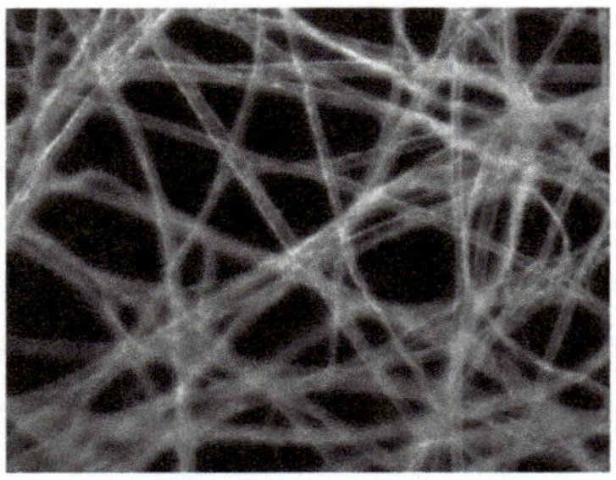

Dark field illumination, sample contrast comes from light scattered by the sample.

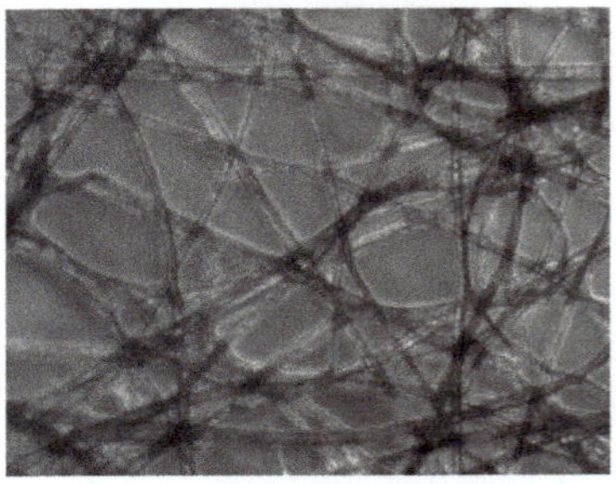

Phase contrast illumination, sample contrast comes from interference of different path lengths of light through the sample.

Bright field

Bright field microscopy is the simplest of all the light microscopy techniques. Sample illumination is via transmitted white light, i.e. illuminated from below and observed from above. Limitations include low contrast of most biological samples and low apparent resolution due to the blur of out of focus material. The simplicity of the technique and the minimal sample preparation required are significant advantages.

Oblique Illumination

The use of oblique (from the side) illumination gives the image a 3-dimensional appearance and can highlight otherwise invisible features. A more recent technique based on this method is *Hoffmann's modulation contrast*, a system found on inverted microscopes for use in cell culture. Oblique illumination suffers from the same limitations as bright field microscopy (low contrast of many biological samples; low apparent resolution due to out of focus objects).

Dark Field

Dark field microscopy is a technique for improving the contrast of unstained, transparent specimens. Dark field illumination uses a carefully aligned light source to minimize the quantity of directly transmitted (unscattered) light entering the image plane, collecting only the light scattered by the sample. Dark field can dramatically improve image contrast – especially of transparent objects – while requiring little equipment setup or sample preparation. However, the technique suffers from low light intensity in final image of many biological samples, and continues to be affected by low apparent resolution.

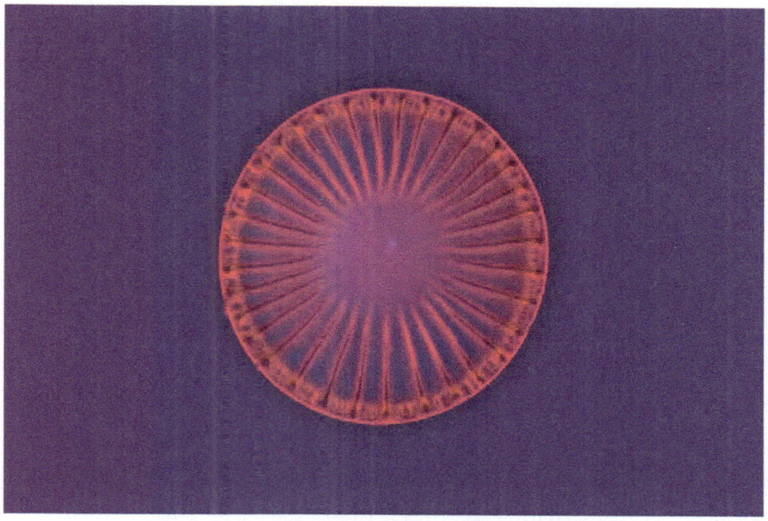

A diatom under Rheinberg illumination

Rheinberg illumination is a special variant of dark field illumination in which transparent, colored filters are inserted just before the condenser so that light rays at high aperture are differently colored than those at low aperture (i.e. the background to the specimen may be blue while the object appears self-luminous red). Other color combinations are possible but their effectiveness is quite variable.

Dispersion staining

Dispersion staining is an optical technique that results in a colored image of a colorless object. This is an optical staining technique and requires no stains or dyes to produce a color effect. There are five different microscope configurations used in the broader technique of dispersion staining. They include brightfield Becke line, oblique, darkfield, phase contrast, and objective stop dispersion staining.

Phase Contrast

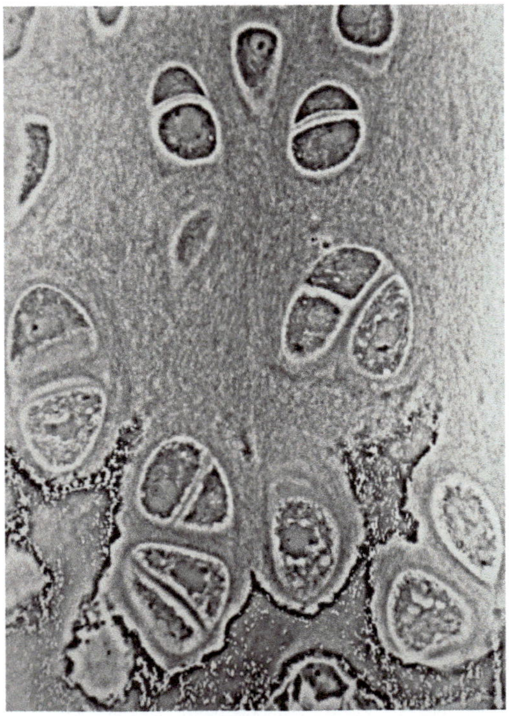

Phase-contrast light micrograph of hyaline cartilage showing chondrocytes and organelles, lacunae and extracellular matrix.

More sophisticated techniques will show proportional differences in optical density. Phase contrast is a widely used technique that shows differences in refractive index as difference in contrast. It was developed by the Dutch physicist Frits Zernike in the 1930s (for which he was awarded the Nobel Prize in 1953). The nucleus in a cell for example will show up darkly against the surrounding cytoplasm. Contrast is excellent; however it is not for use with thick objects. Frequently, a halo is formed even around small objects, which obscures detail. The system consists of a circular annulus in the condenser, which produces a cone of light. This cone is superimposed on a similar sized ring within the phase-objective. Every objective has a different size ring, so for every objective another condenser setting has to be chosen. The ring in the objective has special optical properties: it, first of all, reduces the direct light in intensity, but more importantly, it creates an artificial phase difference of about a quarter wavelength. As the physical properties of this direct light have changed, interference with the diffracted light occurs, resulting in the phase contrast image. One disadvantage of phase-contrast microscopy is halo formation (halo-light ring).

Differential Interference Contrast

Superior and much more expensive is the use of interference contrast. Differences in optical density will show up as differences in relief. A nucleus within a cell will actually show up as a globule in the most often used differential interference contrast system according to Georges Nomarski. However, it has to be kept in mind that this is an *optical effect*, and the relief does not necessarily resemble the true shape. Contrast is very good and the condenser aperture can be used fully open, thereby reducing the depth of field and maximizing resolution.

The system consists of a special prism (Nomarski prism, Wollaston prism) in the condenser that splits light in an ordinary and an extraordinary beam. The spatial difference between the two beams is minimal (less than the maximum resolution of the objective). After passage through the specimen, the beams are reunited by a similar prism in the objective.

In a homogeneous specimen, there is no difference between the two beams, and no contrast is being generated. However, near a refractive boundary (say a nucleus within the cytoplasm), the difference between the ordinary and the extraordinary beam will generate a relief in the image. Differential interference contrast requires a polarized light source to function; two polarizing filters have to be fitted in the light path, one below the condenser (the polarizer), and the other above the objective (the analyzer).

Note: In cases where the optical design of a microscope produces an appreciable lateral separation of the two beams we have the case of classical interference microscopy, which does not result in relief images, but can nevertheless be used for the quantitative determination of mass-thicknesses of microscopic objects.

Interference Reflection

An additional technique using interference is interference reflection microscopy (also known as reflected interference contrast, or RIC). It relies on cell adhesion to the slide to produce an interference signal. If there is no cell attached to the glass, there will be no interference.

Interference reflection microscopy can be obtained by using the same elements used by DIC, but without the prisms. Also, the light that is being detected is reflected and not transmitted as it is when DIC is employed.

Fluorescence

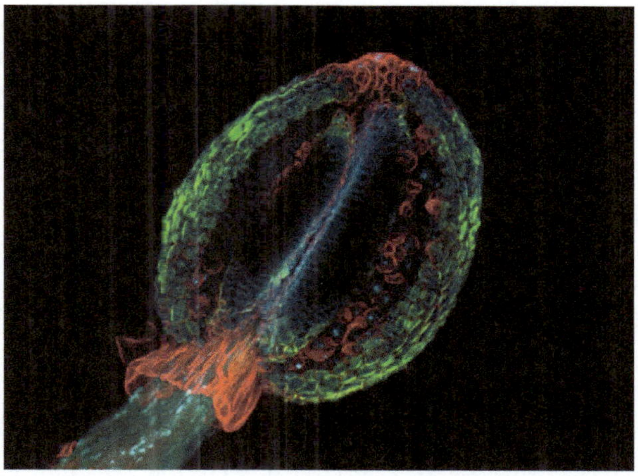

Images may also contain artifacts. This is a confocal laser scanning fluorescence micrograph of thale cress anther (part of stamen). The picture shows among other things a nice red flowing collar-like structure just below the anther. However, an intact thale cress stamen does not have such collar, this is a fixation artifact: the stamen has been cut below the picture frame, and epidermis (upper layer of cells) of stamen stalk has peeled off, forming a non-characteristic structure. Photo: Heiti Paves from Tallinn University of Technology.

When certain compounds are illuminated with high energy light, they emit light of a lower frequency. This effect is known as fluorescence. Often specimens show their characteristic autofluorescence image, based on their chemical makeup.

This method is of critical importance in the modern life sciences, as it can be extremely sensitive, allowing the detection of single molecules. Many different fluorescent dyes can be used to stain different structures or chemical compounds. One particularly powerful method is the combination of antibodies coupled to a fluorophore as in immunostaining. Examples of commonly used fluorophores are fluorescein or rhodamine.

The antibodies can be tailor-made for a chemical compound. For example, one strategy often in use is the artificial production of proteins, based on the genetic code (DNA). These proteins can then be used to immunize rabbits, forming antibodies which bind to the protein. The antibodies are then coupled chemically to a fluorophore and used to trace the proteins in the cells under study.

Highly efficient fluorescent proteins such as the green fluorescent protein (GFP) have been developed using the molecular biology technique of gene fusion, a process that links the expression of the fluorescent compound to that of the target protein. This combined fluorescent protein is, in general, non-toxic to the organism and rarely interferes with the function of the protein under study. Genetically modified cells or organisms directly express the fluorescently tagged proteins, which enables the study of the function of the original protein in vivo.

Growth of protein crystals results in both protein and salt crystals. Both are colorless and microscopic. Recovery of the protein crystals requires imaging which can be done by the intrinsic fluorescence of the protein or by using transmission microscopy. Both methods require an ultraviolet microscope as protein absorbs light at 280 nm. Protein will also fluorescence at approximately 353 nm when excited with 280 nm light.

Since fluorescence emission differs in wavelength (color) from the excitation light, an ideal fluorescent image shows only the structure of interest that was labeled with the fluorescent dye. This high specificity led to the widespread use of fluorescence light microscopy in biomedical research. Different fluorescent dyes can be used to stain different biological structures, which can then be detected simultaneously, while still being specific due to the individual color of the dye.

To block the excitation light from reaching the observer or the detector, filter sets of high quality are needed. These typically consist of an excitation filter selecting the range of excitation wavelengths, a dichroic mirror, and an emission filter blocking the excitation light. Most fluorescence microscopes are operated in the Epi-illumination mode (illumination and detection from one side of the sample) to further decrease the amount of excitation light entering the detector.

An example of fluorescence microscopy today is two-photon or multi-photon imaging. Two photon imaging allows imaging of living tissues up to a very high depth by enabling greater excitation light penetration and reduced background emission signal.

Confocal

Confocal microscopy uses a scanning point of light and a pinhole to prevent out of focus light from reaching the detector. Compared to full sample illumination, confocal microscopy gives slightly higher resolution, and significantly improves optical sectioning. Confocal microscopy is, therefore, commonly used where 3D structure is important.

Single Plane Illumination Microscopy and Light Sheet Fluorescence Microscopy

Using a plane of light formed by focusing light through a cylindrical lens at a narrow angle or by scanning a line of light in a plane perpendicular to the axis of objective, high resolution optical sections can be taken. Single plane illumination, or light sheet illumination, is also accomplished using beam shaping techniques incorporating multiple-prism beam expanders. The images are captured by CCDs. These variants allow very fast and high signal to noise ratio image capture.

Wide-field Multiphoton Microscopy

Wide-field multiphoton microscopy refers to an optical non-linear imaging technique tailored for ultrafast imaging in which a large area of the object is illuminated and imaged without the need for scanning. High intensities are required to induce non-linear optical processes such as two-photon fluorescence or second harmonic generation. In scanning multiphoton microscopes the high intensities are achieved by tightly focusing the light, and the image is obtained by stage- or beam-scanning the sample. In wide-field multiphoton microscopy the high intensities are best achieved using an optically amplified pulsed laser source to attain a large field of view (~100 μm). The image in this case is obtained as a single frame with a CCD without the need of scanning, making the technique particularly useful to visualize dynamic processes simultaneously across the object of interest. With wide-field multiphoton microscopy the frame rate can be increased up to a 1000-fold compared to multiphoton scanning microscopy.

Deconvolution

Fluorescence microscopy is a powerful technique to show specifically labeled structures within a complex environment and to provide three-dimensional information of biological structures. However, this information is blurred by the fact that, upon illumination, all fluorescently labeled structures emit light, irrespective of whether they are in focus or not. So an image of a certain structure is always blurred by the contribution of light from structures that are out of focus. This phenomenon results in a loss of contrast especially when using objectives with a high resolving power, typically oil immersion objectives with a high numerical aperture.

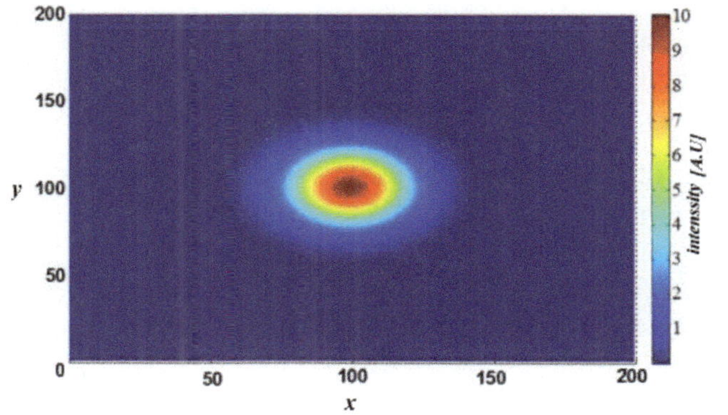

Mathematically modeled Point Spread Function of a pulsed THz laser imaging system.

However, blurring is not caused by random processes, such as light scattering, but can be well defined by the optical properties of the image formation in the microscope imaging system. If one considers a small fluorescent light source (essentially a bright spot), light coming from this spot spreads out further from our perspective as the spot becomes more out of focus. Under ideal conditions, this produces an "hourglass" shape of this point source in the third (axial) dimension. This shape is called the point spread function (PSF) of the microscope imaging system. Since any fluorescence image is made up of a large number of such small fluorescent light sources, the image is said to be "convolved by the point spread function". The mathematically modeled PSF of a terahertz laser pulsed imaging system is shown.

The output of an imaging system can be described using the equation:

$$s(x, y) = PSF(x, y) * o(x, y) + n$$

Where n is the additive noise. Knowing this point spread function means that it is possible to reverse this process to a certain extent by computer-based methods commonly known as deconvolution microscopy. There are various algorithms available for 2D or 3D deconvolution. They can be roughly classified in *nonrestorative* and *restorative* methods. While the nonrestorative methods can improve contrast by removing out-of-focus light from focal planes, only the restorative methods can actually reassign light to its proper place of origin. Processing fluorescent images in this manner can be an advantage over directly acquiring images without out-of-focus light, such as images from confocal microscopy, because light signals otherwise eliminated become useful information. For 3D deconvolution, one typically provides a series of images taken from different focal planes (called a Z-stack) plus the knowledge of the PSF, which can be derived either experimentally or theoretically from knowing all contributing parameters of the microscope.

Sub-diffraction Techniques

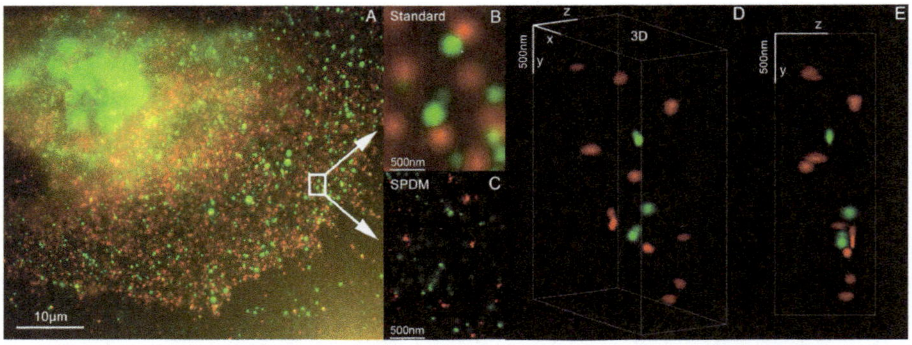

Example of super-resolution microscopy. Image of Her3 and Her2, target of the breast cancer drugTrastuzumab, within a cancer cell.

A multitude of super-resolution microscopy techniques have been developed in recent times which circumvent the diffraction barrier.

This is mostly achieved by imaging a sufficiently static sample multiple times and either modifying the excitation light or observing stochastic changes in the image.

Knowledge of and chemical control over fluorophore photophysics is at the core of these techniques, by which resolutions of ~20 nanometers are regularly obtained.

Serial Time-encoded Amplified Microscopy

Serial time encoded amplified microscopy (STEAM) is an imaging method that provides ultrafast shutter speed and frame rate, by using optical image amplification to circumvent the fundamental trade-off between sensitivity and speed, and a single-pixel photodetector to eliminate the need for a detector array and readout time limitations The method is at least 1000 times faster than the state-of-the-art CCD and CMOS cameras. Consequently, it is potentially useful for a broad range of scientific, industrial, and biomedical applications that require high image acquisition rates, including real-time diagnosis and evaluation of shockwaves, microfluidics, MEMS, and laser surgery.

Extensions

Most modern instruments provide simple solutions for micro-photography and image recording electronically. However such capabilities are not always present and the more experienced microscopist will, in many cases, still prefer a hand drawn image to a photograph. This is because a microscopist with knowledge of the subject can accurately convert a three-dimensional image into a precise two-dimensional drawing. In a photograph or other image capture system however, only one thin plane is ever in good focus.

The creation of careful and accurate micrographs requires a microscopical technique using a monocular eyepiece. It is essential that both eyes are open and that the eye that is not observing down the microscope is instead concentrated on a sheet of paper on the bench besides the microscope. With practice, and without moving the head or eyes, it is possible to accurately record the observed details by tracing round the observed shapes by simultaneously “seeing” the pencil point in the microscopical image.

Practicing this technique also establishes good general microscopical technique. It is always less tiring to observe with the microscope focused so that the image is seen at infinity and with both eyes open at all times.

Other Enhancements

Microspectroscopy:spectroscopy with a microscope.

X-ray

As resolution depends on the wavelength of the light. Electron microscopy has been developed since the 1930s that use electron beams instead of light. Because of the much smaller wavelength of the electron beam, resolution is far higher.

Though less common, X-ray microscopy has also been developed since the late 1940s. The resolution of X-ray microscopy lies between that of light microscopy and electron microscopy.

Electron Microscopy

Until the invention of sub-diffraction microscopy, the wavelength of the light limited the resolution of traditional microscopy to around 0.2 micrometers. In order to gain higher resolution, the use of an electron beam with a far smaller wavelength is used in electron microscopes.

- Transmission electron microscopy (TEM) is quite similar to the compound light microscope, by sending an electron beam through a very thin slice of the specimen. The resolution limit in 2005 was around 0.05 nanometer and has not increased appreciably since that time.
- Scanning electron microscopy (SEM) visualizes details on the surfaces of specimens and gives a very nice 3D view. It gives results much like those of the stereo light microscope. The best resolution for SEM in 2011 was 0.4 nanometer.

Electron microscopes equipped for X-ray spectroscopy can provide qualitative and quantitative elemental analysis.

Scanning probe Microscopy

This is a sub-diffraction technique. Examples of scanning probe microscopes are the atomic force microscope (AFM), the Scanning tunneling microscope, the photonic force microscope and the recurrence tracking microscope. All such methods use the physical contact of a solid probe tip to scan the surface of an object, which is supposed to be almost flat.

Ultrasonic Force

Ultrasonic Force Microscopy (UFM) has been developed in order to improve the details and image contrast on "flat" areas of interest where AFM images are limited in contrast. The combination of AFM-UFM allows a near field acoustic microscopic image to be generated. The AFM tip is used to detect the ultrasonic waves and overcomes the limitation of wavelength that occurs in acoustic microscopy. By using the elastic changes under the AFM tip, an image of much greater detail than the AFM topography can be generated.

Ultrasonic force microscopy allows the local mapping of elasticity in atomic force microscopy by the application of ultrasonic vibration to the cantilever or sample. In an attempt to analyze the results of ultrasonic force microscopy in a quantitative fashion, a force-distance curve measurement is done with ultrasonic vibration applied to the cantilever base, and the results are compared with a model of the cantilever dynamics and tip-sample interaction based on the finite-difference technique.

Ultraviolet Microscopy

Ultraviolet microscopes have two main purposes. The first is to utilize the shorter wavelength of ultraviolet electromagnetic energy to improve the image resolution beyond that of the diffraction limit of standard optical microscopes. This technique is used for non-destructive inspection of devices with very small features such as those found in modern semiconductors. The second application for UV microscopes is contrast enhancement where the response of individual samples is enhanced, relative to their surrounding, due to the interaction of light with the molecules within the sample itself. One example is in the growth of protein crystals. Protein crystals are formed in salt solutions. As salt and protein crystals are both formed in the growth process, and both are commonly transparent to the human eye, they cannot be differentiated with a standard optical microscope. As the tryptophan of protein absorbs light at 280 nm, imaging with a UV microscope with 280 nm bandpass filters makes it simple to differentiate between the two types of crystals. The protein crystals appear dark while the salt crystals are transparent.

Infrared Microscopy

The term *infrared microscopy* refers to microscopy performed at infrared wavelengths. In the typical instrument configuration a Fourier Transform Infrared Spectrometer (FTIR) is combined with an optical microscope and an infrared detector. The infrared detector can be a single point detector, a linear array or a 2D focal plane array. The FTIR provides the ability to perform chemical analysis via infrared spectroscopy and the microscope and point or array detector enable this chemical analysis to be spatially resolved, i.e. performed at different regions of the sample. As such, the technique is also called infrared microspectroscopy (an alternative architecture involves the combination of a tuneable infrared light source and single point detector on a flying objective). This technique is frequently used for infrared chemical imaging, where the image contrast is determined by the response of individual sample regions to particular IR wavelengths selected by the user, usually specific IR absorption bands and associated molecular resonances . A key limitation of conventional infrared microspectroscopy is that the spatial resolution is diffraction-limited. Specifically the spatial resolution is limited to a figure related to the wavelength of the light. For practical IR microscopes, the spatial resolution is limited to 1-3X the wavelength, depending on the specific technique and instrument used. For mid-IR wavelengths, this sets a practical spatial resolution limit of ~3-30 μm.

IR versions of sub-diffraction microscopy also exist. These include IR NSOM, photothermal microspectroscopy, and atomic force microscope based infrared spectroscopy (AFM-IR).

Digital Holographic Microscopy

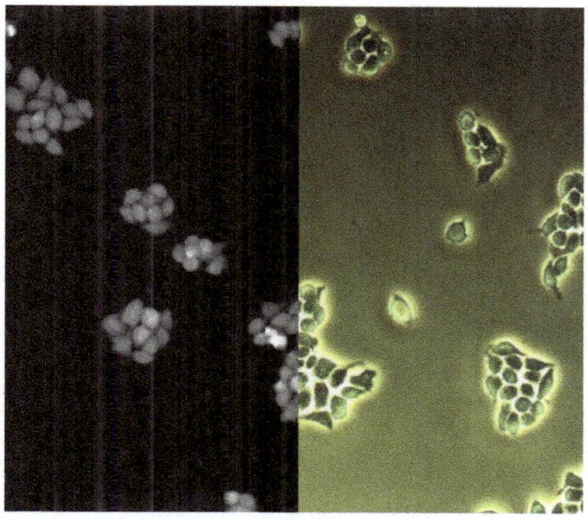

Human cells imaged by DHM phase shift (left) and phase contrast microscopy (right).

In digital holographic microscopy (DHM), interfering wave fronts from a coherent (monochromatic) light-source are recorded on a sensor. The image is digitally reconstructed by a computer from the recorded hologram. Besides the ordinary bright field image, a phase shift image is created.

DHM can operate both in reflection and transmission mode. In reflection mode, the phase shift image provides a relative distance measurement and thus represents a topography map of the reflecting surface. In transmission mode, the phase shift image provides a label-free quantitative measurement of the optical thickness of the specimen. Phase shift images of biological cells are very similar to images of stained cells and have successfully been analyzed by high content analysis software.

A unique feature of DHM is the ability to adjust focus after the image is recorded, since all focus planes are recorded simultaneously by the hologram. This feature makes it possible to image moving particles in a volume or to rapidly scan a surface. Another attractive feature is DHM's ability to use low cost optics by correcting optical aberrations by software.

Digital Pathology (Virtual Microscopy)

Digital pathology is an image-based information environment enabled by computer technology that allows for the management of information generated from a digital slide. Digital pathology is enabled in part by virtual microscopy, which is the practice of converting glass slides into digital slides that can be viewed, managed, and analyzed.

Laser Microscopy

Laser microscopy is a rapidly growing field that uses laser illumination sources in various forms of microscopy. For instance, laser microscopy focused on biological applications uses ultrashort pulse lasers, in a number of techniques labeled as nonlinear microscopy, saturation microscopy, and two-photon excitation microscopy.

High-intensity, short-pulse laboratory x-ray lasers have been under development for several years. When this technology comes to fruition, it will be possible to obtain magnified three-dimensional images of elementary biological structures in the living state at a precisely defined instant. For optimum contrast between water and protein and for best sensitivity and resolution, the laser should be tuned near the nitrogen line at about 0.3 nanometers. Resolution will be limited mainly by the hydrodynamic expansion that occurs while the necessary number of photons is being registered. Thus, while the specimen is destroyed by the exposure, its configuration can be captured before it explodes.

Scientists have been working on practical designs and prototypes for x-ray holographic microscopes, despite the prolonged development of the appropriate laser.

Amateur Microscopy

Amateur Microscopy is the investigation and observation of biological and non-biological specimens for recreational purposes. Collectors of minerals, insects, seashells, and plants may use microscopes as tools to uncover features that help them classify their collected items. Other amateurs may be interested in observing the life found in pond water and of other samples. Microscopes may also prove useful for the water quality assessment for people that keep a home aquarium. Photographic documentation and drawing of the microscopic images are additional tasks that augment the spectrum of tasks of the amateur. There are even competitions for photomicrograph art. Participants of this pastime may either use commercially prepared microscopic slides or engage in the task of specimen preparation.

While microscopy is a central tool in the documentation of biological specimens, it is, in general, insufficient to justify the description of a new species based on microscopic investigations alone. Often genetic and biochemical tests are necessary to confirm the discovery of a new species. A laboratory and access to academic literature is a necessity, which is specialized and, in general, not available to amateurs. There is, however, one huge advantage that amateurs have above professionals: time to explore their surroundings. Often, advanced amateurs team up with professionals to validate their findings and (possibly) describe new species.

In the late 1800s, amateur microscopy became a popular hobby in the United States and Europe. Several 'professional amateurs' were being paid for their sampling trips and microscopic explorations by philanthropists, to keep them amused on the Sunday afternoon (e.g., the diatom specialist A. Grunow, being paid by (among others) a Belgian industrialist). Professor John Phin published "Practical Hints on the Selection and Use of the Microscope," and was also the editor of the "American Journal of Microscopy."

Examples of amateur microscopy images:

 "house bee" Mouth 100X	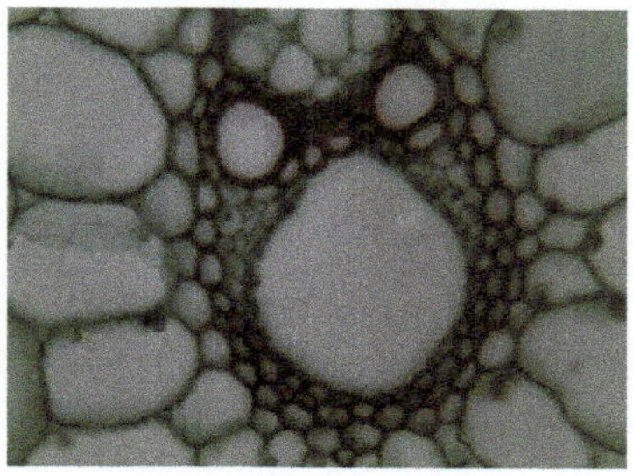 Rice Stem cs 400X
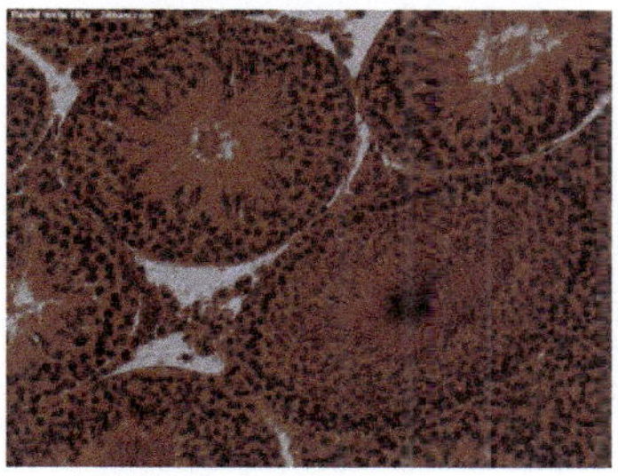 Rabbit Testis 100X	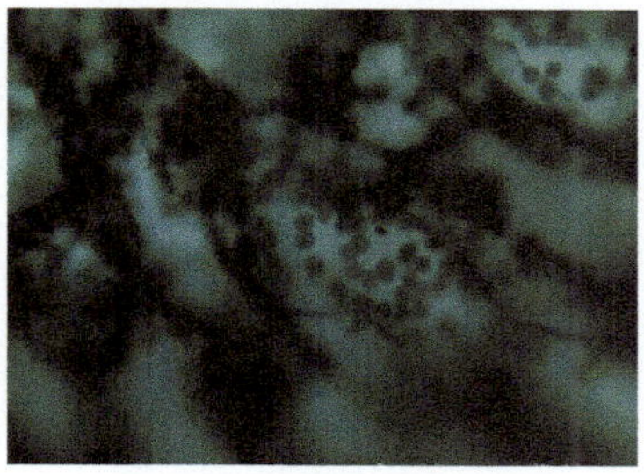 Fern Prothallium 400X

Optical Microscope

The optical microscope, often referred to as light microscope, is a type of microscope which uses visible light and a system of lenses to magnify images of small samples. Optical microscopes are the oldest design of microscope and were possibly invented in their present compound form in the 17th century. Basic optical microscopes can be very simple, although there are many complex designs which aim to improve resolution and sample contrast.

The image from an optical microscope can be captured by normal light-sensitive cameras to generate a micrograph. Originally images were captured by photographic film but modern developments in CMOS and charge-coupled device (CCD) cameras allow the capture of digital images. Purely digital microscopes are now available which use a CCD camera to examine a sample, showing the resulting image directly on a computer screen without the need for eyepieces.

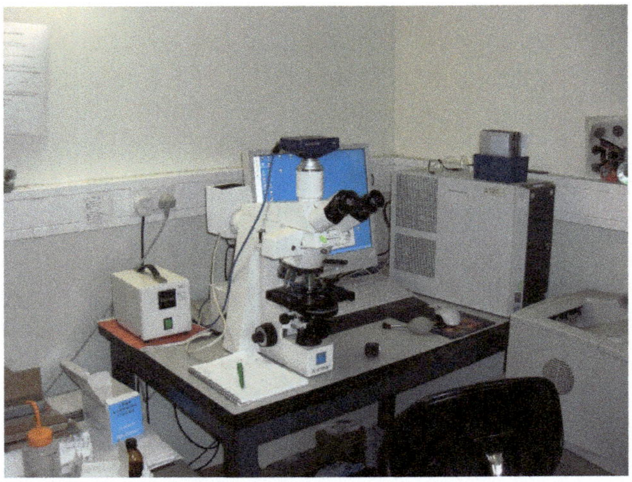

A modern microscope with a mercury bulb for fluorescence microscopy. The microscope has a digital camera, and is attached to a computer.

Alternatives to optical microscopy which do not use visible light include scanning electron microscopy and transmission electron microscopy.

On 8 October 2014, the Nobel Prize in Chemistry was awarded to Eric Betzig, William Moerner and Stefan Hell for "the development of super-resolved fluorescence microscopy," which brings "optical microscopy into the nanodimension".

Types

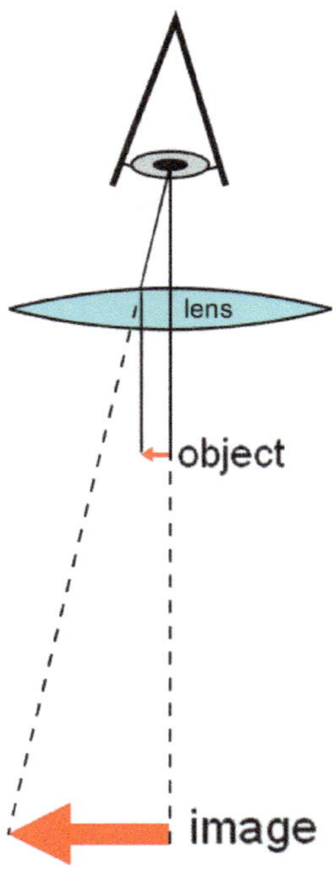

Diagram of a simple microscope

There are two basic types of optical microscopes: simple microscopes and compound microscopes. A simple microscope is one which uses a single lens for magnification, such as a magnifying glass. A compound microscope uses several lenses to enhance the magnification of an object. The vast majority of modern research microscopes are compound microscopes while some cheaper commercial digital microscopes are simple single lens microscopes. Compound microscopes can be further divided into a variety of other types of microscopes which differ in their optical configurations, cost, and intended purposes.

Simple Microscope

A simple microscope uses a lens or set of lenses to enlarge an object through angular magnification alone, giving the viewer an erect enlarged virtual image. The use of a single convex lens or groups of lenses are still found in simple magnification devices such as the magnifying glass, loupes, and eyepieces for telescopes and microscopes.

Compound Microscope

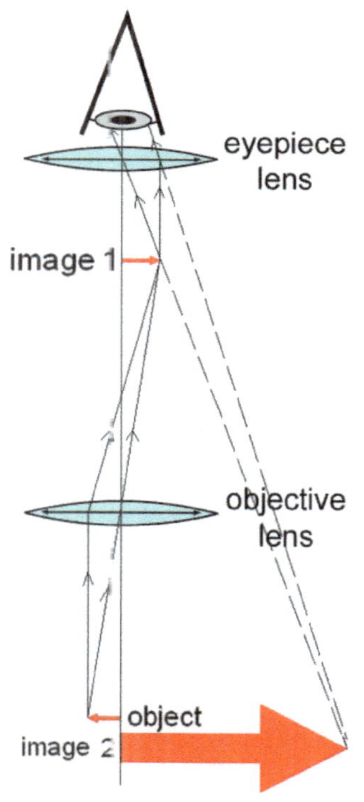

Diagram of a compound microscope

A compound microscope uses a lens close to the object being viewed to collect light (called the objective lens) which focuses a real image of the object inside the microscope (image 1). That image is then magnified by a second lens or group of lenses (called the eyepiece) that gives the viewer an enlarged inverted virtual image of the object (image 2). The use of a compound objective/eyepiece combination allows for much higher magnification. Common compound microscopes often feature exchangeable objective lenses, allowing the user to quickly adjust the magnification. A compound microscope also enables more advanced illumination setups, such as phase contrast.

Other Microscope Variants

There are many variants of the compound optical microscope design for specialized purposes. Some of these are physical design differences allowing specialization for certain purposes:

- Stereo microscope, a low-powered microscope which provides a stereoscopic view of the sample, commonly used for dissection.
- Comparison microscope, which has two separate light paths allowing direct comparison of two samples via one image in each eye.
- Inverted microscope, for studying samples from below; useful for cell cultures in liquid, or for metallography.
- Student microscope, designed for low cost, durability, and ease of use.
- Fiber optic connector inspection microscope, designed for connector end-face inspection

Other microscope variants are designed for different illumination techniques:

- Petrographic microscope, whose design usually includes a polarizing filter, rotating stage and gypsum plate to facilitate the study of minerals or other crystalline materials whose optical properties can vary with orientation.
- Polarizing microscope, similar to the petrographic microscope.
- Phase contrast microscope, which applies the phase contrast illumination method.
- Epifluorescence microscope, designed for analysis of samples which include fluorophores.
- Confocal microscope, a widely used variant of epifluorescent illumination which uses a scanning laser to illuminate a sample for fluorescence.
- Student microscope – an often low-power microscope with simplified controls and sometimes low quality optics designed for school use or as a starter instrument for children.

Digital Microscope

A miniature USB microscope.

A digital microscope is a microscope equipped with a digital camera allowing observation of a sample via a computer. Microscopes can also be partly or wholly computer-controlled with various levels of automation. Digital microscopy allows greater analysis of a microscope image, for example measurements of distances and areas and quantitaton of a fluorescent or histological stain.

Low-powered digital microscopes, USB microscopes, are also commercially available. These are essentially webcams with a high-powered macro lens and generally do not use transillumination. The camera attached directly to the USB port of a computer, so that the images are shown directly on the monitor. They offer modest magnifications (up to about 200×) without the need to use eyepieces, and at very low cost. High power illumination is usually provided by an LED source or sources adjacent to the camera lens.

Digital microscopy with very low light levels to avoid damage to vulnerable biological samples is available using sensitive photon-counting digital cameras. It has been demonstrated that a light source providing pairs of entangled photons may minimize the risk of damage to the most light-sensitive samples. In this application of ghost imaging to photon-sparse microscopy, the sample is illuminated with infrared photons, each of which is spatially correlated with an entangled partner in the visible band for efficient imaging by a photon-counting camera.

History

Invention

It is difficult to say who invented the compound microscope. Cornelis Drebbel has a strong claim, whilst the Dutch spectacle-maker Zacharias Janssen is sometimes claimed to have invented it in 1590 (a claim made by his son and fellow countrymen, in different testimony in 1634 and 1655). Another claim is that Janssen's competitor, Hans Lippershey, invented the compound microscope. Another favourite for the title of 'inventor of the microscope' was Galileo Galilei. He developed an *occhiolino* or compound microscope with a convex and a concave lens in 1609. Galileo's microscope was celebrated in the Accademia dei Lincei in 1624 and was the first such device to be given the name "microscope" a year later by fellow Lincean Giovanni Faber. Faber coined the name from the Greek words (micron) meaning "small", and (skopein) meaning "to look at", a name meant to be analogous with "telescope", another word coined by the Linceans.

Christiaan Huygens, another Dutchman, developed a simple 2-lens ocular system in the late 17th century that was achromatically corrected, and therefore a huge step forward in microscope development. The Huygens ocular is still being produced to this day, but suffers from a small field size, and other minor disadvantages.

Popularization

Antonie van Leeuwenhoek (1632–1724) is credited with bringing the microscope to the attention of biologists, even though simple magnifying lenses were already being produced in the 16th century. Van Leeuwenhoek's home-made microscopes were simple microscopes, with a single very small, yet strong lens. They were awkward in use, but enabled van Leeuwenhoek to see detailed images. It took about 150 years of optical development before the compound microscope was able to provide the same quality image as van Leeuwenhoek's simple microscopes, due to difficulties in configuring multiple lenses.

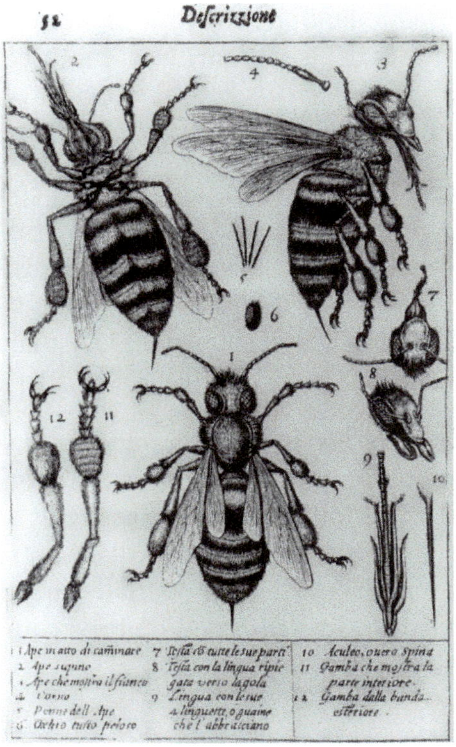

The oldest published image known to have been made with a microscope: bees by Francesco Stelluti, 1630

Lighting Techniques

While basic microscope technology and optics have been available for over 400 years it is much more recently that techniques in sample illumination were developed to generate the high quality images seen today.

In August 1893 August Köhler developed Köhler illumination. This method of sample illumination gives rise to extremely even lighting and overcomes many limitations of older techniques of sample illumination. Before development of Köhler illumination the image of the light source, for example a lightbulb filament, was always visible in the image of the sample.

The Nobel Prize in physics was awarded to Dutch physicist Frits Zernike in 1953 for his development of phase contrast illumination which allows imaging of transparent samples. By using interference rather than absorption of light, extremely transparent samples, such as live mammalian cells, can be imaged without having to use staining techniques. Just two years later, in 1955, Georges Nomarski published the theory for differential interference contrast microscopy, another interference-based imaging technique.

Fluorescence Microscopy

Modern biological microscopy depends heavily on the development of fluorescent probes for specific structures within a cell. In contrast to normal transilluminated light microscopy, in fluorescence microscopy the sample is illuminated through the objective lens with a narrow set of wavelengths of light. This light interacts with fluorophores in the sample which then emit light of a longer wavelength. It is this emitted light which makes up the image.

Since the mid 20th century chemical fluorescent stains, such as DAPI which binds to DNA, have been used to label specific structures within the cell. More recent developments include immunofluorescence, which uses fluorescently labelled antibodies to recognise specific proteins within a sample, and fluorescent proteins like GFP which a live cell can express making it fluorescent.

Components

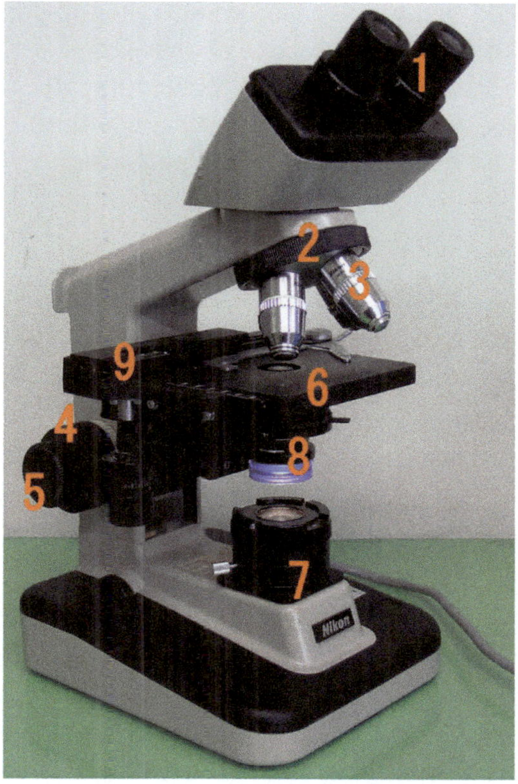

Basic optical transmission microscope elements (1990s)

All modern optical microscopes designed for viewing samples by transmitted light share the same basic components of the light path. In addition, the vast majority of microscopes have the same 'structural' components (numbered below according to the image on the right):

- Eyepiece (ocular lens) (1)
- Objective turret, revolver, or revolving nose piece (to hold multiple objective lenses) (2)
- Objective lenses (3)
- Focus knobs (to move the stage)
    - Coarse adjustment (4)
    - Fine adjustment (5)
- Stage (to hold the specimen) (6)
- Light source (a light or a mirror) (7)

- Diaphragm and condenser (8)
- Mechanical stage (9)

Eyepiece (Ocular Lens)

The eyepiece, or ocular lens, is a cylinder containing two or more lenses; its function is to bring the image into focus for the eye. The eyepiece is inserted into the top end of the body tube. Eyepieces are interchangeable and many different eyepieces can be inserted with different degrees of magnification. Typical magnification values for eyepieces include 5×, 10× (the most common), 15X and 20×. In some high performance microscopes, the optical configuration of the objective lens and eyepiece are matched to give the best possible optical performance. This occurs most commonly with apochromatic objectives.

Objective Turret (Revolver Or Revolving Nose Piece)

Objective turret, revolver, or revolving nose piece is the part that holds the set of objective lenses. It allows the user to switch between objective lenses.

Objective

At the lower end of a typical compound optical microscope, there are one or more objective lenses that collect light from the sample. The objective is usually in a cylinder housing containing a glass single or multi-element compound lens. Typically there will be around three objective lenses screwed into a circular nose piece which may be rotated to select the required objective lens. These arrangements are designed to be parfocal, which means that when one changes from one lens to another on a microscope, the sample stays in focus. Microscope objectives are characterized by two parameters, namely, magnification and numerical aperture. The former typically ranges from 5× to 100× while the latter ranges from 0.14 to 0.7, corresponding to focal lengths of about 40 to 2 mm, respectively. Objective lenses with higher magnifications normally have a higher numerical aperture and a shorter depth of field in the resulting image. Some high performance objective lenses may require matched eyepieces to deliver the best optical performance.

Oil Immersion Objective

Two Leica oil immersion microscope objective lenses: 100× (left) and 40× (right)

Some microscopes make use of oil-immersion objectives or water-immersion objectives for greater resolution at high magnification. These are used with index-matching material such as immersion oil or water and a matched cover slip between the objective lens and the sample. The refractive index of the index-matching material is higher than air allowing the objective lens to have a larger numerical aperture (greater than 1) so that the light is transmitted from the specimen to the outer face of the objective lens with minimal refraction. Numerical apertures as high as 1.6 can be achieved. The larger numerical aperture allows collection of more light making detailed observation of smaller details possible. An oil immersion lens usually has a magnification of 40 to 100×.

Focus Knobs

Adjustment knobs move the stage up and down with separate adjustment for coarse and fine focusing. The same controls enable the microscope to adjust to specimens of different thickness. In older designs of microscopes, the focus adjustment wheels move the microscope tube up or down relative to the stand and had a fixed stage.

Frame

The whole of the optical assembly is traditionally attached to a rigid arm, which in turn is attached to a robust U-shaped foot to provide the necessary rigidity. The arm angle may be adjustable to allow the viewing angle to be adjusted.

The frame provides a mounting point for various microscope controls. Normally this will include controls for focusing, typically a large knurled wheel to adjust coarse focus, together with a smaller knurled wheel to control fine focus. Other features may be lamp controls and/or controls for adjusting the condenser.

Stage

The stage is a platform below the objective which supports the specimen being viewed. In the center of the stage is a hole through which light passes to illuminate the specimen. The stage usually has arms to hold slides (rectangular glass plates with typical dimensions of 25×75 mm, on which the specimen is mounted).

At magnifications higher than 100× moving a slide by hand is not practical. A mechanical stage, typical of medium and higher priced microscopes, allows tiny movements of the slide via control knobs that reposition the sample/slide as desired. If a microscope did not originally have a mechanical stage it may be possible to add one.

All stages move up and down for focus. With a mechanical stage slides move on two horizontal axes for positioning the specimen to examine specimen details.

Focusing starts at lower magnification in order to center the specimen by the user on the stage. Moving to a higher magnification requires the stage to be moved higher vertically for re-focus at the higher magnification and may also require slight horizontal specimen position adjustment. Horizontal specimen position adjustments are the reason for having a mechanical stage.

Due to the difficulty in preparing specimens and mounting them on slides, for children it's best to begin with prepared slides that are centered and focus easily regardless of the focus level used.

Light Source

Many sources of light can be used. At its simplest, daylight is directed via a mirror. Most microscopes, however, have their own adjustable and controllable light source – often a halogen lamp, although illumination using LEDs and lasers are becoming a more common provision.

Condenser

The condenser is a lens designed to focus light from the illumination source onto the sample. The condenser may also include other features, such as a diaphragm and/or filters, to manage the quality and intensity of the illumination. For illumination techniques like dark field, phase contrast and differential interference contrast microscopy additional optical components must be precisely aligned in the light path.

Magnification

The actual power or magnification of a compound optical microscope is the product of the powers of the ocular (eyepiece) and the objective lens. The maximum normal magnifications of the ocular and objective are 10× and 100× respectively, giving a final magnification of 1,000×.

Magnification and Micrographs

When using a camera to capture a micrograph the effective magnification of the image must take into account the size of the image. This is independent of whether it is on a print from a film negative or displayed digitally on a computer screen.

In the case of photographic film cameras the calculation is simple; the final magnification is the product of: the objective lens magnification, the camera optics magnification and the enlargement factor of the film print relative to the negative. A typical value of the enlargement factor is around 5× (for the case of 35mm film and a 15x10 cm (6×4 inch) print).

In the case of digital cameras the size of the pixels in the CMOS or CCD detector and the size of the pixels on the screen have to be known. The enlargement factor from the detector to the pixels on screen can then be calculated. As with a film camera the final magnification is the product of: the objective lens magnification, the camera optics magnification and the enlargement factor.

Operation

The optical components of a modern microscope are very complex and for a microscope to work well, the whole optical path has to be very accurately set up and controlled. Despite this, the basic operating principles of a microscope are quite simple.

The objective lens is, at its simplest, a very high-powered magnifying glass, i.e. a lens with a very short focal length. This is brought very close to the specimen being examined so that the light from the specimen comes to a focus about 160 mm inside the microscope tube. This creates an enlarged image of the subject. This image is inverted and can be seen by removing the eyepiece and placing a piece of tracing paper over the end of the tube. By carefully focusing a brightly lit specimen, a highly enlarged image can be seen. It is this real image that is viewed by the eyepiece lens that provides further enlargement.

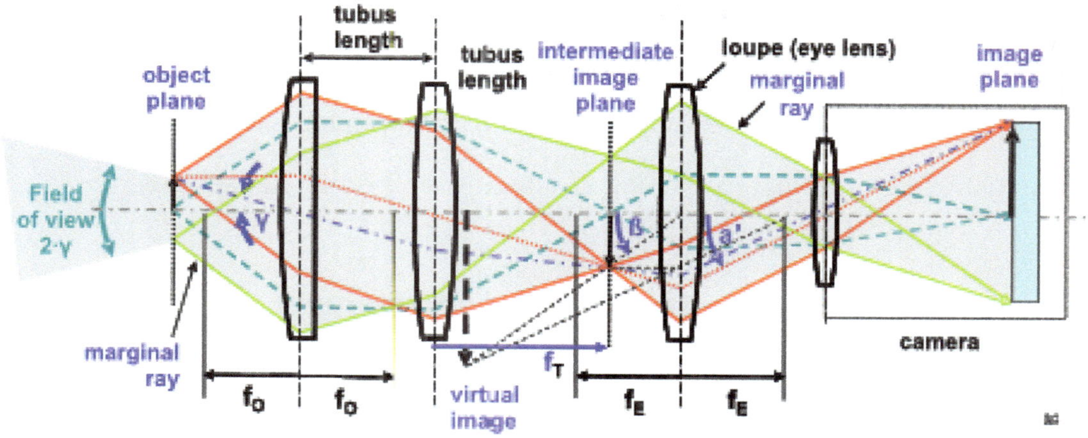

Optical path in a typical microscope

In most microscopes, the eyepiece is a compound lens, with one component lens near the front and one near the back of the eyepiece tube. This forms an air-separated couplet. In many designs, the virtual image comes to a focus between the two lenses of the eyepiece, the first lens bringing the real image to a focus and the second lens enabling the eye to focus on the virtual image.

In all microscopes the image is intended to be viewed with the eyes focused at infinity (mind that the position of the eye in the above figure is determined by the eye's focus). Headaches and tired eyes after using a microscope are usually signs that the eye is being forced to focus at a close distance rather than at infinity.

Illumination Techniques

Many techniques are available which modify the light path to generate an improved contrast image from a sample. Major techniques for generating increased contrast from the sample include cross-polarized light, dark field, phase contrast and differential interference contrast illumination. A recent technique (Sarfus) combines cross-polarized light and specific contrast-enhanced slides for the visualization of nanometric samples.

Other Techniques

Modern microscopes allow more than just observation of transmitted light image of a sample; there are many techniques which can be used to extract other kinds of data. Most of these require additional equipment in addition to a basic compound microscope.

- Reflected light, or incident, illumination (for analysis of surface structures)
- Fluorescence microscopy, both:
    - Epifluorescence microscopy

    - Confocal microscopy
- Microspectroscopy (where a UV-visible spectrophotometer is integrated with an optical microscope)
- Ultraviolet microscopy
- Near-Infrared microscopy
- Multiple transmission microscopy for contrast enhancement and aberration reduction.
- Automation (for automatic scanning of a large sample or image capture)

Applications

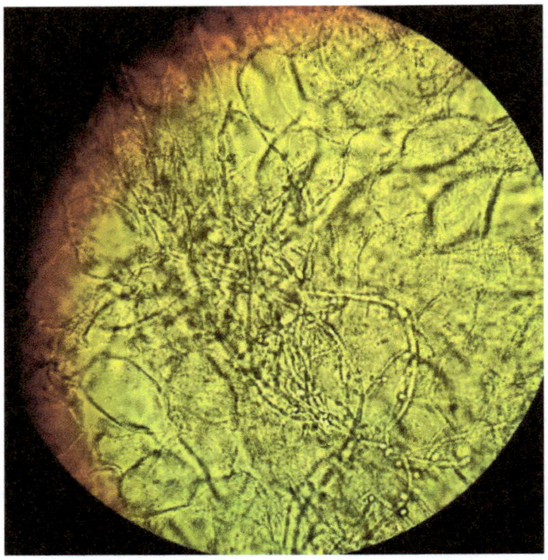

A 40x magnification image of cells in a medical smear test taken through an optical microscope using a wet mount technique, placing the specimen on a glass slide and mixing with a salt solution

Optical microscopy is used extensively in microelectronics, nanophysics, biotechnology, pharmaceutic research, mineralogy and microbiology.

Optical microscopy is used for medical diagnosis, the field being termed histopathology when dealing with tissues, or in smear tests on free cells or tissue fragments.

In industrial use, binocular microscopes are common. Aside from applications needing true depth perception, the use of dual eyepieces reduces eye strain associated with long workdays at a microscopy station. In certain applications, long-working-distance or long-focus microscopes are beneficial. An item may need to be examined behind a window, or industrial subjects may be a hazard to the objective. Such optics resemble telescopes with close-focus capabilities.

Limitations

At very high magnifications with transmitted light, point objects are seen as fuzzy discs surrounded by diffraction rings. These are called Airy disks. The *resolving power* of a microscope is taken as the ability to distinguish between two closely spaced Airy disks (or, in other words the ability of the microscope to reveal adjacent structural detail as distinct and separate). It is these impacts of diffraction that limit the ability to resolve fine details. The extent and magnitude of the diffraction patterns are affected by both the wavelength of light (λ), the refractive materials used to manufacture the objective lens and the numerical aperture (NA) of the objective lens. There is therefore a finite limit beyond which it is impossible to resolve separate points in the objective field, known as the diffraction limit. Assuming that optical aberrations in the whole optical set-up are negligible, the resolution *d*, can be stated as:

$$d = \frac{\lambda}{2NA}$$

Usually a wavelength of 550 nm is assumed, which corresponds to green light. With air as the external medium, the highest practical *NA* is 0.95, and with oil, up to 1.5. In practice the lowest value of *d* obtainable with conventional lenses is about 200 nm. A new type of lens using multiple scattering of light allowed to improve the resolution to below 100 nm.

Surpassing the Resolution Limit

Multiple techniques are available for reaching resolutions higher than the transmitted light limit described above. Holographic techniques, as described by Courjon and Bulabois in 1979, are also capable of breaking this resolution limit, although resolution was restricted in their experimental analysis.

Using fluorescent samples more techniques are available. Examples include Vertico SMI, near field scanning optical microscopy which uses evanescent waves, and stimulated emission depletion. In 2005, a microscope capable of detecting a single molecule was described as a teaching tool.

Despite significant progress in the last decade, techniques for surpassing the diffraction limit remain limited and specialized.

While most techniques focus on increases in lateral resolution there are also some techniques which aim to allow analysis of extremely thin samples. For example, sarfus methods place the thin sample on a contrast-enhancing surface and thereby allows to directly visualize films as thin as 0.3 nanometers.

Structured Illumination SMI

SMI (spatially modulated illumination microscopy) is a light optical process of the so-called point spread function (PSF) engineering. These are processes which modify the PSF of a microscope in a suitable manner to either increase the optical resolution, to maximize the precision of distance measurements of fluorescent objects that are small relative to the wavelength of the illuminating light, or to extract other structural parameters in the nanometer range.

Localization Microscopy SPDMphymod

SPDM (spectral precision distance microscopy), the basic localization microscopy technology is a light optical process of fluorescence microscopy which allows position, distance and angle measurements on "optically isolated" particles (e.g. molecules) well below the theoretical limit of resolution for light microscopy. "Optically isolated" means that at a given point in time, only a single particle/molecule within a region of a size determined by conventional optical resolution (typically approx. 200–250 nm diameter) is being registered. This is possible when molecules within such a region all carry different spectral markers (e.g. different colors or other usable differences in the light emission of different particles).

Many standard fluorescent dyes like GFP, Alexa dyes, Atto dyes, Cy2/Cy3 and fluorescein molecules can be used for localization microscopy, provided certain photo-physical conditions are present. Using this so-called SPDMphymod (physically modifiable fluorophores) technology a single laser wavelength of suitable intensity is sufficient for nanoimaging.

3D Super Resolution Microscopy

3D super resolution microscopy with standard fluorescent dyes can be achieved by combination of localization microscopy for standard fluorescent dyes SPDMphymod and structured illumination SMI.

STED

Stimulated emission depletion (STED) microscopy image of actin filaments within a cell.

Stimulated emission depletion is a simple example of how higher resolution surpassing the diffraction limit is possible, but it has major limitations. STED is a fluorescence microscopy technique which uses a combination of light pulses to induce fluorescence in a small sub-population of fluorescent molecules in a sample. Each molecule produces a diffraction-limited spot of light in the image, and the centre of each of these spots corresponds to the location of the molecule. As the number of fluorescing molecules is low the spots of light are unlikely to overlap and therefore can be placed accurately. This process is then repeated many times to generate the image. Stefan Hell of the Max Planck Institute for Biophysical Chemistry was awarded the 10th German Future Prize in 2006 and Nobel Prize for Chemistry in 2014 for his development of the STED microscope and associated methodologies.

Alternatives

In order to overcome the limitations set by the diffraction limit of visible light other microscopes have been designed which use other waves.

- Atomic force microscope (AFM)
- Scanning electron microscope (SEM)
- Scanning ion-conductance microscopy (SICM)
- Scanning tunneling microscope (STM)
- Transmission electron microscopy (TEM)
- Ultraviolet microscope
- X-ray microscope

It is important to note that higher frequency waves have limited interaction with matter, for example soft tissues are relatively transparent to X-rays resulting in distinct sources of contrast and different target applications.

The use of electrons and X-rays in place of light allows much higher resolution – the wavelength of the radiation is shorter so the diffraction limit is lower. To make the short-wavelength probe non-destructive, the atomic beam imaging system (atomic nanoscope) has been proposed and widely discussed in the literature, but it is not yet competitive with conventional imaging systems.

STM and AFM are scanning probe techniques using a small probe which is scanned over the sample surface. Resolution in these cases is limited by the size of the probe; micromachining techniques can produce probes with tip radii of 5–10 nm.

Additionally, methods such as electron or X-ray microscopy use a vacuum or partial vacuum, which limits their use for live and biological samples (with the exception of an environmental scanning electron microscope). The specimen chambers needed for all such instruments also limits sample size, and sample manipulation is more difficult. Color cannot be seen in images made by these methods, so some information is lost. They are however, essential when investigating molecular or atomic effects, such as age hardening in aluminium alloys, or the microstructure of polymers.

Lenses and the Bending of Light

To understand how a light microscope operates, one should know something about the way in which lenses bend and focus light to form images. When a ray of light passes from one medium to another, refraction occurs, that is, the ray is bent at the interface. Refractive index is the measure of the relative velocity at which light passes through a material. When light passes from air to glass, a medium with a greater refractive index, it is slowed and bent towards the normal. As light leaves glass and returns to air, a medium with a lower refractive index, it accelerates and is bent away from the normal. As a result the prism bends light passing through it. A lens functions somewhat like a collection of prisms. Light rays from a distant source are focused at the focal point F. The focal point lies at a distance f, the focal length, from the lens center. Our eyes cannot focus on objects nearer than about 25 cm or 10 inches and can be overcome by using a convex lens as a simple magnifier (or microscope). Lens strength is related to focal length; a lens with a shorter focal length will magnify an object more than a weaker lens having a longer focal length.

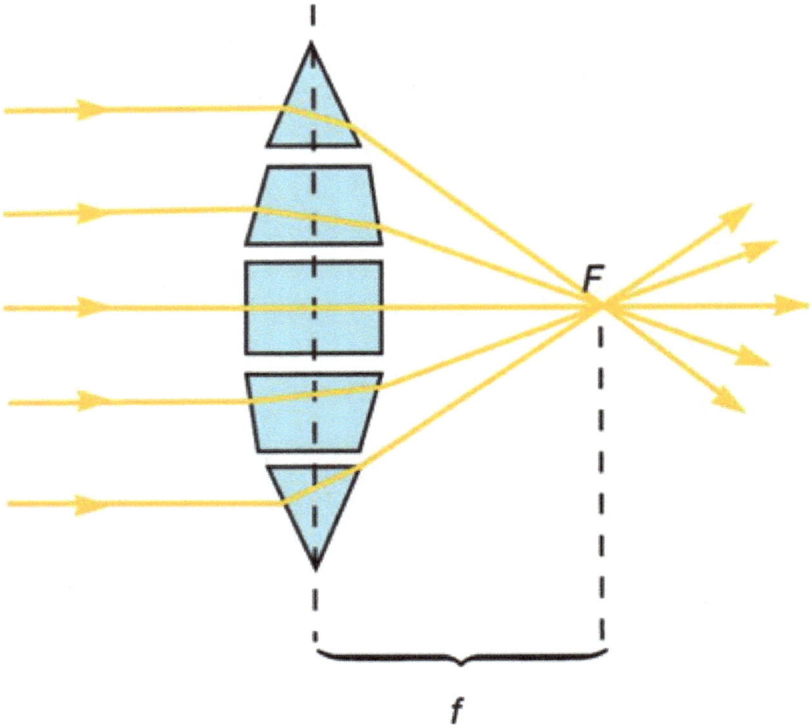

Light refraction and lens function

Microscopic resolution:

Resolution is the ability of a lens to separate or distinguish between small objects that are close to each other. Much of the optical theory underlying microscope design was developed by the German physicist Ernst Abbe in the 1870's.

$d = 0.5\lambda / n\sin\theta$

d = distance between two objects that reveals them as separate entities

λ = wavelength of light

$n\sin\theta$ = numerical aperture (NA)

As'd' becomes smaller, the resolution increases. The wavelength must be shorter than the distance between two objects. Thus greatest resolution is obtained with light of the shortest wavelength, light at the blue end of the visible spectrum (450 to 500nm). The $n\sin\theta$ or NA is more difficult to understand. The θ = 1/2th angle of the cone of light entering an objective. Light that strikes the specimen after passing through condenser is cone-shaped. Cone of light has a narrow angle then resolution is low. Cone of light has a broad angle and spreads out rapidly after passing through a spectrum, closely packed objects appear widely separated and more resolved. The angle of the cone of light that can enter a lens depends on the refractive index (n) of the medium in which the lens works. The refractive index of air is 1.00, since sinθ cannot be greater than 1.00 (the maximum θ is 90° and sin 90° is 1.00), no lens working in air will have a NA greater than 1.00. The practical way to raise the NA above 1.00 is therefore to increase the refractive index with immersion oil which has a refractive index of glass (1.25-1.4, glass and immersion oil). Air is replaced with immersion oil, light rays that did not enter the objective due to reflection and refraction will now do so and there will be an increase in NA and resolution.

d = (0.5) x (530nm) / 1.25 = 212 nm or 0.2μm

The maximum resolution of a bright field microscope with oil immersion is 0.2μm.

Electron Microscope

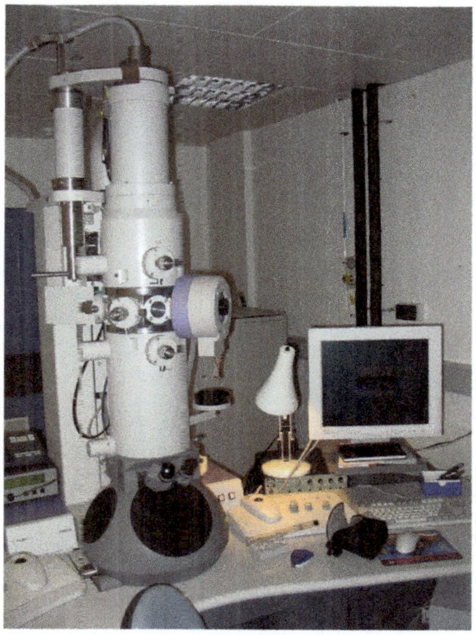

A modern transmission electron microscope

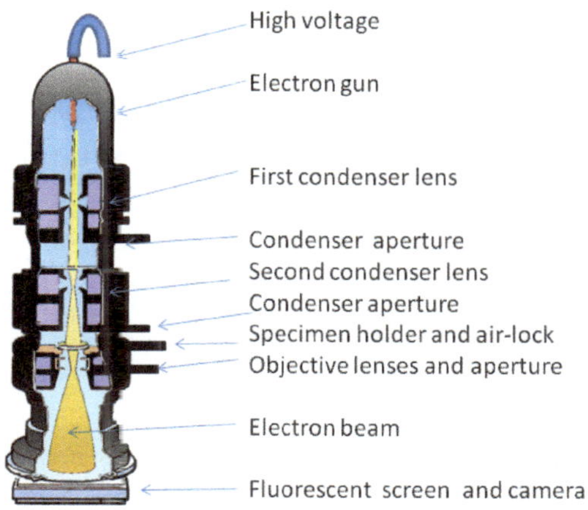

Transmiss on Electron Microscope
Diagram of a transmission electron microscope

Electron microscope constructed by Ernst Ruska in 1933

An electron microscope is a microscope that uses a beam of accelerated electrons as a source of illumination. As the wavelength of an electron can be up to 100,000 times shorter than that of visible light photons, electron microscopes have a higher resolving power than light microscopes and can reveal the structure of smaller objects. A transmission electron microscope can achieve better than 50 pm resolution and magnifications of up to about 10,000,000x whereas most light microscopes are limited by diffraction to about 200 nm resolution and useful magnifications below 2000x.

Transmission electron microscopes use electrostatic and electromagnetic lenses to control the electron beam and focus it to form an image. These electron optical lenses are analogous to the glass lenses of an optical light microscope.

Electron microscopes are used to investigate the ultrastructure of a wide range of biological and inorganic specimens including microorganisms, cells, large molecules, biopsy samples, metals, and crystals. Industrially, electron microscopes are often used for quality control and failure analysis. Modern electron microscopes produce electron micrographs using specialized digital cameras and frame grabbers to capture the image.

History

The first electromagnetic lens was developed in 1926 by Hans Busch.

According to Dennis Gabor, the physicist Leó Szilárd tried in 1928 to convince Busch to build an electron microscope, for which he had filed a patent.

The physicist Ernst Ruska and the electrical engineer Max Knoll constructed the prototype electron microscope in 1931, capable of four-hundred-power magnification; the apparatus was the first demonstration of the principles of electron microscopy. Two years later, in 1933, Ruska built an electron microscope that exceeded the resolution attainable with an optical (light) microscope. Moreover, Reinhold Rudenberg, the scientific director of Siemens-Schuckertwerke, obtained the patent for the electron microscope in May 1931.

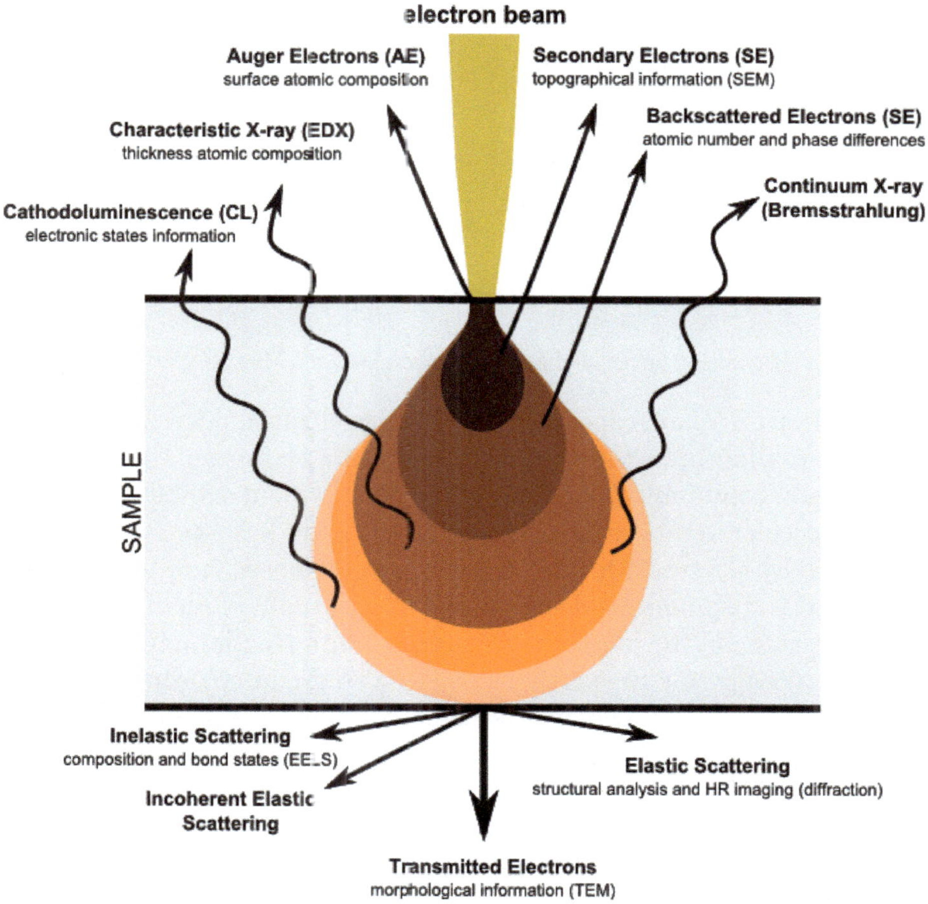

Diagram illustrating the phenomena resulting of the interaction of highly energetic electrons with matter.

In 1932, Ernst Lubcke of Siemens & Halske built and obtained images from a prototype electron microscope, applying concepts described in the Rudenberg patent applications. Five years later (1937), the firm financed the work of Ernst Ruska and Bodo von Borries, and employed Helmut Ruska (Ernst's brother) to develop applications for the microscope, especially with biological specimens. Also in 1937, Manfred von Ardenne pioneered the scanning electron microscope. The first commercial electron microscope was produced in 1938 by Siemens. The first North American electron microscope was constructed in 1938, at the University of Toronto, by Eli Franklin Burton and students Cecil Hall, James Hillier, and Albert Prebus; and Siemens produced a transmission electron microscope (TEM) in 1939. Although contemporary transmission electron microscopes are capable of two million-power magnification, as scientific instruments, they remain based upon Ruska's prototype.

Types

Transmission Electron Microscope (TEM)

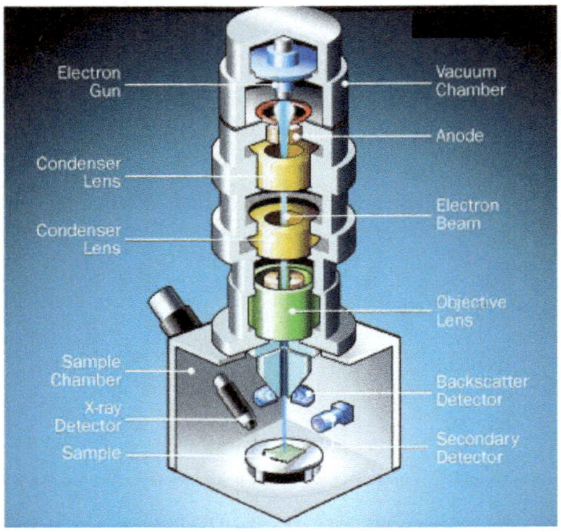

Operating principe of a Transmission Electron Microscope

The original form of electron microscope, the transmission electron microscope (TEM) uses a high voltage electron beam to illuminate the specimen and create an image. The electron beam is produced by an electron gun, commonly fitted with a tungsten filament cathode as the electron source. The electron beam is accelerated by an anode typically at +100 keV (40 to 400 keV) with respect to the cathode, focused by electrostatic and electromagnetic lenses, and transmitted through the specimen that is in part transparent to electrons and in part scatters them out of the beam. When it emerges from the specimen, the electron beam carries information about the structure of the specimen that is magnified by the objective lens system of the microscope. The spatial variation in this information (the "image") may be viewed by projecting the magnified electron image onto a fluorescent viewing screen coated with a phosphor or scintillator material such as zinc sulfide. Alternatively, the image can be photographically recorded by exposing a photographic film or plate directly to the electron beam, or a high-resolution phosphor may be coupled by means of a lens optical system or a fibre optic light-guide to the sensor of a digital camera. The image detected by the digital camera may be displayed on a monitor or computer.

The resolution of TEMs is limited primarily by spherical aberration, but a new generation of aberration correctors have been able to partially overcome spherical aberration to increase resolution. Hardware correction of spherical aberration for the high-resolution transmission electron microscopy (HRTEM) has allowed the production of images with resolution below 0.5 angstrom (50 picometres) and magnifications above 50 million times. The ability to determine the positions of atoms within materials has made the HRTEM an important tool for nano-technologies research and development.

Transmission electron microscopes are often used in electron diffraction mode. The advantages of electron diffraction over X-ray crystallography are that the specimen need not be a single crystal or even a polycrystalline powder, and also that the Fourier transform reconstruction of the object's magnified structure occurs physically and thus avoids the need for solving the phase problem faced by the X-ray crystallographers after obtaining their X-ray diffraction patterns of a single crystal or polycrystalline powder.

The major disadvantage of the transmission electron microscope is the need for extremely thin sections of the specimens, typically about 100 nanometers. Biological specimens are typically required to be chemically fixed, dehydrated and embedded in a polymer resin to stabilize them sufficiently to allow ultrathin sectioning. Sections of biological specimens, organic polymers and similar materials may require special treatment with heavy atom labels in order to achieve the required image contrast.

Scanning Electron Microscope (SEM)

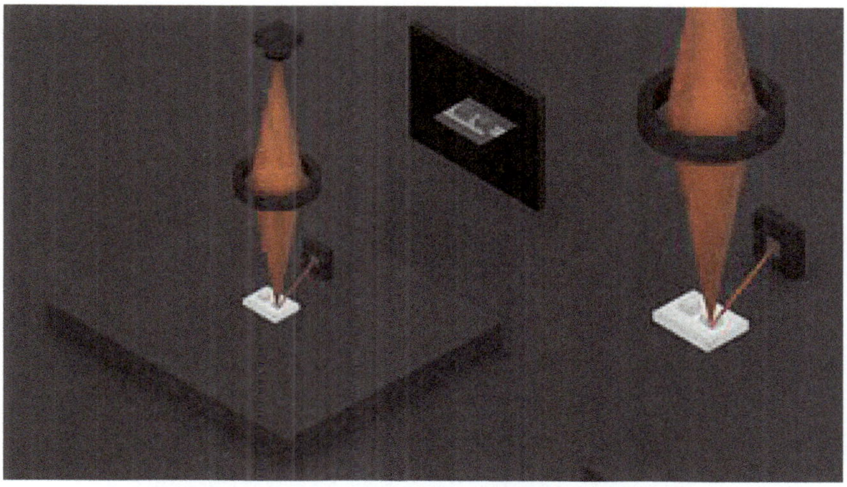

Operating principe of a Scanning Electron Microscope

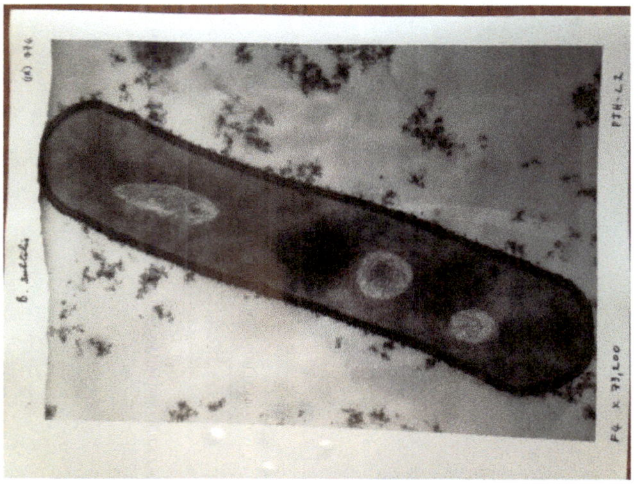

Image of bacillus subtilis taken with a 1960s electron microscope.

The SEM produces images by probing the specimen with a focused electron beam that is scanned across a rectangular area of the specimen (raster scanning). When the electron beam interacts with the specimen, it loses energy by a variety of mechanisms. The lost energy is converted into alternative forms such as heat, emission of low-energy secondary electrons and high-energy backscattered electrons, light emission (cathodoluminescence) or X-ray emission, all of which provide signals carrying information about the properties of the specimen surface, such as its topography and composition. The image displayed by an SEM maps the varying intensity of any of these signals into the image in a position corresponding to the position of the beam on the specimen when the signal was generated. In the SEM image of an ant shown below and to the right, the image was constructed from signals produced by a secondary electron detector, the normal or conventional imaging mode in most SEMs.

Generally, the image resolution of an SEM is at least an order of magnitude poorer than that of a TEM. However, because the SEM image relies on surface processes rather than transmission, it is able to image bulk samples up to many centimeters in size and (depending on instrument design and settings) has a great depth of field, and so can produce images that are good representations of the three-dimensional shape of the sample. Another advantage of SEM is its variety called environmental scanning electron microscope (ESEM) can produce images of sufficient quality and resolution with the samples being wet or contained in low vacuum or gas. This greatly facilitates imaging biological samples that are unstable in the high vacuum of conventional electron microscopes.

An image of an ant in a scanning electron microscope

Color

In their most common configurations, electron microscopes produce images with a single brightness value per pixel, with the results usually rendered in grayscale. However, often these images are then colorized through the use of feature-detection software, or simply by hand-editing using a graphics editor. This may be done to clarify structure or for aesthetic effect and generally does not add new information about the specimen.

In some configurations information about several specimen properties is gathered per pixel, usually by the use of multiple detectors. In SEM, the attributes of topography and material contrast can be obtained by a pair of backscattered electron detectors and such attributes can be superimposed in a single color image by assigning a different primary color to each attribute. Similarly, a combination of backscattered and secondary electron signals can be assigned to different colors and superimposed on a single color micrograph displaying simultaneously the properties of the specimen.

Some types of detectors used in SEM have analytical capabilities, and can provide several items of data at each pixel. Examples are the Energy-dispersive X-ray spectroscopy (EDS) detectors used in elemental analysis and Cathodoluminescence microscope (CL) systems that analyse the intensity and spectrum of electron-induced luminescence in (for example) geological specimens. In SEM systems using these detectors it is common to color code the signals and superimpose them in a single color image, so that differences in the distribution of the various components of the specimen can be seen clearly and compared. Optionally, the standard secondary electron image can be merged with the one or more compositional channels, so that the specimen's structure and composition can be compared. Such images can be made while maintaining the full integrity of the original signal, which is not modified in any way.

Reflection Electron Microscope (REM)

In the reflection electron microscope (REM) as in the TEM, an electron beam is incident on a surface but instead of using the transmission (TEM) or secondary electrons (SEM), the reflected beam of elastically scattered electrons is detected. This technique is typically coupled with reflection high energy electron diffraction (RHEED) and *reflection high-energy loss spectroscopy (RHELS)*. Another variation is spin-polarized low-energy electron microscopy (SPLEEM), which is used for looking at the microstructure of magnetic domains.

Scanning Transmission Electron Microscope (STEM)

The STEM rasters a focused incident probe across a specimen that (as with the TEM) has been thinned to facilitate detection of electrons scattered through the specimen. The high resolution of the TEM is thus possible in STEM. The focusing action (and aberrations) occur before the electrons hit the specimen in the STEM, but afterward in the TEM. The STEMs use of SEM-like beam rastering simplifies annular dark-field imaging, and other analytical techniques, but also means that image data is acquired in serial rather than in parallel fashion. Often TEM can be equipped with the scanning option and then it can function both as TEM and STEM.

Sample Preparation

An insect coated in gold for viewing with a scanning electron microscope

Materials to be viewed under an electron microscope may require processing to produce a suitable sample. The technique required varies depending on the specimen and the analysis required:

- *Chemical fixation* – for biological specimens aims to stabilize the specimen's mobile macromolecular structure by chemical crosslinking of proteins with aldehydes such as formaldehyde and glutaraldehyde, and lipids with osmium tetroxide.
- *Negative stain* – suspensions containing nanoparticles or fine biological material (such as viruses and bacteria) are briefly mixed with a dilute solution of an electron-opaque solution such as ammonium molybdate, uranyl acetate (or formate), or phosphotungstic acid. This mixture is applied to a suitably coated EM grid, blotted, then allowed to dry. Viewing of this preparation in the TEM should be carried out without delay for best results. The method is important in microbiology for fast but crude morphological identification, but can also be used as the basis for high resolution 3D reconstruction using EM tomography methodology when carbon films are used for support. Negative staining is also used for observation of nanoparticles.
- *Cryofixation* – freezing a specimen so rapidly, in liquid ethane, and maintained at liquid nitrogen or even liquid helium temperatures, so that the water forms vitreous (non-crystalline) ice. This preserves the specimen in a snapshot of its solution state. An entire field called cryo-electron microscopy has branched from this technique. With the development of cryo-electron microscopy of vitreous sections (CEMOVIS), it is now possible to observe samples from virtually any biological specimen close to its native state.
- *Dehydration* – or replacement of water with organic solvents such as ethanol or acetone, followed by critical point drying or infiltration with embedding resins. Also freeze drying.
- *Embedding, biological specimens* – after dehydration, tissue for observation in the transmission electron microscope is embedded so it can be sectioned ready for viewing. To do this the tissue is passed through a 'transition solvent' such as propylene oxide (epoxypropane) or acetone and then infiltrated with an epoxy resin such as Araldite, Epon, or Durcupan; tissues may also be embedded directly in water-miscible acrylic resin. After the resin has been polymerized (hardened) the sample is thin sectioned (ultrathin sections) and stained – it is then ready for viewing.
- *Embedding, materials* – after embedding in resin, the specimen is usually ground and polished to a mirror-like finish using ultra-fine abrasives. The polishing process must be performed carefully to minimize scratches and other polishing artifacts that reduce image quality.
- *Metal shadowing* – Metal (e.g. platinum) is evaporated from an overhead electrode and applied to the surface of a biological sample at an angle. The surface topography results in variations in the thickness of the metal that are seen as variations in brightness and contrast in the electron microscope image.
- *Replication* – A surface shadowed with metal (e.g. platinum, or a mixture of carbon and platinum) at an angle is coated with pure carbon evaporated from carbon electrodes at right angles to the surface. This is followed by removal of the specimen material (e.g. in an acid

bath, using enzymes or by mechanical separation) to produce a surface replica that records the surface ultrastructure and can be examined using transmission electron microscopy.

- *Sectioning* – produces thin slices of specimen, semitransparent to electrons. These can be cut on an ultramicrotome with a diamond knife to produce ultra-thin sections about 60–90 nm thick. Disposable glass knives are also used because they can be made in the lab and are much cheaper.

- *Staining* – uses heavy metals such as lead, uranium or tungsten to scatter imaging electrons and thus give contrast between different structures, since many (especially biological) materials are nearly "transparent" to electrons (weak phase objects). In biology, specimens can be stained "en bloc" before embedding and also later after sectioning. Typically thin sections are stained for several minutes with an aqueous or alcoholic solution of uranyl acetate followed by aqueous lead citrate.

- *Freeze-fracture or freeze-etch* – a preparation method particularly useful for examining lipid membranes and their incorporated proteins in "face on" view. The fresh tissue or cell suspension is frozen rapidly (cryofixation), then fractured by breaking or by using a microtome while maintained at liquid nitrogen temperature. The cold fractured surface (sometimes "etched" by increasing the temperature to about −100 °C for several minutes to let some ice sublime) is then shadowed with evaporated platinum or gold at an average angle of 45° in a high vacuum evaporator. A second coat of carbon, evaporated perpendicular to the average surface plane is often performed to improve stability of the replica coating. The specimen is returned to room temperature and pressure, then the extremely fragile "pre-shadowed" metal replica of the fracture surface is released from the underlying biological material by careful chemical digestion with acids, hypochlorite solution or SDS detergent. The still-floating replica is thoroughly washed free from residual chemicals, carefully fished up on fine grids, dried then viewed in the TEM.

- *Ion beam milling* – thins samples until they are transparent to electrons by firing ions (typically argon) at the surface from an angle and sputtering material from the surface. A subclass of this is focused ion beam milling, where gallium ions are used to produce an electron transparent membrane in a specific region of the sample, for example through a device within a microprocessor. Ion beam milling may also be used for cross-section polishing prior to SEM analysis of materials that are difficult to prepare using mechanical polishing.

- *Conductive coating* – an ultrathin coating of electrically conducting material, deposited either by high vacuum evaporation or by low vacuum sputter coating of the sample. This is done to prevent the accumulation of static electric fields at the specimen due to the electron irradiation required during imaging. The coating materials include gold, gold/palladium, platinum, tungsten, graphite, etc.

- *Earthing* – to avoid electrical charge accumulation on a conductive coated sample, it is usually electrically connected to the metal sample holder. Often an electrically conductive adhesive is used for this purpose.

Disadvantages

Electron microscopes are expensive to build and maintain, but the capital and running costs of confocal light microscope systems now overlaps with those of basic electron microscopes. Microscopes designed to achieve high resolutions must be housed in stable buildings (sometimes underground) with special services such as magnetic field cancelling systems.

The samples largely have to be viewed in vacuum, as the molecules that make up air would scatter the electrons. An exception is the environmental scanning electron microscope, which allows hydrated samples to be viewed in a low-pressure (up to 20 Torr or 2.7 kPa) and/or wet environment.

Scanning electron microscopes operating in conventional high-vacuum mode usually image conductive specimens; therefore non-conductive materials require conductive coating (gold/palladium alloy, carbon, osmium, etc.). The low-voltage mode of modern microscopes makes possible the observation of non-conductive specimens without coating. Non-conductive materials can be imaged also by a variable pressure (or environmental) scanning electron microscope.

Small, stable specimens such as carbon nanotubes, diatom frustules and small mineral crystals (asbestos fibres, for example) require no special treatment before being examined in the electron microscope. Samples of hydrated materials, including almost all biological specimens have to be prepared in various ways to stabilize them, reduce their thickness (ultrathin sectioning) and increase their electron optical contrast (staining). These processes may result in *artifacts*, but these can usually be identified by comparing the results obtained by using radically different specimen preparation methods. Since the 1980s, analysis of cryofixed, vitrified specimens has also become increasingly used by scientists, further confirming the validity of this technique.

Transmission Electron Microscope (TEM)

Transmission electron microscopy (TEM) involves a high voltage electron beam emitted by a cathode and formed by magnetic lenses. The electron beam that has been partially transmitted through the very thin (and so semitransparent for electrons) specimen carries information about the structure of the specimen. The spatial variation in this information (the "image") is then magnified by a series of magnetic lenses until it is recorded by hitting a fluorescent screen, photographic plate, or light sensitive sensor such as a CCD (charge-coupled device) camera. The image detected by the CCD may be displayed in real time on a monitor or computer. Transmission electron microscopes produce two-dimensional, black and white images. Resolution of the TEM is also limited by spherical and chromatic aberration, but a new generation of aberration correctors has been able to overcome or limit these aberrations. Software correction of spherical aberration has allowed the production of images with sufficient resolution.

Specimens are prepared by negative staining, shadowing with metal or free-etching and can resolve objects as close as 2.5 nm.

Preparation of specimens for TEM:

- Specimens must be around 20-100 nm thick and should be able to maintain its structure when bombarded with electrons under high vacuum.

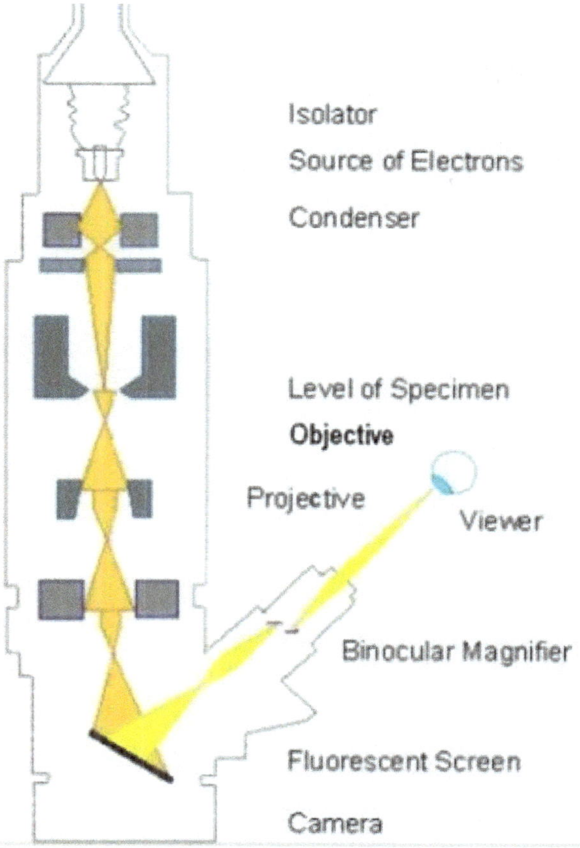

Transmission electron microscope

- Thin slice can be cut with the necessary support like plastic
- After fixate with chemicals like glutaraldehyde or osmium tetroxide to stabilize cell structures, the specimen is dehydrated with organic solvents (e.g., acetone or ethanol)
- Specimen is soaked in un-polymerized, liquid epoxy plastic until it is completely permeated, and then the plastic is hardened to form a solid block. Thin sections are cut from this block with a glass or diamond knife using a special instrument called an ultra microtome
- Specimens are soaked with solutions of heavy metal salts like lead citrate and Uranyl acetate (make them more electron opaque), thus increasing the contrast in the material
- Stained thin sections are then mounted on tiny copper grids and viewed.

Three types:

1. Negative staining:
   - Specimen is spread out in a thin film with either phosphotungstic acid or uranyl acetate
   - Heavy metals do not penetrate the specimen but render the background dark, whereas the specimen appearslight in photographs

- Excellent way to study the structure of viruses, bacterial gas vacuoles and other similar objects

2. Shadowing:

- It is coated with a thin film of platinum or other heavy metals by evaporation at an angle of about 45° from horizontal so that the metal strikes the microorganism on only one side
- The area coated with metal scatters electrons and appears light in photographs. Whereas the uncoated side and the shadow region created by the object is dark
- Particularly useful in studying virus morphology, bacterial flagella and plasmids

3. Freeze-etching technique:

- Disclose the shape of organelles within microorganism by this technique
- Cells are rapidly frozen in liquid nitrogen and then warmed to -100°C in a vacuum chamber
- A knife that has been pre cooled with liquid nitrogen (-196°C) fractures the frozen cells, which are very brittle and break along lines by greatest weakness
- Specimen is left in the high vacuum for a minute or more so that some ice can sublimate away and uncover more structural detail
- Finally, the exposed surfaces are shadowed and coated with layers of platinum and carbon to form a replica of the surface
- After the specimen has been removed chemically, this replica is studied I the TEM and provides a detailed, three-dimensional view of intracellular structure.

Table: Features and differences between light and electron microscope

Feature	Light Microscope	Electron Microscope
Highest practical magnification	About 1000-1500	Over 100,000
Best resolution	0.2 µm	0.5 nm
Radiation source	Visible light	Electron beam
Medium of travel	Air	High vacuum
Type of lens	Glass	Electromagnet
Source of contrast	Differential light absorption	Scattering of electrons
Focusing mechanism	Adjust lens position mechanically	Adjust current to the magnetic lens
Method of changing magnification	Switch the objective lens or eye piece	Adjust current to the magnetic lens
Specimen mount	Glass slide	Metal grid (usually copper)

Scanning Electron Microscope (SEM)

The Scanning Electron Microscope (SEM) produces images by detecting secondary electrons which are emitted from the surface due to excitation by the primary electron beam. In the SEM, the electron beam is scanned across the surface of the sample in a raster pattern, with detectors building up an image by mapping the detected signals with beam position. The SEM image relies on electron interactions at the surface rather than transmission it is able to image bulk samples and has a much greater depth of view, and so can produce images that are a good representation of the 3D structure of the sample. SEM images are therefore considered to provide us with 3D, topographical information about the sample surface but will still always be only in black and white. This is specially used to study the external surface features of microorganisms.

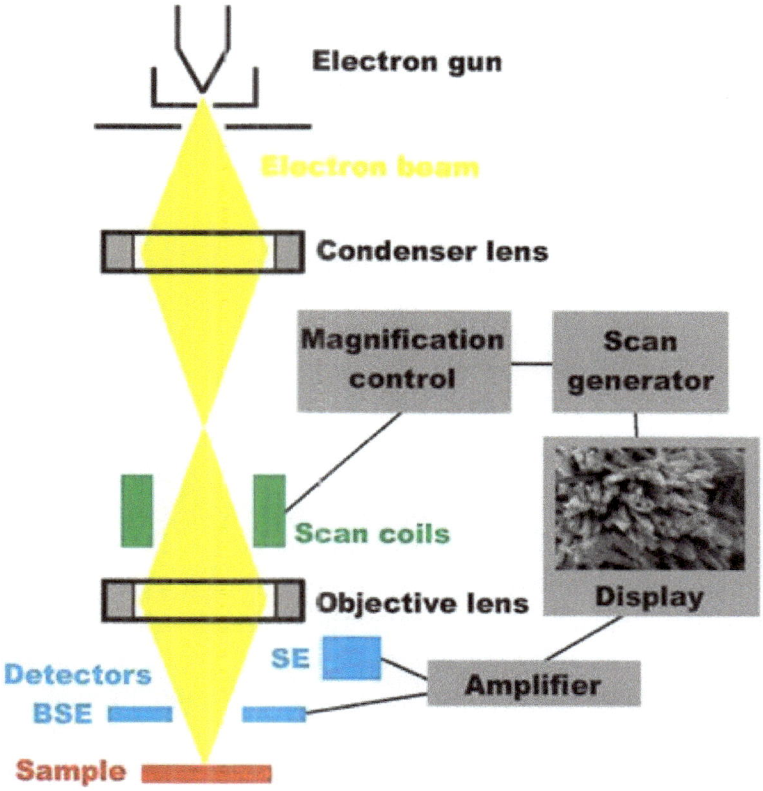

Scanning electron microscope

Disadvantages of Electron Microscopy

Electron microscopes are very expensive to buy and maintain. They are dynamic rather than static in their operation: requiring extremely stable high voltage supplies, extremely stable currents to each electromagnetic coil/lens, continuously-pumped high/ultra-high vacuum systems and a cooling water supply circulation through the lenses and pumps. As they are very sensitive to vibration and external magnetic fields, microscopes aimed at achieving high resolutions must be housed in buildings with special services.

New Microscopy Technique

Confocal Microscopy

Confocal microscopy is an optical imaging technique used to increase optical resolution and of a micrograph by using point illumination and a spatial pinhole to eliminate out-of-focus light in specimens that are thicker than the focal plane. It enables the reconstruction of three-dimensional structures from the obtained images. This technique has gained popularity in the scientific and industrial communities and typical applications are in life sciences, semiconductor inspection and materials science

Three types of confocal microscopes are commercially available.

Confocal laser scanning microscopes use a pair of mirrors (one for the x and the other for the y axis) to scan the laser across the sample and «descan» the image across a fixed pinhole and detector.

- Spinning-disk (Nipkow disk) confocal microscopes use a series of moving pinholes on a disc to scan spot of light.
- Programmable Array Microscopes (PAM) uses an electronically controlled spatial light modulator (SLM) that produces a set of moving pinholes. The SLM is a device containing an array of pixels with some property (opacity, reflectivity or optical rotation) of the individual pixels that can be adjusted electronically. The SLM contains micro-electromechanical mirrors or liquid crystal components. The image is usually acquired by a CCD camera.

Each of these classes of confocal microscope has particular advantages and disadvantages. Most systems are either optimized for either recording speed (i.e. video capture) or high spatial resolution. Confocal laser scanning microscopes can have a programmable sampling density and very high resolutions while Nipkow and PAM use a fixed sampling density defined by the camera's resolution. Imaging frame rates are typically very slow for laser scanning systems (e.g. less than 3 frames/second). Commercial spinning-disk confocal microscopes achieve frame rates of over 50 per second – a desirable feature for dynamic observations such as live cell imaging. In practice, Nipkow and PAM allow multiple pinholes scanning the same are in parallel as long as the pinholes are sufficiently far apart. Cutting-edge development of confocal laser scanning microscopy now allows better than standard video rate (60 frames/second) imaging by using multiple micro-electromechanical systems -based scanning mirrors. Confocal X-ray fluorescence imaging is a newer technique that allows control over depth, in addition to horizontal and vertical aiming, for example, when analyzing buried layers in a painting

Scanning Probe Microscope:

Scanning tunneling Microscope (1980):

A scanning tunneling microscope (STM) is an instrument for imaging surfaces at the atomic level. Its development in 1981 earned its inventors, Gerd Binnig and Heinrich Rohrer (at IBM Zürich), the Nobel Prize in Physics in 1986. For an STM, good resolution is considered to be 0.1 nm lateral resolution and 0.01 nm depth resolution. With this resolution, individual atoms within materials are routinely imaged and manipulated. The STM can be used not only in ultra-high vacuum but also in air, water, and various other liquid or gas ambients, and at temperatures ranging from near zero kelvin to a few hundred degrees Celsius. The STM is based on the concept of quantum tunneling . When a conducting tip is brought very near to the surface to be examined, a bias (voltage difference) applied between the two can allow electrons to tunnel through the vacuum between them. The resulting *tunneling current* is a function of tip position, applied voltage, and the local density of states (LDOS) of the sample. Information is acquired by monitoring the current as the tip's position scans across the surface, and is usually displayed in image form. STM can be a challenging technique, as it requires extremely clean and stable surfaces, sharp tips, excellent vibration control , and sophisticated electronics, but nonetheless many hobbyists have built their own. Recently, it has been used to directly view DNA. This microscope can examine objects when they are immersed in water, it may be particularly useful in studying biological molecules. It has a needle like probe with a point so sharp that often there is only one atom at its tip. The arrangement of atoms on the specimen surface is determined by moving the probe tip back and forth over the surface (3-D image of the surface atoms is made as its motion os recorded and analyzed by a computer)

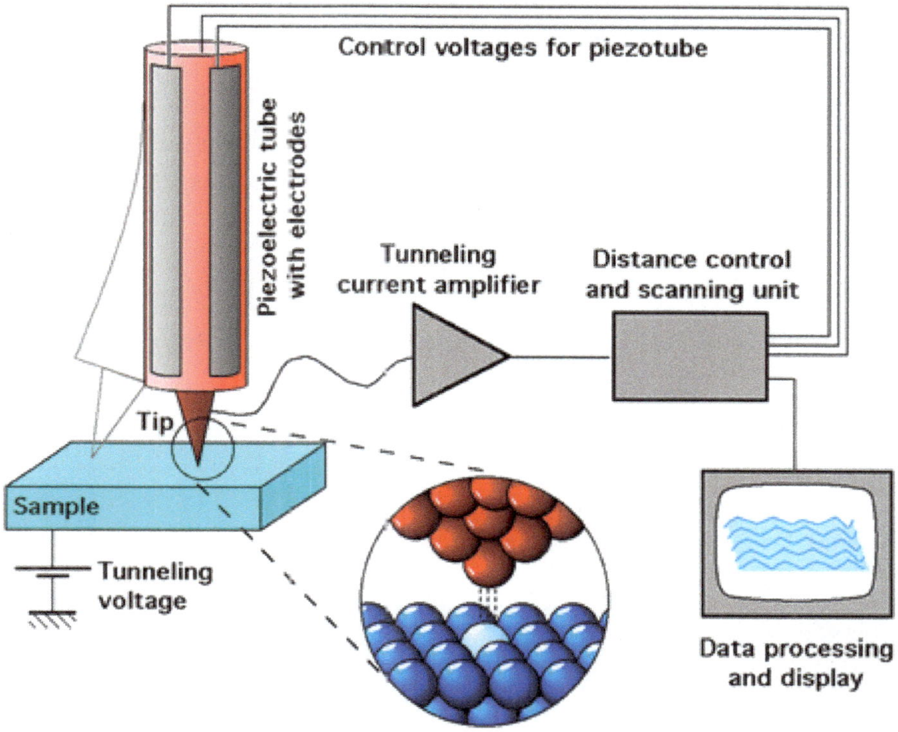

Scanning tunneling microscope

Atomic force microscope:

Atomic force microscopy (AFM) or scanning force microscopy (SFM) is a very high-resolution type of scanning probe microscopy, with demonstrated resolution on the order of fractions of a nanometer, more than 1000 times better than the optical diffraction limit. The first commercially available atomic force microscope was introduced in 1989. The AFM is one of the foremost tools for imaging, measuring, and manipulating matter at the nanoscale. The information is gathered by "feeling" the surface with a mechanical probe. Piezoelectric elements that facilitate tiny but accurate and precise movements on (electronic) command enable the very precise scanning. In some variations, electric potentials can also be scanned using conducting cantilevers. In newer more advanced versions, currents can even be passed through the tip to probe the electrical conductivity or transport of the underlying surface, but this is much more challenging with very few research groups reporting consistent data.The primary modes of operation for an AFM are static mode and dynamic mode. In static mode, the cantilever is "dragged" across the surface of the sample and the contours of the surface are measured directly using the deflection of the cantilever. In the dynamic mode, the cantilever is externally oscillated at or close to its fundamental resonance frequency or a harmonic. The oscillation amplitude, phase and resonance frequency are modified by tip-sample interaction forces. These changes in oscillation with respect to the external reference oscillation provide information about the sample's characteristics. The AFM can be used to image and manipulate atoms and structures on a variety of surfaces. The atom at the apex of the tip "senses" individual atoms on the underlying surface when it forms incipient chemical bonds with each atom. Because these chemical interactions subtly alter the tip's vibration frequency, they can be detected and mapped. This principle was used to distinguish between atoms of silicon, tin and lead on an alloy surface, by comparing these 'atomic fingerprints' to values obtained from large-scale density functional theory (DFT) simulations.The trick is to first measure these forces precisely for each type of atom expected in the sample, and then to compare with forces given by DFT simulations. The team found that the tip interacted most strongly with silicon atoms, and interacted 23% and 41% less strongly with tin and lead atoms, respectively. Thus, each different type of atom can be identified in the matrix as the tip is moved across the surface.

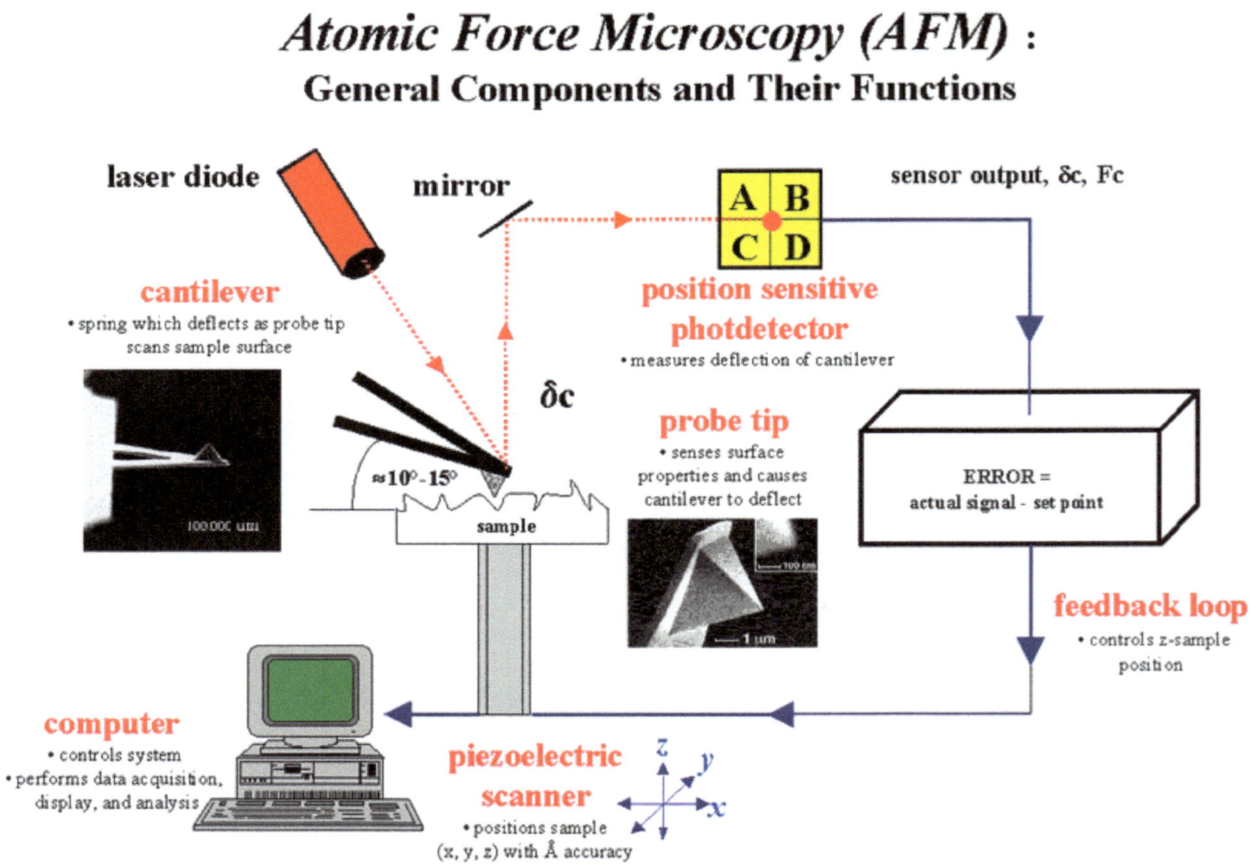

Atomic force microscopy

STED (stimulated emission depletion) microscopy has the potential to overcome the diffraction limit of optical microscopy. The techniques based on molecular photo activation and structured illumination. New techniques aimed at increasing the depth penetration in tissue. Other approaches have concentrated on phase imaging, such as interference microscopy, digital holography, and polarization imaging and continue to improve the image contrast based on nonlinear mechanisms, such as multi-photon excitation, multi-harmonic generation, coherent Raman scattering, and stimulated emission.

Preparation of Specimen

Although living microorganisms can be directly examined with light microscopy, they often must be fixed and stained to increase visibility, accentuate specific morphological features and preserve them for future study.

Fixation:

- Is a process by which the internal and external structures of cells and microorganisms are preserved and fixed in position

Two different types of fixation:

- Heat-fix bacterial smears by gently flame heating an air-dried film of bacteria. This preserves overall morphology but not structures within cells
- Chemical fixatives such as ethanol, acetic acid, mercuric chloride, formaldehyde and glutaraldehyde penetrate the cells and react with cellular components, usually proteins and lipids to render them inactive, insoluble and immobile

Dyes:

Dyes have two features in common:

- They have chromophore groups (groups with conjugate double bonds that give the dye its color)
- They can bind with cells by ionic, covalent, or hydrophobic bonding. Ex. +vely charged dye binds to –vely charged structures of the cell

Ionizable dyes are divided into two classes based on their charged group.

Basic dyes – methylene blue, basic fuchsin,crystal violet, safranin, malachite green – have +vely charged groups. These bind to –vely charged molecules like nucleic acids and many proteins. Surfaces of bacterial cells are –vely charged, so most often used in bacteriology.

Acid dyes – eosin, rose Bengal and acid fuchsin – posses –vely charged groups such as carboxyls (-COOH) and phenolic hydroxyls (-OH). These bind to +vely charged cell structures.

Staining Methods:

Preparation of bacterial smears:

All the methods start by preparing 'fixed' smears on the microscope slides. An aqueous suspension is first made, in the case of a non-liquid specimen, by taking a small amount of the sample and suspending it in a single drop of distilled water on a 25 mm x 75 mm slide. Do not make the smear too thick. If you have a liquid sample, a single drop is used directly from the culture container. The suspension made by either method is air dried, then "fixed" by passing rapidly through the Bunsen flame two or three times. Cover the smears with 95% methyl alcohol for 1 min, and then allow the smear to cool before staining.

Staining of Specimens

Staining technique used in microscopy to enhance contrast in the microscopic image. Stains and dyes are frequently used in biology and medicine to highlight structures in biological tissues for viewing, often with the aid of different microscopes. Stains may be used to define and examine bulk tissues, cell populations, or organelles within individual cells.

Stains

A stain is a substance that adheres to a cell, giving the cell color. The presence of color gives the cells significant contrast. Different stains have different affinities for different organisms, or different parts of organisms. They thus may be used to differentiate different types of organisms or to view specific parts of organism.

Simple Stains

Is a stain using only a single dye that does not differentiate between different types of organisms. There is a single staining step and everything that stains is stained the same color.

Simple staining:

- A simple stain is an aqueous or alcohol solution of a single basic dye
- Simple stains usually highlight the entire microorganism so that cellular shapes and structures are visible.
- Stain is applied to the fixed smear for a certain length of time and then washed off, and the slide is dried and examined.
- Basic dyes like crystal violet, methylene blue and carbolfuchsin are frequently used to determine the size, shape and arrangement of bacteria

Differential Stain

A differential stain uses more than one dye and stains different kinds of organisms with different colors. The differential stains are Gram stain and Ziehl - Neelsen acid fast stains.

Gram Staining

Gram staining or Gram stain, also called Gram's method, is a method of staining used to differentiate bacterial species into two large groups (gram-positive and gram-negative). The name comes from the Danish bacteriologist Hans Christian Gram, who developed the technique.

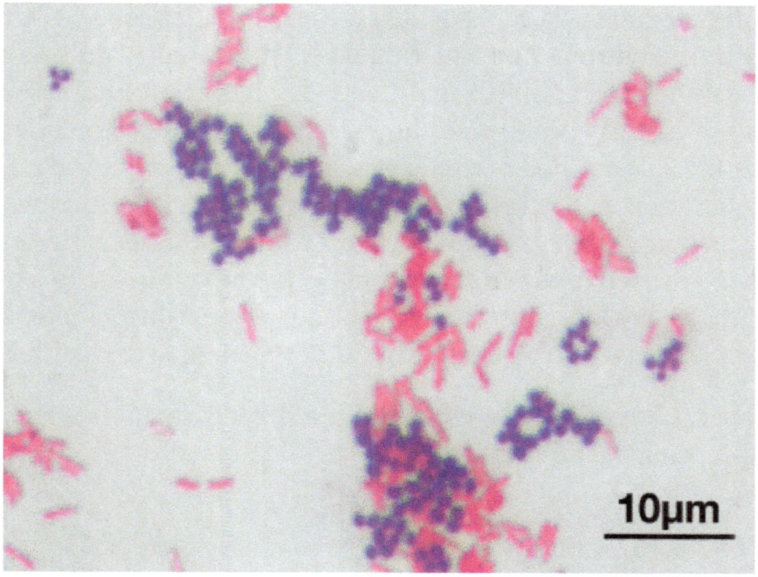

A Gram stain of mixed *Staphylococcus aureus* (*S. aureus* ATCC 25923, gram-positive cocci, in purple) and *Escherichia coli* (*E. coli* ATCC 11775, gram-negative bacilli, in red), the most common Gram stain reference bacteria

Gram staining differentiates bacteria by the chemical and physical properties of their cell walls by detecting peptidoglycan, which is present in the cell wall of Gram-positive bacteria. Gram-positive bacteria retain the crystal violet dye, and thus are stained violet, while the Gram-negative bacteria do not; after washing, a counterstain is added (commonly safranin or fuchsine) that will stain these Gram-negative bacteria a pink color. Both Gram-positive bacteria and Gram-negative bacteria pick up the counterstain. The counterstain, however, is unseen on Gram-positive bacteria because of the darker crystal violet stain.

The Gram stain is almost always the first step in the preliminary identification of a bacterial organism. While Gram staining is a valuable diagnostic tool in both clinical and research settings, not all bacteria can be definitively classified by this technique. This gives rise to *gram-variable* and *gram-indeterminate* groups.

History

The method is named after its inventor, the Danish scientist Hans Christian Gram (1853–1938), who developed the technique while working with Carl Friedländer in the morgue of the city hospital in Berlin in 1884. Gram devised his technique not for the purpose of distinguishing one type of bacterium from another but to make bacteria more visible in stained sections of lung tissue. He published his method in 1884, and included in his short report the observation that the typhus bacillus did not retain the stain.

Uses

Gram staining is a bacteriological laboratory technique used to differentiate bacterial species into two large groups (gram-positive and gram-negative) based on the physical properties of their cell walls. Gram staining is not used to classify archaea, formerly archaeabacteria, since these microorganisms yield widely varying responses that do not follow their phylogenetic groups.

The Gram stain is not an infallible tool for diagnosis, identification, or phylogeny, and it is of extremely limited use in environmental microbiology. It is used mainly to make a preliminary morphologic identification or to establish that there are significant numbers of bacteria in a clinical specimen. It cannot identify bacteria to the species level, and for most medical conditions, it should not be used as the sole method of bacterial identification. In clinical microbiology laboratories, it is used in combination with other traditional and molecular techniques to identify bacteria. Some organisms are gram-variable (meaning they may stain either negative or positive); some are not stained with either dye used in the Gram technique and are not seen. In a modern environmental or molecular microbiology lab, most identification is done using genetic sequences and other molecular techniques, which are far more specific and informative than differential staining.

Gram staining has been suggested to be as effective a diagnostic tool as PCR in one primary research report regarding gonococcal urethritis.

Medical

Gram stains are performed on body fluid or biopsy when infection is suspected. Gram stains yield results much more quickly than culturing, and is especially important when infection would make an important difference in the patient's treatment and prognosis; examples are cerebrospinal fluid for meningitis and synovial fluid for septic arthritis.

Staining Mechanism

Gram-positive bacteria have a thick mesh-like cell wall made of peptidoglycan (50–90% of cell envelope), and as a result are stained purple by crystal violet, whereas gram-negative bacteria have a thinner layer (10% of cell envelope), so do not retain the purple stain and are counter-stained pink by safranin. There are four basic steps of the Gram stain:

- Applying a primary stain (crystal violet) to a heat-fixed smear of a bacterial culture. Heat fixation kills some bacteria but is mostly used to affix the bacteria to the slide so that they don't rinse out during the staining procedure
- The addition of iodide, which binds to crystal violet and traps it in the cell
- Rapid decolorization with ethanol or acetone
- *Counterstaining* with safranin. Carbol fuchsin is sometimes substituted for safranin since it more intensely stains anaerobic bacteria, but it is less commonly used as a counterstain.

Crystal violet (CV) dissociates in aqueous solutions into CV^+ and chloride (Cl^-) ions. These ions penetrate through the cell wall and cell membrane of both gram-positive and gram-negative cells. The CV^+ ion interacts with negatively charged components of bacterial cells and stains the cells purple.

Iodide (I–or I–3) interacts with CV^+ and forms large complexes of crystal violet and iodine (CVI) within the inner and outer layers of the cell. Iodine is often referred to as a mordant, but is a trapping agent that prevents the removal of the CV–I complex and, therefore, color the cell.

When a decolorizer such as alcohol or acetone is added, it interacts with the lipids of the cell membrane. A gram-negative cell loses its outer lipopolysaccharide membrane, and the inner peptidoglycan layer is left exposed. The CV–I complexes are washed from the gram-negative cell along with the outer membrane. In contrast, a gram-positive cell becomes dehydrated from an ethanol treatment. The large CV–I complexes become trapped within the gram-positive cell due to the multilayered nature of its peptidoglycan. The decolorization step is critical and must be timed correctly; the crystal violet stain is removed from both gram-positive and negative cells if the decolorizing agent is left on too long (a matter of seconds).

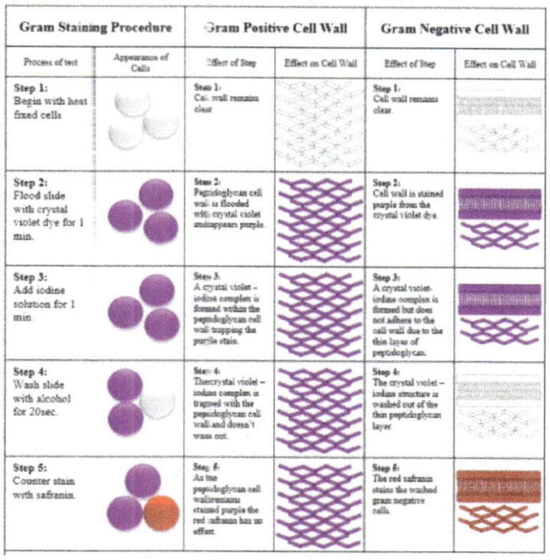

Gram stain procedure and effect on Gram-negative and Gram-positive cell walls.

After decolorization, the gram-positive cell remains purple and the gram-negative cell loses its purple color. Counterstain, which is usually positively charged safranin or basic fuchsine, is applied last to give decolorized gram-negative bacteria a pink or red color.

Examples

Gram-positive Bacteria

Gram-positive bacteria generally have a single membrane surrounded by a thick peptidoglycan. This rule is followed by two phyla: *Firmicutes* (except for the classes Mollicutes and Negativicutes) and the *Actinobacteria*. In contrast, members of the Chloroflexi (green non-sulfur bacteria) are monoderms but possess a thin or absent (class Dehalococcoidetes) peptidoglycan and can stain negative, positive or indeterminate; members of the Deinococcus-Thermus group, stain positive but are diderms with a thick peptidoglycan.

Historically, the gram-positive forms made up the phylum Firmicutes, a name now used for the largest group. It includes many well-known genera such as *Bacillus*, *Listeria*, *Staphylococcus*, *Streptococcus*, *Enterococcus*, and *Clostridium*. It has also been expanded to include the Mollicutes, bacteria like *Mycoplasma* that lack cell walls and so cannot be stained by Gram, but are derived from such forms.

Some bacteria have cell walls which are particularly adept at retaining stains. These will appear positive by Gram stain even though they are not closely related to other gram-positive bacteria. These are called acid fast bacteria, and can only be differentiated from other gram-positive bacteria by special staining procedures.

Gram-negative Bacteria

Gram-negative bacteria generally possess a thin layer of peptidoglycan between two membranes (*diderms*). Most bacterial phyla are gram-negative, including the cyanobacteria, spirochaetes, and green sulfur bacteria, and most Proteobacteria (exceptions being some members of the *Rickettsiales* and the insect-endosymbionts of the *Enterobacteriales*).

Gram Variable and Gram-indeterminate Bacteria

Some bacteria, after staining with the Gram stain, yield a gram-variable pattern: a mix of pink and purple cells are seen. In cultures of Bacillus, Butyrivibrio, and Clostridium, a decrease in peptidoglycan thickness during growth coincides with an increase in the number of cells that stain gram-negative. In addition, in all bacteria stained using the Gram stain, the age of the culture may influence the results of the stain.

Gram-indeterminate bacteria do not respond predictably to Gram staining and, therefore, cannot be determined as either gram-positive or gram-negative. Examples include many species of *Mycobacterium*, including *M. tuberculosis* and *M. leprae.*

Orthographic Note

The term Gram staining is derived from the surname of Hans Christian Gram, the eponym (Gram) is therefore capitalized but not the common noun (stain) as is usual for scientific terms. The adjectives 'gram-positive' and 'gram-negative'; as eponymous adjectives, their initial letter can be either lowercase 'g' or capital 'G', depending on whose style guide (if any) governs the document being written. Lowercase style is used by the US Centers for Disease Control and Prevention and other style regimens such as the AMA style. Dictionaries may use lowercase, uppercase, or both. Uppercase 'Gram-positive' or 'Gram-negative' usage is also common in many scientific journal articles and publications. When articles are submitted to journals, each journal may or may not apply house style to the postprint version. Preprint versions contain whichever style the author happened to use. Even style regimens that use lowercase for the adjectives 'gram-positive' and 'gram-negative' still use capital for 'Gram stain'.

Differential Staining

Differential staining requires the use of chemical reagents that are applied sequentially to a heat-fixed smear. The first reagent is called the primary stain. Its function is to impart color to all cells. In order to establish a color contrast, the second reagent used is the decolorizing agent. Based on the chemical composition of the cellular components, the decolorizing agent may or may not remove the primary stain from the entire cell or only from certain cell structures. The final reagent, counter stain , has a contrasting color to that of the primary stain. Following decolorization, if the primary stain is not washed out, the counter stain cannot be absorbed and the cell or its components will retain the color of the primary stain. If the primary stain is removed, the decolorized cellular components will accept and assume the contrasting color of the counterstain. The most important differential stain used in bacteriology is the Gram stain. It divides bacterial cells into two major groups, gram-positive and gram-negative, which makes it an essential tool for classification and differentiation of microorganisms.

Primary Stain

Crystal Violet

This is used first and stains all cells purple.

Mordant

Gram's Iodine

This serves as a substance that forms an insoluble complex by binding to primary stain. The resultant crystal violet-iodine (CV-I) complex serves to intensify the color of the stain, and all the cells will appear purple-black at this point.

Decolorizing Agent

Ethyl Alcohol, 95%

The reagent serves a dual function as a lipid solvent and as a protein dehydrating agent. Its action is determined by the lipid concentration of the microbial cell wall. In gram-positive cells, the low lipid concentration is important for retention of the CV-I complex. Therefore, the small amount of lipid content is readily dissolved by the action of the alcohol, forming minute cell wall pores. These are then closed by alcohol's dehydrating effect. Consequently, the tightly bound primary stain is difficult to remove, and the cells remain purple. In gram-negative cells, the high lipid concentration found in outer layers of the cell wall is dissolved by the alcohol, creating large pores in the cell wall that do not close appreciably on dehydration of cell wall proteins. This facilitates release of the unbound CV-I complex, leaving those cells colorless or unstained.

Counterstain

Safranin

This is the final reagent, used to stain red those cells that have previously decolorized. Since only gram-negative cells undergo decolorization, they may now absorb the counterstain. Gram-positive cells retain the purple color of the primary stain.

Procedure

1. One or two loopful of suspended cells was applied directly to the glass slide with a sterile inoculating loop and spread evenly.
2. The smear was then heat fixed by rapid passage of the air-dried smear two-three times over the flame of the Bunsen burner.
3. The heat fixed smear was flooded with crystal violet and incubated for 1 minute.
4. Crystal violet was washed away with distilled water.
5. The smear was flooded with Gram's iodine mordant and incubated for 1 minute.

6. The mordant was washed away with distilled water.
7. Ethyl alcohol (95%) was added drop wise to the smear until crystal violet fails to wash off.
8. The alcohol was washed away with distilled water.
9. The smear was counterstained with safranin for 45 seconds.
10. The counterstain was washed away with distilled water.
11. The slide was dried (either blot dried or in an incubator) and examined in a compound microscope.

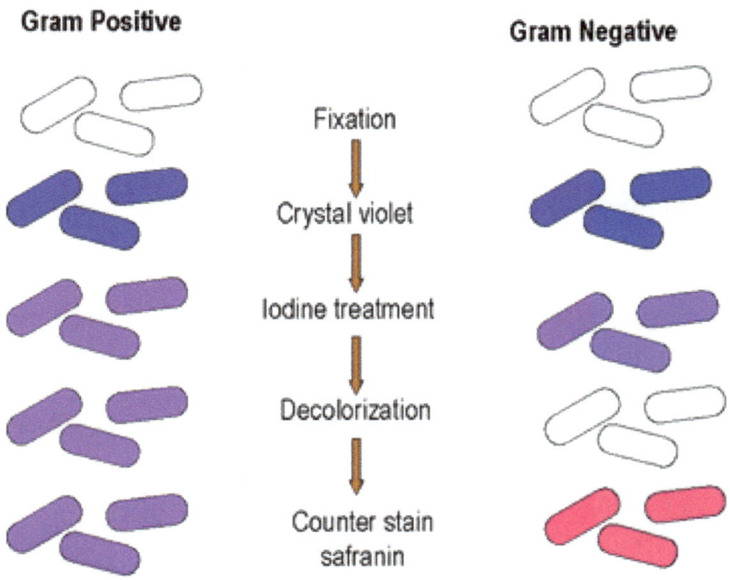

Procedure of Gram staining

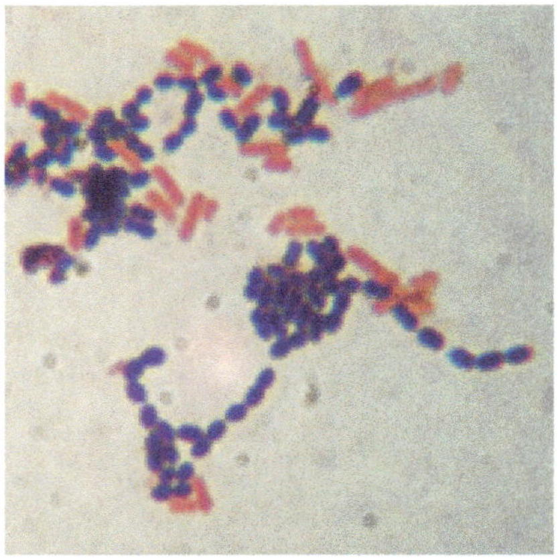
Gram positive and negative cells when viewed under microscope

Acid fast staining techniques principle :

Most of the bacteria can be stained with aqueous solutions of para rosaniline dyes but certain organisms especially which belongs to family mycobacterium resist entry of these weaker dyes and hence these weak solutions are unsuitable for staining them. They can be stained with strong solutions which contain phenols which help in penetration of the in to the cell. Heating further enhances entry of the dye in to the cell. These organisms once stained resist decolorization with acids such as H_2SO_4, HCL, HNO_3 etc. and even with alcohol. Those organisms which resist decolorize with acid retain primary stain and are called acid fast organisms and those which are decolorized by acid will take up counter stain and are called non acid fast organisms. Reason for acid fastness is due to the presence of long chain fatty acid, mycolic acid in the cell wall of these organisms.

References

- Appendix A, Journal of the Royal Microscopical Society, 1906; page 716. A discussion of Zeiss measuring microscopes
- Chang, Kenneth (8 October 2014). "2 Americans and a German Are Awarded Nobel Prize in Chemistry". New York Times. Retrieved 8 October 2014
- Manuel Gunkel; et al. (2009). "Dual color localization microscopy of cellular nanostructures". Biotechnology journal. 4 (6): 927–38. doi:10.1002/biot.200900005. PMID 19548231
- Albert Van Helden; Sven Dupré; Rob van Gent (2010). The Origins of the Telescope. Amsterdam University Press. p. 24. ISBN 978-90-6984-615-6
- Ritter, Karl; Rising, Malin (8 October 2014). "2 Americans, 1 German win chemistry Nobel". AP News. Retrieved 8 October 2014
- Luft, J.H. (1961). "Improvements in epoxy resin embedding methods". The Journal of biophysical and biochemical cytology. 9 (2). p. 409. PMC 2224998. PMID 13764136
- Albert Van Helden; Sven Dupré; Rob van Gent (2010). The Origins of the Telescope. Amsterdam University Press. pp. 32–36, 43. ISBN 978-90-6984-615-6
- Kenneth, Spring; Keller, H. Ernst; Davidson, Michael W. "Microscope objectives". Olympus Microscopy Resource Center. Retrieved 29 Oct 2008
- Juniper, B.E.; Bradley, D.E. (1958). "The carbon replica technique in the study of the ultrastructure of leaf surfaces". Journal of ultrastructure research. 2 (1): 16–27. doi:10.1016/s0022-5320(58)90045-5
- Gould, Stephen Jay (2000). "Chapter 2: The Sharp-Eyed Lynx, Outfoxed by Nature". The Lying Stones of Marrakech: Penultimate Reflections in Natural History. New York, N.Y: Harmony. ISBN 0-224-05044-3
- Reynolds, E. S. (1963). "The use of lead citrate at high pH as an electron-opaque stain in electron microscopy". Journal of Cell Biology. 17: 208–212. doi:10.1083/jcb.17.1.208. PMC 2106263. PMID 13986422
- "The Scale of Things". Office of Basic Energy Sciences, U.S. Department of Energy. 2006-05-26. Archived from the original on 2010-02-01. Retrieved 2010-01-31
- John G. Holt; Noel R. Krieg; Peter H.A. Sneath; James T. Staley; Stanley T. Williams (1994). Bergey's Manual of Determinative Bacteriology (9th ed.). Lippincott Williams & Wilkins. p. 11. ISBN 0-683-00603-7
- "SPLEEM". National Center for Electron Microscopy (NCEM). Archived from the original on 2010-05-29. Retrieved 2010-01-31
- Ryan K.J., Ray C.G. (editors) (2004). Sherris Medical Microbiology (4th ed.). McGraw Hill. pp. 232f. ISBN 0838585299
- Madigan, M.T.; Martinko J.; Parker J. (2004). Brock Biology of Microorganisms (10th ed.). Lippincott Williams & Wilkins. ISBN 0-13-066271-2

5

Microbial Nutrition and Microbial Growth

Microbes require nutrition to conduct their functions. Microbial nutritional requirements are divided into two parts, i.e. macronutrients and micronutrients. Microbes require macronutrients in huge amounts, which include elements such as carbon, oxygen, nitrogen, and components of carbohydrates, such as lipids and proteins. The requirement for macronutrients is comparatively less. These include cobalt, molybdenum, nickel and copper. The aspects elucidated in this chapter are of vital importance, and provide a better understanding of microbiology.

Microbial Nutrient Requirements

Microbial cell composition shows that 95% of cell dry weight is made up of a few major elements: Carbon, oxygen, hydrogen, nitrogen, sulfur, phosphorous, potassium, calcium, magnesium and iron.

Macronutrients or macro elements:

These are required by microorganisms in relatively large amounts. Carbon, oxygen, hydrogen nitrogen, sulfurs and phosphorous are components of carbohydrates, lipids, proteins and nucleic acids. The remaining four macro elements (K, Ca, Mg and Fe) exist in the cell as cations.

K^{+} - is required for the activity by a number of enzymes, including those involved in protein synthesis.

Ca^{2+} - contributes to the heat resistance of bacterial endospores. 15% of spore contains dipicolinic acid and calcium.

Mg^{2+} - serves as a cofactor for many enzymes, complexes with ATP and stabilizes ribosomes and cell membranes.

Fe^{2+} and Fe^{2+} - part of cytochromes and a cofactor for enzymes and electron-carrying proteins.

Micronutrients or Trace elements:

These are manganese, zinc, cobalt, molybdenum, nickel and copper. These are normally part of enzymes and cofactors, and they aid in the catalysis of reactions and maintenance of protein structure.

Zn^{2+} - is present at the active site of some enzymes but is also involved in the association of regulatory and catalytic subunits in *E. coli* aspartate carbomoyl transferase.

Mn^{2+} - aids many enzymes catalyzing the transfer of phosphate groups.

Mo^{2+} - required for nitrogen fixation.

Co^{2+} - is a component of Vitamin B12.

Besides macro and micro nutrients, some microorganisms may have particular requirements that reflect the special nature of their morphology or environment. Diatoms need silicic acid to construct their beautiful cell walls of silica. Bacteria growing in saline lakes and oceans depend on the presence of high concentrations of sodium ion. Microorganisms require a balanced mixture of all the above nutrients for proper growth.

Requirements for carbon, hydrogen and oxygen:

Carbon is needed for the skeleton or backbone of all organic molecules and molecules serving as carbon sources normally also contribute both oxygen and hydrogen atoms. One important carbon source that does not supply hydrogen or energy is CO_2 . Autotrophs – can use CO_2 as their sole or principal source of carbon. Many microorganisms are autotrophic, and most of these carry out photosynthesis and use light as their energy source. Some autotrophs oxidize inorganic molecules and derive energy from electron transfer. Heterotrophs – are organisms that use reduced pre-formed organic molecules as carbon sources. Ex. Glycolytic pathway produces carbon skeleton for use in biosynthesis and also releases energy as ATP and NADH. Actinomycetes will degrade amyl alcohol, paraffin and even rubber. *Burkholderia cepacia* can use over 100 different carbon compounds. Some microorganisms can metabolize even relatively indigestible human-made substances such as pesticides. Indigestible molecules can be oxidized and degraded in the presence of a growth promoting nutrient that is metabolized at the same time, a process called Co-metabolism. The products of this breakdown can then be used as nutrients by other microorganisms.

Nutritional types of microorganisms:

In addition to Carbon, hydrogen and oxygen all organisms require sources of energy and electrons for growth.

Carbon sources:

Autotrophs - CO_2 sole or principal biosynthetic carbon source

Heterotrophs – reduced, preformed organic molecules from other organisms.

Energy sources:

Phototrophs – use light as their energy source.

Chemotrophs – obtain energy from the oxidation of chemical compounds (either organic or in organic)

Electron sources:

Lithotrophs – use reduced inorganic substances as their electron source.

Organotrophs – extract electrons from organic compounds.

Four major nutritional classes based on their primary sources of carbon, energy and electrons is known.

Phtotolithotrophic autotrophs or photoautotrophs or photolithoautotrophs:

Source of energy – light energy

Source of electrons – Inorganic hydrogen/ electron

Carbon source - CO_2

Example: Algae, purple and green sulfur bacteria and cyanobacteria.

Photoorganotrophic heterotrophy or photoorganoheterotrophy:

Source of energy – light energy

Source of electrons – organic hydrogen/ electron

Carbon source – organic carbon sources (CO_2 may also be used)

Example: Purple and green nonsulfur bacteria (common inhabitants of lakes and streams)

Chemolithotrophic autotrophs or chemolithoautotrophy:

Source of energy – Chemical energy source (inorganic)

Source of electrons – Inorganic hydrogen/ electron donor

Carbon source - CO_2

Example: Sulfur-oxidizing bacteria, hydrogen bacteria, nitrifying bacteria, iron-oxidizing bacteria.

Chemoorganotrophic heterotrophs or chemoorganoheterotrophy:

Source of energy – Chemical energy source (organic)

Source of electrons – Inorganic hydrogen/ electron donor

Carbon source – organic carbon source

Example: Protozoan, fungi, most non-photosynthetic bacteria (including most pathogens)

The most common nutritional types are photolithoautotrophs and chemoorganoheterotrophs. Bacteria *Beggiatoa* rely on inorganic energy sources and organic (or sometimes CO_2) carbon sources. These microbes are sometimes called Mixotrophic because they combine chemolithoautotrophic and heterotrophic metabolic processes.

Requirements for nitrogen, phosphorous and sulfur:

Nitrogen is needed for the synthesis of amino acids, purines, pyramidines, some carbohydrates and lipids, enzyme cofactors and other substances. Most phototrophs and many nonphotosynthetic microorganisms reduce nitrate to ammonia and incorporate the ammonia in assimilatory nitrate reduction. A variety of bacteria like many Cyanobacteria and *Rhizobiium* can reduce and assimilate atmospheric nitrogen using the nitrogenase systems. Phosphorous is present in nucleic acids, phospholipids, ATP, several cofactors, some proteins and other cell components. All microorganisms use inorganic phosphate as their phosphorous source and incorporate it directly. *E. coli* can use both organic and inorganic phosphate. Organophosphates such as hexose 6- phosphate can be taken up directly by transport proteins. Other organophosphates are often hydrolyzed in the periplasm by the enzyme alkaline phosphatase to produce inorganic phosphate which is then transported across the plasma membrane. When inorganic phosphate is outside the bacterium, it crosses the outer membrane by the use of a porin protein channel. Sulfur is needed for the synthesis of substances like the amino acids cysteine and methionine, some carbohydrates biotin and thiamine. Most of them use sulfate as a source of sulfur and reduce it by assimilatory sulfate reduction; a few require a reduced form of sulfur such as cysteine.

Growth factors:

Many microorganisms have the enzymes and pathways necessary to synthesize all cell components. Many lack one or more enzymes and hence require organic compounds because they are essential cell components or precursors of such components and cannot be synthesized by the organisms are called – growth factors. There are three major classes of growth factors:

Amino acids – needed for protein synthesis.

Purines and Pyramidines – for nucleic acid synthesis

Vitamins – small organic molecules that usually make up all or part of enzyme cofactors, and only very small amounts sustain growth.

Knowledge of the specific growth factor requirements of many microorganisms makes possible quantitative growth response assays for a variety of substances. The observation that many microorganisms can synthesize large quantities of vitamins has led to their use in industry. Several water-soluble and fat-soluble vitamins are produced using industrial fermentations.

Ribofalvin – *Clostridium, Candida, Ashbya, Eremothecium*

Coenzyme A – *Brevibacterium*

Vitamin B_{12} – *Streptomyces, Propionibacterium, Pseudomonas*

Vitamin C – *Gluconobacter, Erwinia, Corynebacterium*

β- Carotene – *Dunaliella*

Vitamin D - *Saccharomyces*

Uptake of Nutrients

Uptake of the required nutrients by the microbial cell is important. Since microorganisms live in nutrient poor habitats, they must be able to transport nutrients from dilute solutions into the cell against concentration gradient. Finally, they must pass through a selectively permeable plasma membrane. Microorganisms use different transport mechanisms like facilitated diffusion, active transport and group translocation. Eukaryotic microorganisms do not employ group translocation but take up nutrients by endocytosis.

Movement of materials across the plasmamembrane is mostly done by two processes:

Passive processes : Substances cross the area from an area of high concentration to an area of low concentration without any expenditure of energy (ATP). Example, simple diffusion, osmosis and facilitated diffusion.

Active process: The cell must use energy (ATP) to move substances from areas of low concentration to areas of high concentration. Example, Group translocation.

Passive processes:

Passive or simple diffusion :Often called diffusion, is the process in which molecules move from a region of higher concentration to one of lower concentration. The rate is dependent on the size of the concentration gradient between a cell's exterior and its interior. Very small molecules such as water and oxygen and carbon dioxide move across membranes by simple or passive diffusion. Larger molecules, ions, and polar substances do not cross membranes by this method.

Osmosis: Is the net movement of solvent molecules across a selectively permeable membrane from an area in which the solvent molecules are highly concentrated to an area of low concentration until equilibrium is reached. In living systems the chief solvent is water. The three types of solutions which are normally found are isotonic, hypotonic and hypertonic.

Facilitated diffusion: The rate of diffusion across selectively permeable membrane is greatly increased by using carrier proteins, sometimes called permeases which are embedded in the plasma membrane. Because a carrier aids the diffusion process, it is called as facilitated diffusion. Carrier proteins also resemble enzymes in their specificity for the substances to be transported; each carrier is selective and will transport only closely related solutes. Because there is no energy input, molecules will continue to enter only as long as their concentration is greater on the outside. Two widespread major intrinsic protein channels in bacteria are aquaporins that transport water and glycerol facilitators which aid glycerol diffusion. The carrier protein complex spans the membrane. After the solute molecule binds to the outside, the carrier may change conformation and release the molecule on the cell interior. The carrier would subsequently change back to its original shape and be ready to pick up another molecule. The mechanism is driven by concentration gradients and therefore is reversible.

Examples. Glycerol is transported by facilitated diffusion in *E. coli, Salmonella typhimurium, Pseudomonas, Bacillus* and many other bacteria. This is prominent in eukaryotes where it is used to transport a variety of sugars and amino acids.

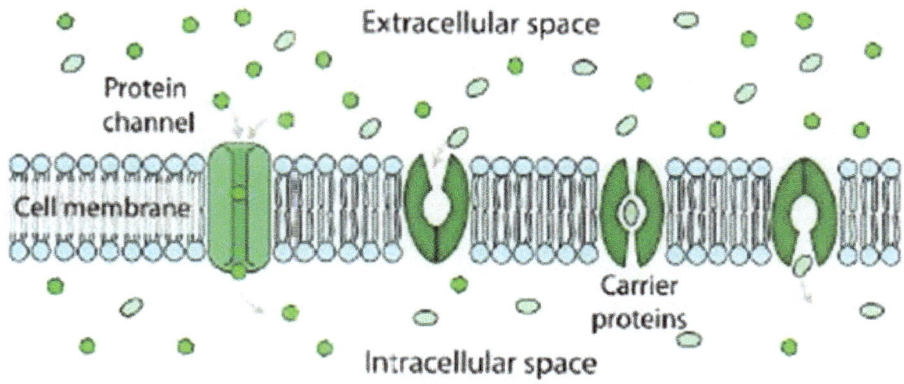

Facilitated diffusion. The carrier proteins aid in the release of solute molecules from extracellular space to the intracellular space.

Active Transport

Active transport is the transport of solute molecules to higher concentrations or against a concentration gradient, with the use of metabolic energy input. It resembles facilitated diffusion in the involvement of protein carrier activity, but differs in its use of metabolic energy and in its ability to concentrate substances. One example of active transport is binding protein transport systems or ATP-binding cassette transporters (ABC transporters) are active in bacteria, archaea and eukaryotes.

ABC transporters: These transporters consist of two hydrophobic membrane spanning domains associated on their cytoplasmic surfaces with two nucleotide-binding domains. The membrane spanning domains from a pore in the membrane and the nucleotide binding domains bind and hydrolyze ATP to drive uptake. ABC transporters employ special substrate binding proteins, which are located in the periplasmic space of gram-negative bacteria or are attached to membrane lipids on the external face of the gram positive plasma membrane. These binding proteins, bind the molecule to be transported and then interact with the membrane transport proteins to move the solute molecule inside the cell. *E. coli* transports a variety of sugars (arabinose, maltose, galactose, and ribose) and amino acids (glutamate, histidine, leucine) by this mechanism.

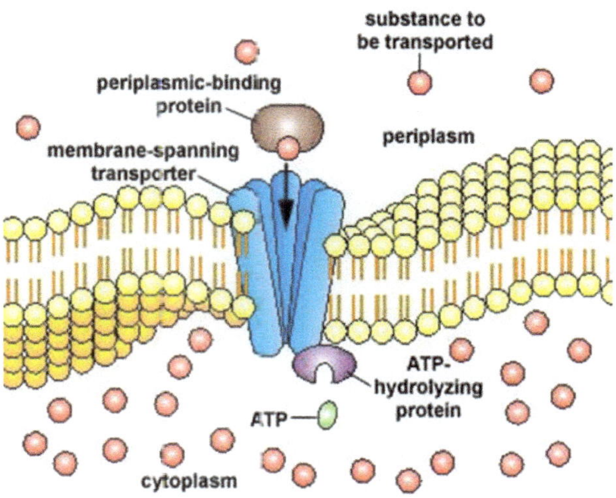

ABC transporter system.

Eukaryotic ABC transporters are sometimes of great medical importance. Some tumor cells pump drugs out using these transporters. Cystic fibrosis results from a mutation that inactivates an ABC transporter that acts as a chloride ion channel in lungs.

Bacteria also use Proton gradients generated during electron transport to drive active transport. The lactose permease of *E. coli* transports a lactose molecule inward as a proton simultaneously enters the cell. Such linked transport of two substances in the same direction is called Symport. *E. coli* also uses proton symport to take up amino acids and organic acids like succinate and malate.

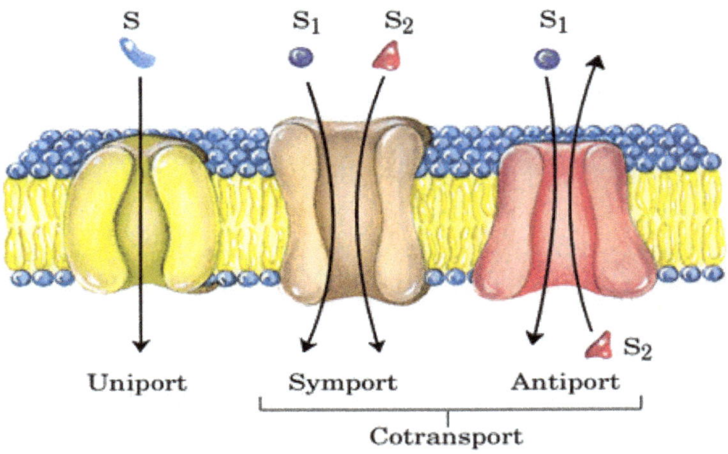

Comparison of the three different types of transport mechanisms.

A proton gradient also can power active transport indirectly, often through the formation of a sodium ion gradient. In *E. coli* , sodium transport system pumps sodium outward in response to the inward movement of protons. Such linked transport in which the transported substances move in opposite directions is called Anitport. The sodium gradient generated by this proton anitport system then drives the uptake of sugars and amino acids. *E. coli* has at least transport systems for the sugar galactose.

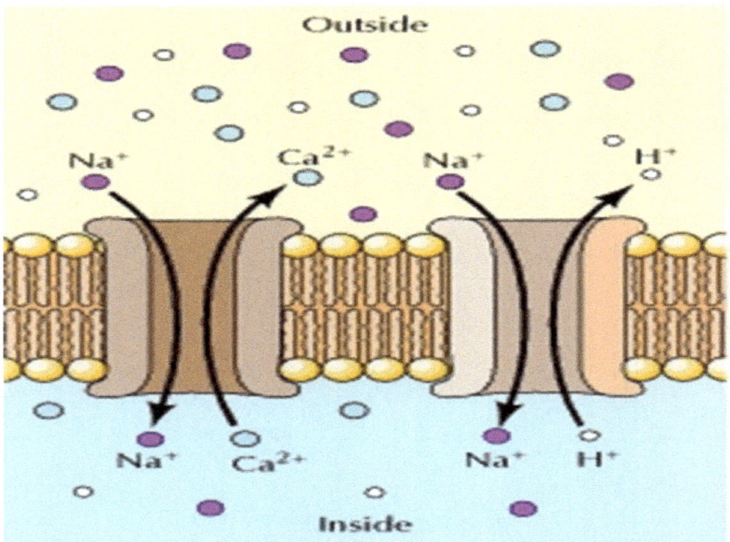

Antiport mechanism - linked transport of two substances in opposite direction.

Group Translocation: In active transport, solute molecules move across a membrane without modification. Many prokaryotes also take up molecules by group translocation, a process in which a molecule is transported into the cell while being chemically altered. For example, Phosphoenolpyruvate: Sugar phosphotransferease system (PTS). It transports a variety of sugars while phosphorylating them using phosphoenolpyruvate (PEP) as the phosphate donor.

PEP + Sugar (outside) → Pyruvate + Sugar-P (inside)

In *E. coli* and *Salmonella typhimurium*, it consists of two enzymes and a low molecular weight heat stable protein (HPr). HPr and enzyme I (EI) are cytoplasmic. Enzyme II (EII) is more variable in structure and often composed of three subunits or domains. EIIA is cytoplasmic and soluble. EIIB also is hydrophilic but frequently attached to EIIC, a hydrophobic protein that is embedded in the membrane. A high energy phosphate is transferred from PEP to enzyme II (EII) with the aid of enzyme I (EI) and HPr. Then a sugar molecule is phosphorylated as it is carried across the membrane by enzyme II (EII). Enzyme II (EII) transports only specific sugars and varies with PTS, whereas enzyme I (EI) and HPr are common to all PTS's.

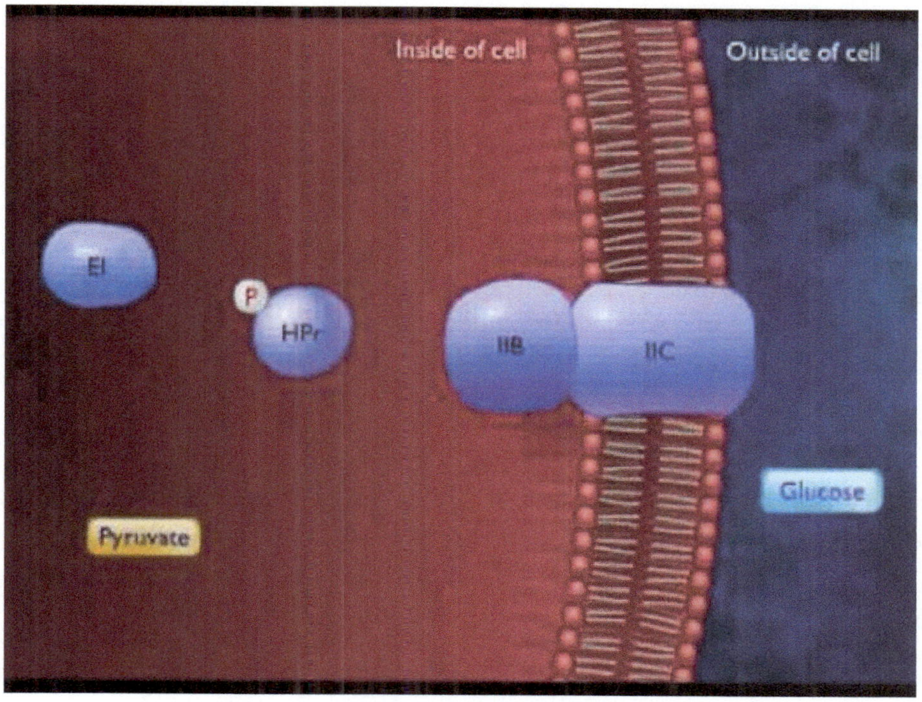

Group translocation: Bacterial PTS transport.

PTS's are widely distributed in prokaryotes. Aerobic bacteria lack PTS's. Genera *Escherichia, Salmonella, Staphylococcus* and other facultative anaerobic bacteria have phosphotransferase systems; some obligate anaerobic bacteria (*Clostridium*) also have PTS's. Many carbohydrates are transported by these systems. *E. coli* takes up glucose, fructose, mannitol, sucrose, N-acetylglucosamine, cellobiose and other carbohydrates by group translocation.

Iron Uptake:

All microorganisms require iron for use in cytochromes and many enzymes. Iron uptake is made difficult by the extreme insolubility of ferric ion (Fe^{3+}) and its derivatives, which leave little free, iron available for transport. Many bacteria and fungi have overcome this difficulty by secreting siderophores. Siderophores – are low molecular weight molecules that are able to complex with ferric ion and supply it to the cell. These are either hydroxamates or phenolates-catecholates. Ferrichrome is a hydroxamate produced by many fungi; enterobactin is the catecholate formed by *E. coli*. Microorganisms secrete sidereophores when little iron is available in the medium. Once the iron-siderophore complex has reached the cell surface, it binds to a siderophore receptor protein. The iron is either released to enter the cell directly or the whole iron-siderophore complex is transported inside by an ABC transporter. In *E. coli*, the siderophore receptor is in the outer membrane of the cell envelope; when the iron reaches the periplasmic space, it moves through the plasma membrane with the aid of the transporter. After the iron has entered the cell, it is reduced to the ferrous form (Fe^{2+}). Iron is so crucial to microorganisms that many use more than one route of iron uptake to ensure an adequate supply.

Culture Media

In the laboratory, bacteria are normally grown or cultured in either liquid medium, in flasks, bottles or large culture vessels called fermenters or solid medium in Petri dishes which are round, normally plastic or glass dishes. Introduction of microbes into or onto these media is called inoculation. The nature of the medium depends on the microorganism's natural environment because its nutrient requirements reflect its natural surroundings. The culture medium must contain all the nutrients that the microorganism requires for growth like water, a source of energy, carbon, nitrogen, essential inorganic ions and a number of trace elements. Bacteria that can synthesize all they require from the basic ingredients are called Prototrophs – and most microorganisms that survive in the outside environment can do this. Microbes that have become adapted to life in a situation rich with nutrients such as human body may require other growth factors like vitamins, amino acids or nitrogenous bases to be provided. These are called auxotrophs.

The media that are used in microbiology laboratories to culture bacteria are referred to as artificial media or synthetic media, because they do not occur naturally; rather they are prepared in the laboratory. There are number of ways of categorizing the media, one way is to classify based on whether the exact contents of the media are known or not.

Chemically defined media : Is one in which all the ingredients are known; and was prepared in the laboratory by adding a certain number of grams of each of the components (Carbohydrates, amino acids, salts etc). Particularly photolithoautotrophs such as Cyanobacteria and eukaryotic algae can be grown on relatively simple media containing CO_2 as a carbon source (often added as sodium carbonate or bicarbonate), nitrate or ammonia as a nitrogen source, sulfate, phosphate and a variety of minerals. Chemoorganoheterotrophs can be grown in a defined media with glucose as a carbon source and an ammonium salt as a nitrogen source.

Complex media : contains undefined ingredients and the exact contents are not known. Complex medium may be sufficiently rich and complete to meet the nutritional requirements of many different microorganisms. They contain undefined components like peptones, meat extract and yeast extract. Peptones – are proteins hydrolysates prepared by partial proteolytic digestion of meat, casein, soyameal, gelatin and other protein sources (Carbon, energy and nitrogen). Beef extract – aqueous extracts from lean beef and contain amino acids, peptides, nucleotides, organic acids and minerals and vitamins.Yeast extract – from brewer's yeast and contain an excellent source of B vitamins, nitrogen and carbon compounds. Three commonly used complex media are:

1. Nutrient broth – Peptone (5.0g/L), Beef Extract (3.0g/L)

2. Tryptic soya broth – Tryptone (enzyme digest of casein), Peptone (enzyme digest of soybean meal), glucose, NaCl and K_2HPO_4

3. MacConkey Agar – Pancreatic digest of gelatin, casein, peptic digest of animal tissue, bile salts mixture, NaCl, neutral red, crystal red and agar.

These media are routinely used for cultivation of bacteria in the laboratory and particularly useful for cultivation of bacteria whose growth requirements have not been defined.

Culture media can also be categorized as liquid or solid. Liquid media (also called broths) are contained in tubes, flasks and fermenters. Solid media are prepared by adding agar to liquid medium and then pouring the media into tubes or Petridishes where the media will solidify. Agar is a complex polysaccharide that is obtained from red marine algae. Other solidifying agents are gelatin, silica. A 1% or 2% agar can be used to solidify, but 1.5% is the most commonly used.

Enriched Medium : Is a broth or solid medium containing a rich supply of special nutrients that promotes the growth of fastidious organisms (that have complex nutrition and environmental requirements). It is usually prepared by adding extra nutrients to a medium called nutrient agar. Blood agar (nutrient agar + 5% sheep red blood cells): It is bright red in color and distinguishes between the hemolytic and non-hemolytic bacteria (*Streptococci* and other pathogens)

Chocolate agar (nutrient agar + powdered hemoglobin): It is brown in color and is considered more enriched than blood agar as hemoglobin is more readily accessible in chocolate agar. Pathogens like *Neisseria gonorrhoeae* and *Haemophillus influenza,* which will not grow on blood agar, can be cultured.

Types of Media:

Selective Media: Are designed to suppress the growth of unwanted bacteria and encourage the growth of the desired microbes.

Examples: MacConkey agar inhibits growth of gram positive bacteria and thus is selective for gram-negative bacteria. Phenylethyl alcohol (PEA) agar and Colistinalidixic acid (CAN) agar are selective for gram positive bacteria.

Manitol salt agar contains a concentration of 7.5% NaCl that is quite inhibitory to most human pathogens. Bile salts, a component of fecus, inhibit most gram positive bacteria while permitting many gram negative rods to grow. This media is used for selecting intestinal pathogens which contain bile salts. Dyes such as methylene blue and crystal violet also inhibit certain gram positive bacteria (Staphylococcus can produce acid from mannitol and turn the phenol red dye to bright yellow.

A medium containing acetate as a carbon source would be selective for organisms that grow on acetate. Sabaraud's dextrose agar is used to isolate fungi and cellulose for cellulose digesting bacteria.

Differential media : Makes it easier to distinguish colonies of the desired organism from other colonies growing on the same plate.

MacConkey agar is also a differential and selective medium to distinguish between lactose fermenting organisms (e.g, *E. coli*) and other non-fermenters (e.g, *Shigella* sp.). Lactose in the medium is fermented by *E. coli* producing acid which causes an indicator dye to change color to red and colonies that do not ferment lactose are white. Dyes can be used as differential agents because many of them are pH indicators that change color in response to the production of an acid or base. MacConkey agar contains neutral red, a dye that is yellow when neutral and pink or red when acidic.

Mannitol salt agar is used to screen for *Staphylococcus aureus,* and it turns the originally pink medium to yellow due to its ability to ferment mannitol. In a sense, blood agar is also differential because it is used to determine the hemolytic and non-hemolytic bacteria.

Blood agar is both differential and an enriched one. It distinguishes between hemolytic and non-hemolytic bacteria. Hemolytic bacteria (eg., many *Streptococci* and *Staphylococci*) produce clear zones around their colonies because of red blood cell destruction.

Isolation of pure cultures:

In natural habitats, microorganisms usually grow in complex, mixed populations containing several species. One needs a pure culture, a population of cells arising from a single cell, to characterize an individual species. Robert Koch introduced pure culture techniques.

Spread plate and streak plate:

Spread plate - Mixture of cells is spread out on an agar surface so that every cell grows into a completely separate colony, each colony representing a pure culture. A small volume of around 30 to 300 cells (mixed) is transferred to the center of the plate and spread evenly over the surface with a sterile bent-glass rod.

Streak plate – the microbial mixture is transferred to the edge of an agar plate with an inoculating loop or swab and then streaked out over the surface in one of several patterns. In both these techniques, successful isolation depends on spatial separation of single cells.

The pour plate : Extensively used with bacteria and fungi. The original sample is diluted several times to reduce the microbial population sufficiently to obtain separate colonies when plating. The microbes are mixed with molten agar which has been cooled t 45° Cand then poured into pertidishes. All these techniques require petridishes (special culture dishes) after their inventor Julius Richard Petri (1887). They consist of two round halves, the top half overlapping the bottom. Each bacterium replicates to form a colony that is visible to the naked eye. A colony contains up to 109 copies of the original bacterium.

Colony morphology and growth:

Colony development on agar surfaces aids the microbiologist in identifying bacteria because individual species often form colonies of characteristic size and appearance. Structure of bacterial colonies also has been examined with the scanning electron microscope. Generally the most rapid cell growth occurs at the colony edges. Growth is much slower in the center and cell autolysis takes place in the older central portion of some colonies. These differences in growth appear due to gradients of oxygen, nutrients and toxic products within the colony. At the colony edge, oxygen and nutrients are plentiful. Bacteria growing on solid surfaces such as agar can form complex and intricate colony shapes. Nutrient diffusion and availability, bacterial chemotaxis and the presence of liquid on the surface all appear to play a role in pattern formation. These patterns vary with nutrient availability and the hardness of the agar surface.

Growth of Bacterial Cultures

Growth is defined as an increase in cellular constituents which leads to a rise in cell number. As we are aware, microorganisms reproduce by binary fission or by budding. In order to study growth, normally one follows the changes in the total population number. The cells copy their DNA almost continuously and divide again and again by the process called binary fission . Binary fission which has been described earlier a s the form of asexual reproduction in single-celled organisms by which one cell divides into two cells of the same size. Fortunately, few prokaryotic populations can sustain exponential growth for long. Environments are usually limiting in resources such as food and space. Prokaryotes also produce metabolic waste products that may eventually pollute the colony's environment. Still, you can understand why certain bacteria can make you sick so soon after infection or why food can spoil so rapidly. Refrigeration retards food spoilage not because the cold kills the bacteria on food but because most microorganisms reproduce very slowly at such low temperatures.

Fission (Biology)

In biology, fission is the division of a single entity into two or more parts and the regeneration of those parts into separate entities resembling the original. The object experiencing fission is usually a cell, but the term may also refer to how organisms, bodies, populations, or species split into discrete parts. The fission may be *binary fission*, in which a single entity produces two parts, or *multiple fission*, in which a single entity produces multiple parts.

Binary Fission

Organisms in the domains of Archaea and Bacteria reproduce with binary fission. This form of asexual reproduction and cell division is also used by some organelles within eukaryotic organisms (e.g., mitochondria). Binary fission results in the reproduction of a living prokaryotic cell (or organelle) by dividing the cell into two parts, each with the potential to grow to the size of the original.

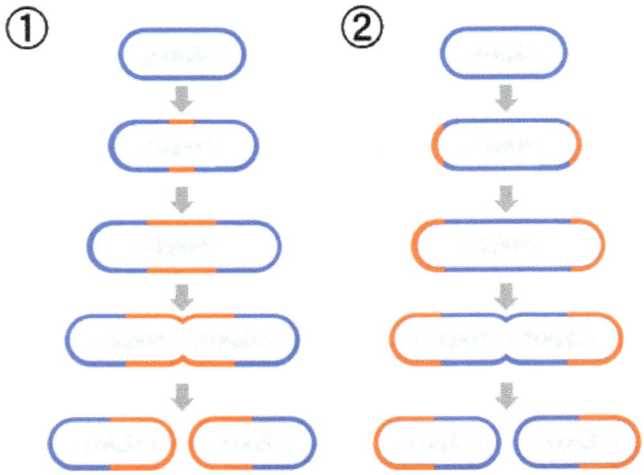

Schematic diagram of cellular growth (longation) and binary fission of bacilli. Blue and red lines indicate old and newly generated bacterial cell wall, respectively. (1) growth at the center of bacterial body. e.g. *Bacillus subtilis*, *E. coli*, and others. (2) apical growth. e.g. *Corynebacterium diphtheriae*.

Fission of Prokaryotes

Unlike the process of mitosis used by eukaryotic cells, binary fission takes place without the formation of a spindle apparatus in the cell. The single DNA molecule first replicates, then attaches each copy to a different part of the cell membrane. When the cell begins to pull apart, the replicated and original chromosomes are separated. The consequence of this asexual method of reproduction is that all the cells are genetically identical, meaning that they have the same genetic material (barring random mutations).

Process of Bacterial Fission

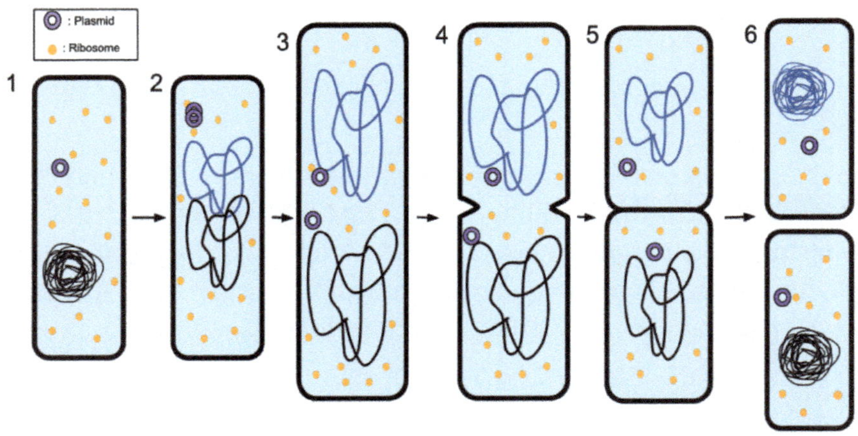

Binary fission in a prokaryote

The process of binary fission in bacteria involves the following steps. First, the cell's DNA is replicated. The replicated DNA copies then move to opposite poles of the cell in an energy-dependent process. The cell lengthens. Then, the equatorial plate of the cell constricts and separates the plasma membrane such that each new cell has exactly the same genetic material.

More specifically, the following steps occur:

1. The bacterium before binary fission is when the DNA is tightly coiled.
2. The DNA of the bacterium has uncoiled and duplicated.
3. The DNA is pulled to the separate poles of the bacterium as it increases size to prepare for splitting.
4. The growth of a new cell wall begins to separate the bacterium.
5. The new cell wall fully develops, resulting in the complete split of the bacterium.
6. The new daughter cells have tightly coiled DNA rods, ribosomes, and plasmids; these are now brand new organisms.

Speed of Bacterial Fission

Binary fission is generally rapid though its speed varies between species. For *E. coli*, cells typically divide about every 20 minutes at 37 °C. Because the new cells will, in turn, undergo binary fission on their own, the time binary fission requires is also the time the bacterial culture requires to double in the number of cells it contains. This time period can therefore be referred to as the doubling time. Some species other than *E. coli* may have faster or slower doubling times: some strains of *Mycobacterium tuberculosis* may have doubling times of nearly 100 hours. Bacterial growth is limited by factors including nutrient availability and available space, so binary fission occurs at much lower rates in bacterial cultures once they enter the stationary phase of growth.

Fission of Organelles

Some organelles in eukaryotic cells reproduce using binary fission. Mitochondrial fission occurs frequently within the cell, even when the cell is not actively undergoing mitosis, and is necessary to regulate the cell's metabolism.

Types of Binary Fission

Binary fission in organisms can occur in four ways, irregular,longitudinal, transverse, oblique.

1. Irregular: In this fission, cytokinesis may take place along any plane but it is always perpendicular to the plane of karyokinesis. e.g. amoeba
2. Longitudinal: Here cytokinesis takes place along the longitudinal axis. e.g. in flagellates like Euglena.
3. Transverse: Here cytokinesis takes place along the transverse axis. e.g. in *ciliate* protzoans like paramecium.
4. Oblique: In this type of binary fission cytokinesis occurs obliquely. e.g. ceratium

Multiple Fission

Fission of Protists

Multiple fission at the cellular level occurs in many protists, e.g. sporozoans and algae. The nucleus of the parent cell divides several times by mitosis, producing several nuclei. The cytoplasm then separates, creating multiple daughter cells.

Some parasitic, single-celled organisms undergo a multiple fission-like process to produce numerous daughter cells from a single parent cell. Isolates of the human parasite *Blastocystis hominis* were observed to begin such a process within 4 to 6 days. Cells of the fish parasite *Trypanoplasma borreli* have also been observed participating in both binary and multiple fission.

Fission of Apicomplexans

In the apicomplexans, a phylum of parasitic protists, multiple fission, or schizogony, is manifested either as merogony, sporogony or gametogony. Merogony results in merozoites, which are multiple daughter cells, that originate within the same cell membrane, sporogony results in sporozoites, and gametogony results in microgametes.

Fission of Green Algae

Green algae can divide into more than two daughter cells. The exact number of daughter cells depends on the species of algae and is an effect of temperature and light.

Multiple Fission of Bacteria

Most species of bacteria primarily undergo binary reproduction. Some species and groups of bacteria may undergo multiple fission as well, sometimes beginning or ending with the production of spores. The species *Metabacterium polyspora*, a symbiont of guinea pigs, has been found to produce multiple endospores in each division. Some species of cyanobacteria have also been found to reproduce through multiple fission.

Plasmotomy

Some protozoans reproduce by yet another mechanism of fission called as plasmotomy. In this type of fission, a multinucleate adult parent undergoes cytokinesis to form two multinucleate(or coenocytic) daughter cells. The daughter cells so produced undergo karyokinesis, further.

Such type of reproduction are shown by opalina and pelomyxa.

Clonal Fragmentation

Fragmentation in multicellular or colonial organisms is a form of asexual reproduction or cloning where an organism is split into fragments. Each of these fragments develop into mature, fully grown individuals that are clones of the original organism. In echinoderms, this method of reproduction is usually known as fissiparity.

Population Fission

Any splitting of a single population of individuals into discrete parts may be considered fission. A population may undergo fission for a variety of reasons, including migration or geographic isolation. Because the fission leads to genetic variance in the newly isolated, smaller populations, population fission is a precursor to speciation.

Growth Curve (Biology)

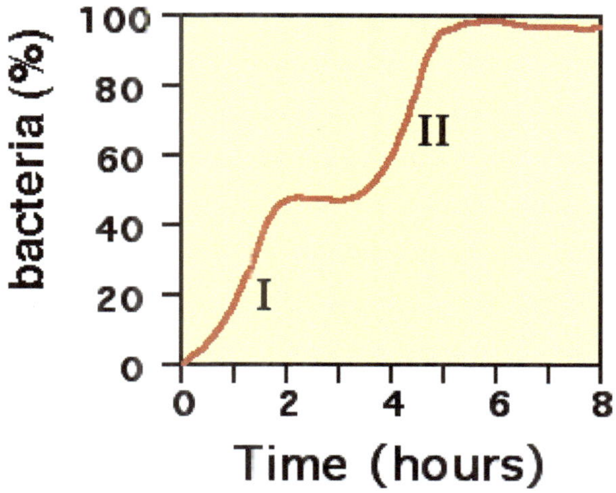

Figure: A bi-phasic bacterial growth curve.

A growth curve is an empirical model of the evolution of a quantity over time. Growth curves are widely used in biology for quantities such as population size or biomass (in population ecology and demography, for population growth analysis), individual body height or biomass (in physiology, for growth analysis of individuals). Values for the measured property can be plotted on a graph as a function of time.

Bacterial Growth

In this example the number of bacteria present in a nutrient-containing broth was measured during the course of an 8-hour cell growth experiment. The observed pattern of bacterial growth is bi-phasic because two different sugars were present, glucose and lactose. The bacteria prefer to consume glucose (Phase I) and only use the lactose (Phase II) after the glucose has been depleted. Analysis of the molecular basis for this bi-phasic growth curve led to the discovery of the basic mechanisms that control gene expression.

Cancer Cell Growth

Cancer research is an area of biology where growth curve analysis plays an important role. In many types of cancer, the rate at which tumors shrink following chemotherapy is related to the rate of tumor growth before treatment. Tumors that grow rapidly are generally more sensitive to the toxic effects that conventional anticancer drugs have on the cancer cells. Many conventional anticancer drugs (for example, 5-Fluorouracil) interfere with DNA replication and can cause the death of cells that attempt to replicate their DNA and divide. A rapidly growing tumor will have more actively dividing cells and more cell death upon exposure to such anticancer drugs.

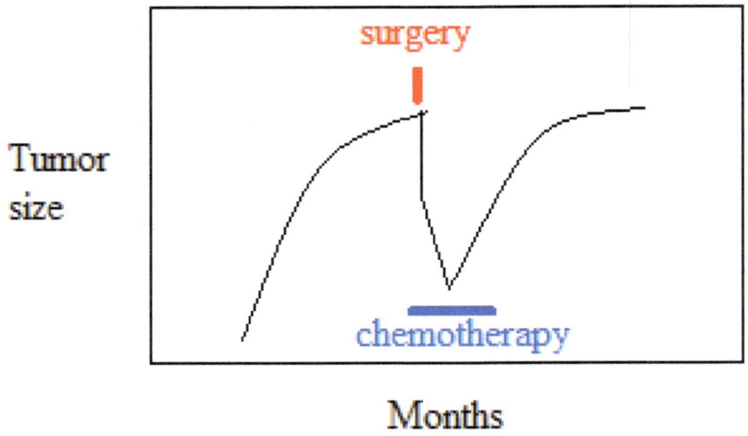

Figure. Tumor growth curve.

In the example shown in Figure, a tumor is found after the cell growth rate has slowed. Most of the cancer cells are removed by surgery. The remaining cancer cells begin to proliferate rapidly and cancer chemotherapy is started. Many tumor cells are killed by the chemotherapy, but eventually some cancer cells that are resistant to the chemotherapy drug begin to grow rapidly. The chemotherapy is no longer useful and is discontinued.

The Growth of Children

Children who fall significantly below the normal range of growth curves for body height can be tested for growth hormone deficiency and might be treatable with hormone injections.

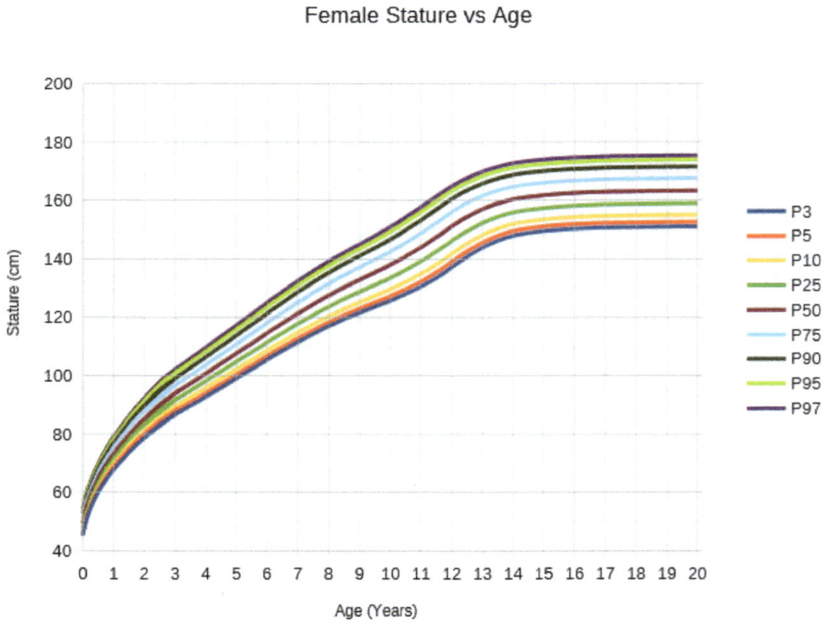

Female Stature vs Age (US CDC)

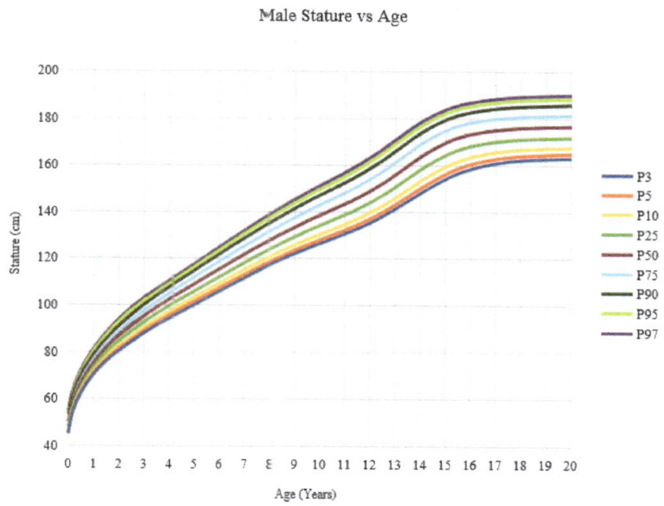

Male Stature vs Age (US CDC)

Exponential Growth

Some growth curves for certain biological systems display periods of exponential growth. Typically, periods of exponential growth are of limited duration due to depletion of some rate-limiting resource.

Lag Phase

Lag Phase: As the cells are introduced into the new medium, no immediate increase in cell number occurs. During this phase, the cells are undergoing a period of intense metabolic activity involving synthesis of enzymes and various other molecules required to divide in the coming phase. This phase can vary considerably in length depending on the nature of the medium and the microorganism. The medium may be different from the one the microorganism was growing in previously. The cells may be old and depleted of ATP, essential cofactors and ribosomes; these must be synthesized before growth can begin. So, the microorganism requires time to recover and young, vigorously growing cultures and fresh medium are to be used for the lag phase to be short.

Log Phase

Log Phase: In this phase the cell starts dividing in a logarithmic way and this is also called as exponential phase and the growth is balanced. Cellular reproduction is high during this period and the plot during this phase is a straight line. The cells are most active metabolically during this phase and the population is most uniform during this phase; therefore exponential phase cultures are usually used in biochemical and physiological studies. But during this phase, the microorganisms may be particularly sensitive to adverse conditions. On the whole, in this phase the cells are growing and dividing and increasing in cell number.The rate of exponential growth of a bacterial culture is expressed as generation time , also the doubling time of the bacterial population. Generation time (G) is defined as the time (t) per generation/ n (n = number of generations). Hence, G=t/n is the equation from which calculations of generation time can be derived.

Exponential phase or log phase is balanced growth. That is, all cellular constituents are manufactured at constant rates relative to each other. If nutrient levels or other environmental conditions change, unbalanced growth results. The individual cells may take slightly longer than others to go from lag phase to the log phase, and they do not all divided precisely together. If they divided together and the generation time is same, the number of cells in a culture would increase in a stair – step pattern, exactly doubling every 20 min or a particular time – a hypothetical situation called Synchronous growth. In an actual culture, each cell divides sometime during the 20 min generation time, with about 1/20 cells dividing each minute – a natural situation called nonsynchronous growth or asynchronous growth which appears as a smooth line, not as steps.

Organisms in a tube of culture medium can maintain logarithmic growth for only a limited time. As the number of organisms' increases, nutrients are used up, metabolic wastes accumulate, living space may become limiting factor and aerobes suffer from oxygen depletion.

Stationary Phase

Stationary Phase: Exponential growth cannot be continued forever in a batch culture (e.g. a closed system such as a test tube or flask). Population growth is limited by one of three factors: 1. exhaustion of available nutrients; 2. accumulation of inhibitory metabolites or end products; 3. exhaustion of space, in this case called a lack of "biological space".During the stationary phase, if viable cells are being counted, it cannot be determined whether some cells are dying and an equal number of cells are dividing, or the population of cells has simply stopped growing and dividing. The stationary phase, like the lag phase, is not necessarily a period of quiescence. Bacteria that produce secondary metabolites, such as antibiotics, do so during the stationary phase of the growth cycle (Secondary metabolites are defined as metabolites produced after the active stage of growth). It is during the stationary phase that spore-forming bacteria have to induce or unmask the activity of dozens of genes that may be involved in sporulation process. Starving bacteria frequently produce a variety of starvation proteins, which make the cell much more resistant to damage. They increase peptidoglycan cross-linking and cell wall strength. The Dps (DNA- binding proteins from starved cells) protein protects the DNA. Bacterial pathogens like *Salmonella typhymurium* become more virulent when starved.

Death Phase

Death Phase: Due to the conditions during the stationary phase, the death phase is seen as there is a decline in the number of viable cells. This phase also is like the log phase where the cell number is declining in a logarithmic way. The cell is said to be dead if it does not revive itself and reproduce when incubated again in a fresh medium. In this phase, the number of live cells decreases at a logarithmic rate, as indicated by the straight downward sloping diagonal line. The duration of this phase is as highly variable as the duration of log phase. Both depend primarily on the genetic characteristics of the organism.

Mathematics of Growth:

Microbial growth during the exponential phase is very important and of interest to microbiologists and the analysis applies to microorganisms dividing by binary fission. The time required by a cell to divide is called the generation time or doubling time. In the laboratory, under favorable conditions, a growing bacterial population doubles at regular intervals. Growth is by geometric progression: 1, 2, 4, 8, etc. or 2^0, 2^1, 2^2, 2^3.........2^n (where n = the number of generations). This is called exponential growth. In reality, exponential growth is only part of the bacterial life cycle, and not representative of the normal pattern of growth of bacteria in Nature. This might vary from organism to organism depending upon the environmental conditions etc. For example in *E. coli* the generation time is 20 min and hence after 20 generations a single initial cell would increase to over 1 million cells. This would require a little less than 7 hours. The population is doubling every generation; hence the increase in population is always $2^n$where n is the number of generations. The resulting population increase is exponential or logarithmic.

When growing exponentially by binary fission, the increase in a bacterial population is by geometric progression. If we start with one cell, when it divides, there are 2 cells in the first generation, 4 cells in the second generation, 8 cells in the third generation, and so on. The generation time is the time interval required for the cells (or population) to divide.

G (generation time) = (time, in minutes or hours)/n(number of generations)

$G = t/n$

t = time interval in hours or minutes

B = number of bacteria at the beginning of a time interval

b = number of bacteria at the end of the time interval

n = number of generations (number of times the cell population doubles during the time interval)

$b = B \times 2^n$ (This equation is an expression of growth by binary fission)

Solve for n:

$\log b = \log B + n\log 2$

$n=\log b-\log B/\log 2$

$n= \log b-\log B/ .301$

$n = 3.3 \log b/B$

$G = t/n$

Solve for G

$G= t/3.3\log b/B$

Example: What is the generation time of a bacterial population that increases from 10,000 cells to 10,000,000 cells in four hours of growth?

$G= t/3.3\log b/B$

$G= 240\text{minutes}/3.3\log 10^7 /10^4$

$G= 240\text{minutes}/3.3\times 3$

$G = 24$ minutes

Measurement of Microbial Growth:

A number of techniques are available in order to measure growth of microbial populations. Either population number of mass may be calculated ad growth leads to increase in both.

Direct measurement of cell numbers:

Bacteria or microorganisms can be counted directly on the plate and also called as plate counting. Advantage of this method is that it measures the number of viable cells. Disadvantage is that, it is time consuming and expensive as one needs media and other conditions need to be maintained. Bacteria counted on plate counts are referred to as colony forming units as a single cell or a clump of bacterial cells can lead to a colony which contains many cells. The colonies when they are counted in plate count method are to be present sparsely for accurate counting as overcrowding can lead to incorrect counting. To solve this, one has to adapt the serial dilution method in order to get an accurate count.

Serial dilution and pour and spread plate: Supposing one has to accurately count the number of cells given in a solution, then serial dilution needs to be performed. A 1ml of the sample is taken and transferred to a tube containing 9ml of sterile water and this process can be repeated until we reach a considerable dilution (say 10^6 to 10^7). Once the original inoculum is diluted one needs to perform a pour plate or a spread plate technique in order to count the number of bacteria present in the diluted sample and then the original sample. In pour plate method the diluted sample is poured into the petriplate and then the medium which is at nearly 50°C is poured over the inoculum and mixed by gentle agitation. With this method, colonies grow within the nutrient agar as well as on the surface of the agar plate. As certain disadvantages are encountered in this method like heat sensitive microorganisms might not grow and also bacteria when they grow within the nutrient medium might not be useful for diagnostic purposes. In order to avoid these problems, spread plate method is mostly used. A 0.1ml of the diluted sample is added to the surface of the nutrient medium and spread uniformly with the help of a glass spreader and after incubation, the colonies can be counted and the concentration of the bacterial cells in the original sample is calculated as follows:

Number of bacteria/ml = Number of colonies on plate x reciprocal of dilution of sample

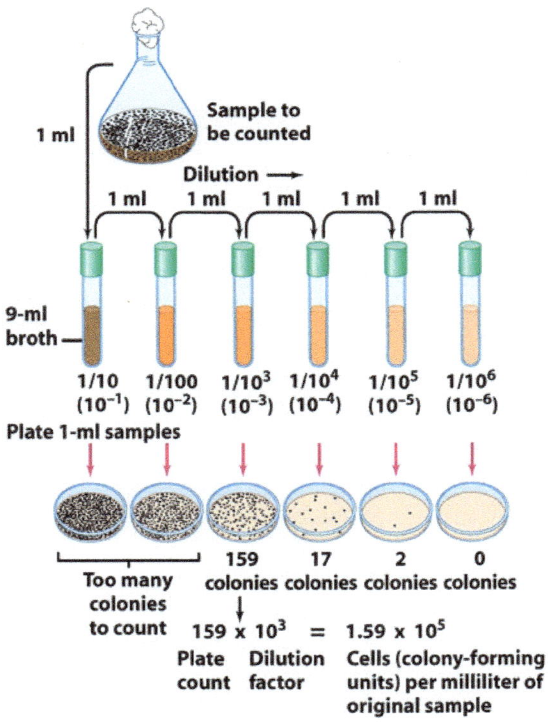

Serial dilution methodology

Membrane Filtration: This method can be used in order to study if the quantity of the bacteria is very small as in aquatic samples like lakes, streams etc. Membranes with different pore sizes are used to trap different microorganisms. The sample is drawn through these special membrane filters and placed on an agar medium or on a pad soaked with liquid media. After incubation, the number of colonies can be counted and the number determined in the original sample. Selective media or differential media can be used for specific microorganisms. This is mostly used for analyzing aquatic samples.

Microscopic count: The Petroff-Hausser counting chamber or slide is easy, inexpensive and relatively quick method and also gives information about the size and morphology of the microorganisms. These specially designed slides have chamber of known depth with an etched grid on the chamber bottom.

Bacteria/mm^3 = (bacteria/square) (25 squares)/ (50)

Bacteria can be counted by taking into account the chamber's volume and any sample dilution. The disadvantage encountered in this method is that fairly large volume is required and also it is difficult to distinguish between living and dead cells. Microorganisms of larger sizes can be counted by using electronic counters such as coulter counter; where in the number of cells in a measured volume of liquid is counted. This method gives accurate results with larger cells and is extensively used in hospital laboratories to count red and white blood cells.

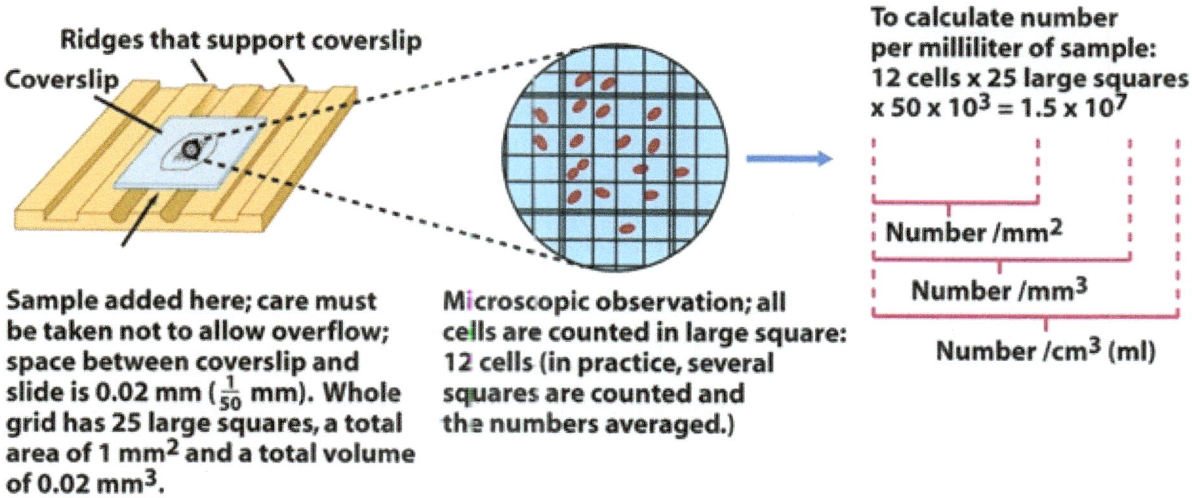

Direct microscopic count of bacterial cells

Indirect methods of measurement of cell mass:

Population growth leads to increase in the total cell mass, as well as in cell numbers. The following methods can be used.

Turbidity: As bacteria grow/multiply in a liquid medium, the medium becomes turbid. Spectrophotometer is used in order to measure the turbidity. A beam of light is transmitted through a bacterial suspension to a light-sensitive detector. The fact that microbial cells scatter light striking them, the amount of scattering is directly proportional to the biomass of cells present and indirectly related to cell number. The extent of light scattering can be measured and is almost linearly related to bacterial concentration at low absorbance levels.

Broth culture showing turbidity

Dry weight: This method is mostly used for filamentous bacteria and moulds. The microorganism is grown in liquid medium, filtered or centrifuged to remove extraneous material, and dried in an oven and then weighted. It is time consuming and hence not very sensitive.

Continuous culture of Microorganisms:

Batch cultures : Nutrient supplies are not renewed nor wastes removed.

Continuous cultures : continuous provision of nutrients and removal of wastes takes place. The population can be maintained in the exponential phase and at a constant biomass concentration for extended periods.

These can again be categorized into two types:

Chemostat : Where sterile medium is fed into the culture vessel at the same rate as the media containing microorganisms is removed.

Turbidostat : It has a photocell that measures the absorbance or turbidity of the cell culture in the growth vessel. The flow rate of media through the vessel is automatically regulated to maintain a predetermined turbidity or cell density.

Used in food and industrial microbiology.

Environmental Factors on Microbial Growth

The growth of microorganisms is greatly affected by the chemical and physical nature of their surroundings. An understanding of these influences aids in the control of microbial growth and the study of the ecological distribution of microorganisms. Prokaryotes are present or grow anywhere life can exist. The environments in which some prokaryotes grow would kill most other organisms. For example *Bacillus infernus* is able to live over 1.5 miles below the earth's surface without O_2 and 60°C temperature. These microorganisms which can thrive and grow in such harsh conditions are often called extremophiles.

The major physical factors which affect microbial growth are solutes and water activity, pH, temperature, oxygen level, pressure and radiation.

Solutes and Water activity: Changes in osmotic concentration of the surroundings can affect microbial growth as a selectively permeable plasma membrane separates the microorganisms from their surroundings. Microorganisms need to keep the osmotic concentration of their cytoplasm somewhat above that of the habitat by the use of compatible solutes, so that the plasma membrane is always pressed firmly against their cell wall. In a hypertonic environment, the prokaryotes increase their internal osmotic concentration through the synthesis or uptake of choline, proline, glutamic acid and other amino acids. A few prokaryotes like *Halobacterium salinarium* raise their osmotic concentration with potassium ions. The enzymes of these bacteria are altered for the requirement of high salt concentrations for normal activity. Halophiles grow optimally in the presence of NaCl or other salts at a concentration above about 0.2M. These have extensively modified the structure of their proteins and membranes rather than simply increasing the intracellular concentrations of solutes. They require higher potassium levels for stability and activity. The plasma membrane of halophiles is also stabilized by high concentration of sodium ions.

Water activity (a_w) is the amount of water available to microorganisms and this can be reduced by interaction with solute molecules (osmotic effect). Water activity is inversely related to osmotic pressure; if a solution has high osmotic pressure, it's a w is low. Microorganisms differ greatly in their ability to adapt to habitats with low water activity. In a low a w habitat, the microorganisms must expend extra effort to grow as it should maintain a high solute concentration to retain water. Such microorganisms are osmotolerant or can grow over wide range of water activity or osmotic concentration. Most of the microorganisms grow at a w =0.98 or higher.

pH: It refers to the acidity or alkalinity of a solution. It is a measure of the hydrogen ion activity of a solution and is defined as the negative logarithm of the hydrogen ion concentration.

$$pH = -\log [H^+] = \log (1/H^+)$$

The pH scale ranges from 1.0 to 14.0 and most microorganisms grow vary widely from pH 0 to 2.0 at the acid end to alkaline lakes and soil that may have pH values between 9.0 and 10. The pH can affect the growth of microorganisms and each species has a definite pH growth range and pH growth optimum. Acidophiles have their growth optimum between pH 0 and 5.5; neutrophiles between 5.5 and 8.0 and alkalophiles prefer pH range of 8.5 to 11.5. Most bacteria and protozoans are neutrophiles, fungi prefer acid surroundings about pH 4 to 6; algae also seem to favour slight acidity. *Cyanidium caldarium* (algae) and archaeon *Sulfolobus acidocaldarium* are inhabitants of acidic hot springs; both grow well around pH 1 to 3 and at high temperature. Drastic changes/variations in cytoplasmic pH can harm microorganisms by disrupting the plasma membrane or inhibiting the activity of enzymes and membrane transport proteins. Prokaryotes die if the internal pH drops much below 5.0 to 5.5. External pH alterations also might alter the ionization of nutrient molecules and thus reduce their availability to the organism. The microorganism needs to maintain a neutral cytoplasmic pH and for this the plasma membrane may be relatively impermeable to protons. Neutrophiles appear to exchange potassium for protons using an antiport transport system. Extreme alkalophiles maintain their internal pH closer to neutrality by exchanging internal sodium ions for external protons. The antiport systems probably correct small variations in pH. In case of too much acidity (below 5.5 to 6.0) *S. typhimurium* and *E. coli* synthesize an array of new proteins as part of what has been called as their acidic tolerance response. If the external pH decreases to 4.5 or lower, chaperones such as acid shock proteins and heat shock proteins are synthesized. Microorganisms can change the pH of their own habitat by producing acidic or basic metabolic waste products. In order to maintain the pH, buffers are often included in the media to prevent growth inhibition. Phosphate is commonly used buffer and a good example of buffering agent. Peptides and amino acids in complex media also have a strong buffering effect.

Temperature: Temperature profoundly affects microorganisms as the most important factor influencing the effect is temperature sensitivity of enzyme-catalyzed reactions. Beyond a certain point of higher temperature, slow growth takes place and damages the microorganisms by denaturing enzymes, transport carriers and other proteins. The plasma membrane also is disrupted as lipid bilayer simply melts and the damage is such an extent that it cannot be repaired. At very low temperature, membranes solidify and enzymes don't work rapidly. In summary, when organisms are above their optimum temperature, both the function and cell structure is affected at low temperature, function is affected. The cardinal temperatures vary greatly between microorganisms. Optimum usually range from 0°C to as high as 75°C, where as microbial growth occurs at temperature extending from -20°C to over 120°C. Archaen *Geogemma barossii* grows anaerobically at 121°C. The major microbial groups differ from one another regarding their maximum growth temperature. Upper limit for protozoans is around 50°C, some algae and fungi can grow at temperatures as high as 55°C to 60°C.

Did you know: Prokaryotes have been found growing at or close of 100°C. Now thermophilic prokaryotes have been reported growing in surface of chimneys or black smokers located along rifts and ridges on the ocean floor that spew sulphide-rich super-heated vent water with temperatures above 350°C. These microbes can grow and reproduce at or above 112°C. The proteins, membranes and nucleic acids of these prokaryotes are remarkably temperature stable and provide ideal subjects for studying the ways in which macromolecules and membranes are stabilized. Some thermostable enzymes from these organisms have important industrial and scientific uses. For ex., the Taq polymerase from the thermophilic *Thermus aquaticus* is used extensively in the polymerase chain reaction.

Microorganisms are classified into five classes based on their temperature ranges for growth.

1. Psychrophiles: Microorganisms grow well at 0°C and the optimum growth temperature of 15°C or lower and maximum at around 20°C. These microorganisms are isolated from Arctic and Antarctic habitats. They have adapted to their environment in several ways. Their enzymes, transport systems and protein synthetic mechanisms function well at low temperatures. The cell membranes have high levels of unsaturated fatty acids and remain semifluid when cold. At higher than 20°C, the psychrophiles begin to leak cellular constituents because of cell membrane disruption. Microorganisms such as *Pseudomonas, Vibrio, Alcaligenes, Bacillus, Arthrobacter, Moritella, Photo-*

bacterium belong to this group. The psychrophilic *Chlamydomonasnivalis* turns a snowfield or glacier pink with its bright red spores.

2. Psychrotrophs or Facultative Psychrophiles: In this group many s pecies can grow ato o C to 7°C, optimum between 20°C and 30°C. The spoilage of refrigerated foods is mainly caused by microorganisms belonging to this group.

3. Mesophiles: Growth optimum around 20°C to 40°C, minimum at 15°C to 20°C and maximum at 45°C or lower. Most of the organisms fall under or within this category including human pathogens.

4. Thermophiles: The microorganisms in this group can grow at temperature of 55°C or higher, minimum is usually around 45°C and growth optima at around 55°C to 65°C. Mostly prokaryotes and a few algae and fungi belong to this group. The habitats in which they grow include, composts, self-heating haystacks, hot water lines and hot springs. Microorganisms have more heat-stable enzymes and proteins synthesis systems, which function at high temperature. Heat stable proteins have high organized, hydrophobic interiors, more hydrogen bonds and other non-covalent bonds strengthen the structure. Amino acids like proline make the polypeptide chain less flexible and chaperones also aid in folding of proteins to stabilize them. DNA also is stabilized by specific histone like proteins. The membrane lipids are also stable and tend to be more saturated, more branched and of higher molecular weight. Archaeal thermophiles have membrane lipids with ether linkages, which protect the lipids from hydrolysis at high temperatures.

5. Hyperthermophiles: Few microorganisms can grow at 96°C or above and have maximum at 100°C; and growth optima between 80°C and about 113°C. *Pyrococcus* and *Pyrpdictiumoccultum* are examples of marine hyperthermophiles found in hot floors of the sea floor.

Oxygen Concentration

Oxygen Concentration: An aerobe is an organism able to grow in the presence of atmospheric $O_2$and the ones that grow in its absence is an anaerobe. Organisms which completely are dependent on atmospheric O_2 for growth are obligate aerobes, and it serves as the terminal electron acceptor for the electron transport chain in aerobic respiration and employs it in the synthesis of sterols and unsaturated fatty acids. Organisms which do not require O_2 for growth but do grow better in its presence are called facultative anaerobes . Aerotelerant anaerobes such as Enterococcus faecalis simply ignore O2 and grow equally well whether it is present or not. Obligate anaerobes like Bacteroides, Fusobacterium, Clostridium pasteurianum, *Methanococcus, Neocallimastix*, do not tolerate O_2 at all and die in its presence. Aerotelerant and obligate anaerobes cannot generate energy through respiration and must employ fermentation or anaerobic respiration pathways for the purpose. Microaerophiles are those organisms that are damaged by the normal atmospheric levels of O_2 (20%) and require O_2 levels between the range of 2% to 16% for growth. The nature of bacterial O_2 responses can be readily determined by growing the bacteria in culture tubes filled with a solid culture medium or a special medium like thioglycollate broth, which contains a reducing agent to lower O_2 levels. Aerobic microorganisms are cultured, either the culture vessel is shaken to aerate the medium or sterile air is pumped. Anaerobic microorganisms require special anaerobic media containing reducing agents such as thioglycollate or cysteine may be used. Removing air with a vacuum pump and flushing out residual oxygen with nitrogen gas is also preferred. Co_2 and nitrogen is added to the chamber since many anaerobes require a small amount of Co_2 for best growth. The technique in which gas pak jar is used can be used.

Prokaryotes and protozoa are found arranged among all the 5 types of microorganisms. Fungi are normally aerobic, but species particularly among yeasts, are facultative anaerobes. Algae are almost always obligate aerobes. The different relationships with O_2 appear due to several factors, including the inactivation of proteins and the effect of toxic O_2 derivatives. Enzymes can be inactivated when sensitive groups like sulfydryls are oxidised. A notable example is the nitrogen fixation enzymes nitrogenise which is very O_2 sensitive.

Oxygen accepts electrons and is readily reduced because its two outer orbital electrons are unpaired. The reduction products such as superoxide radical, hydrogen peroxide and hydroxyl radical can be resulted by flavoproteins, several other cell constituents and radiation.

$$O_2 + e^- \rightarrow O_{2^-}^{\bullet}$$
$$O_{2^-}^{\bullet} + e^- + 2H^+ \rightarrow H_2O_2$$
$$H_2O_2 + e^- + H^+ \rightarrow H_2O + OH^{\bullet}$$

These are extremely toxic because they are powerful oxidizing agents and rapidly destroy cellular constituents. As microorganisms can be killed, they need to protect themselves from such oxygen products. Microorganisms possess enzymes that afford protection against toxic O_2 products. Obligate aerobes and facultative anaerobes usually contain the enzymes superoxide dismutase (SOD) and catalase, which catalyze the destruction of superoxide radical and hydrogen peroxide respectively. Peroxidase also can be used to destroy hydrogen peroxide.

$$2O_{2^-}^{\bullet} + 2H^+ \rightarrow O_2 + H_2O_2 \text{ (Superoxide dismutase)}$$
$$2H_2O_2 \rightarrow 2H_2O \rightarrow O_2 \text{ (Catalase)}$$
$$H_2O_2 + NADH + H+ \rightarrow 2H_2O + NAD^+ \text{ (Peroxidase)}$$

Aerotolerant microorganisms may lack catalase but almost always have superoxide dismutase. Strict anaerobes lack either of the enzymes or have them in very low concentrations and therefore cannot tolerate O_2. Aerobic microorganisms can be grown in an aerated medium which is aerated by shaking the vessel or sterile air is pumped. But for anaerobes, O_2 must be excluded from the medium. Reducing agents such as thioglycollate or cysteine can be used to flush out O_2. Nitrogen gas can also be used to eliminate O_2 with a vacuum pump, and sometimes CO_2 and nitrogen is added to the chamber, as anaerobes require a small amount of CO_2 for best growth and also gas pack jar can also be used.

Pressure: Most microorganisms always are subjected to pressure of 10 atmospheres (atm). The hydrostatic pressure can reach to 600 to 1100 atm in the deep sea with temperature about 2°C to 3°C. Organisms can survive and adapt at these extreme conditions and many are barotolerant, increased pressure does adversely affect them but not as much as it does to nontolerant bacteria. The barophillic organisms are those growing in the guts of deep sea invertebrates such as amphipods and holothurians and grow more rapidly at high pressures. These bacteria may play an important role in nutrient recycling in the deep sea. Bacterial genera of *Photobacteria, Shewanella, Colwellia* are barophiles. Some members of the Archaea are thermophiles for example *Pyrococcus spp., Methanococcus janaschii*.

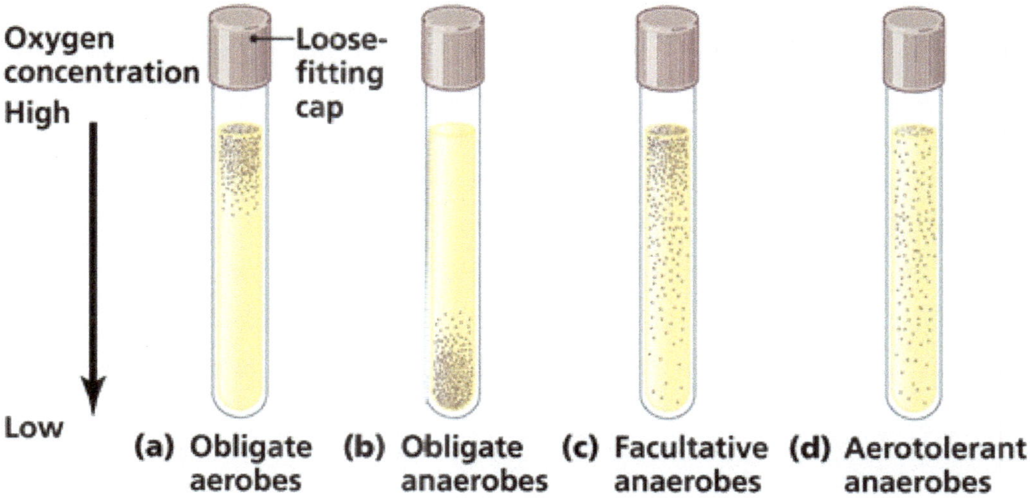

Oxygen requirements in bacteria

Pressure:

Most organisms on land or on the surface of water is always subjected to a pressure of 1 atm. The hydrostatic pressure can reach 600 to 1100 atm in the deep sea. Despite these extremes, bacteria survive and adapt. Many are barotolerant. Some bacteria in the gut of deep sea invertebrates such as amphipods and holothurians are truly barophilic and grow more rapidly at high pressures (Ex. *Photobacterium, shewanella, Colwellia*).

Radiation:

Electromagnetic radiation of various types bombards our world. As the wavelength of electromagnetic radiation decreases, the energy of the radiation increases – gamma rays and X rays are much more energetic than visible light or infrared waves. Sunlight is the major source of radiation on the earth. It includes visible light, ultraviolet radiation, infrared rays and radio waves. Most life is dependent on the ability of photosynthetic organisms to trap the light energy of the sun as visible light. Many forms of electromagnetic radiation are very harmful to microorganisms. Ionizing radiation, radiation of very short wavelength or high energy can cause atoms to lose electrons or ionize. The two major forms of ionizing radiation, X rays which are artificially produced and gamma rays which are emitted during radioisotope decay. Low levels of ionizing radiation will produce mutations, higher levels are directly lethal. Some prokaryotes like *Deinococcus radiodurans* and bacterial endospores are resistant and can cause a variety of changes in cells like; it breaks hydrogen bonds, oxidises double bonds, destroys ring structures and polymerizes some molecules. Oxygen enhances these destructive e effects, probably through the generation of hydroxyl radicals (OH.). Destruction of DNA is the most important cause of death of microorganisms. Ultraviolet radiation kills all kinds of microorganisms due to its short wavelength (approximately 10 to 400 nm) and high energy. The most lethal UV radiation has a wavelength of 260 nm, the wavelength most effectively absorbed by DNA. Formation of thymine dimmers in DNA is the primary mechanism of UV damage; these dimmers inhibit DNA replication and function. This damage is repaired by photo reactivation, where blue light is used by a photo reactivating enzyme (photolyase) to split the thymine dimmers. Dark reactivation, where a short sequence containing the thymine dimmers can also be excised and replaced in the absence of light. Damage can also be repaired by the recA protein in recombination repair and SOS repair.

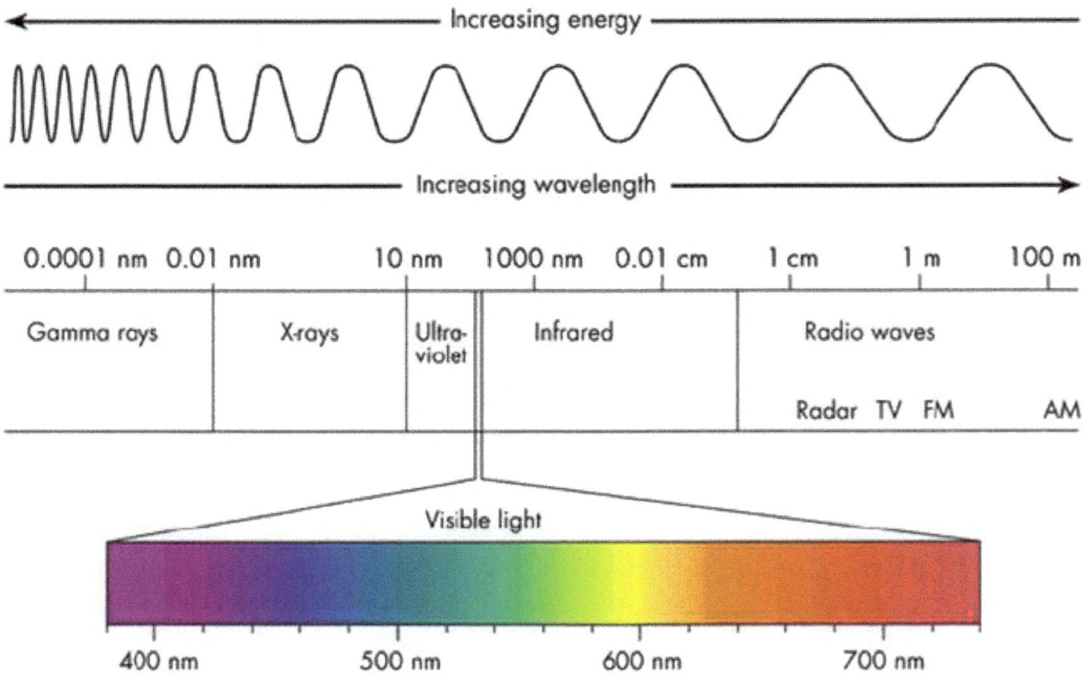

Electromagnetic spectrum

Visible light – immensely beneficial becuas eit is the source of energy for photosynthesis. Visible light when present in sufficieint intensity can damage or kill microbial cells. Pigments called photosensitizers and O_2 are required. All microorganisms possess pigments like chlorophyll, bacteriochlorophull, cytochromes and flavins which absorb light energy, become exited or activated and act as photosensitizers. The excited photosensitizer (P) transfers its energy to O_2 generating singlet oxygen (‘O_2).

$$\text{P (light)} \rightarrow \text{P(activated)}$$

$$\text{P(activated)} + O_2 \rightarrow \text{P} + {}^{`}O_2$$

Singlet oxygen is very reactive, powerful oxidizing agent that will quickly destroy a cell. Many microorganisms that are airborne or live on exposed surface use carotenoid pigments for protection against photooxidation. Carotenoids effectively quench singlet oxygen that is absorbing energy from singlet oxygen and convert it back into the unexcited ground state.

Microbial Growth in Natural Environments:

Microorganisms in a particular location are exposed to many overlapping gradients of nutrients and various other environmental factors. Microorganisms will grow in ‘microenvironments’ until an environmental or nutritional factor limits growth. Liebig’s law of minimum states that the total biomass of an organism will be determined by the nutrient present in the lowest concentration relative to the organisms requirements. Shelford's law of tolerance states that there are limits to environmental factors, below and above which a microorganism cannot survive and grow, regardless of the nutrient supply. Each organism has a specific temperature, pH, oxygen level and hydrostatic pressure to grow. The growth of a microorganism depends on both the nutrient supply and its tolerance of the environmental conditions. In response to low nutrient levels (oligotrophic environments) and intense competition, many microorganisms become more competitive in nutrient capture and exploitation of available resources. Morphology of the organism can change in order to increase its surface area and ability to absorb nutrients. Microorganisms can also undergo step by step shut down of metabolism except for housekeeping genes. Natural substances can also directly inhibit microbial growth and reproduction in low-nutrient environments. These agents include phenolics, tannins, ammonia, ethylene etc. This may be a means by which microorganisms avoid expending limited energy resources until an adequate supply of nutrients becomes available.

Growth of natural prokaryotic populations outside the laboratory can be determined by counting the number of viable microorganisms present. A viable microorganism is one that is able to grow actively, resulting in the formation of a colony or visible turbidity in liquid medium. John R. Postgate of University of Sussex in England was one of the first to note that microorganisms stressed by survival in natural habitats were sensitive to secondary stresses. Such stresses can produce viable microorganisms that have lost the ability to grow on media normally used for their cultivation. To determine the growth potential of such microorganism, Postgate micro viability assay, in which microorganisms are cultured in the thin agar film under a cover slip. The ability of a cell to change its morphology, even if it does not grow beyond the single stage indicates that the microorganism does show life signs.

The situation in natural environments with mixed populations is much more complex. Often only 1 to 10% of observable cells are able to form colonies. The microbiologist is attempting to grow microorganisms that perhaps never have been cultured or characterized. At present, molecular techniques involving PCR amplification and small subunit ribosomal RNA analysis are increasingly used to analyze diversity of uncultured microbial population.

Quorum sensing:

Bacteria can communicate with one another and behave cooperatively. Quorum sensing or autoinduction is a phenomenon in which bacteria monitor their own population density through sensing the levels of signal molecules, sometimes called autoinducers because they can stimulate the cell that releases them. The concentration of these signal molecules increases along with the bacterial population until it rises to a specific threshold and signals the bacteria that the population density has reached a critical level or quorum. The bacteria then begin expressing sets of quorum-dependent genes. Quorum sensing has been found among both gram-negative and gram-positive bacteria. Quorum sensing was first discovered in gram-negative bacteria and is best understood in these microorganisms. The most common signals in gram-negative bacteria are acyl-homoserine lactones (HSLs). These diffuse into the target cell and once they reach a sufficiently high level, acyl HSLs bind to specific receptor proteins and trigger conformational change, they bind to target sites on the DNA and stimulate transcription of quorum sensitive genes. The genes needed to synthesize acyl HSLs is also produced frequently, thus amplifying the effect by the production and release of more auto inducer molecules.

Examples in gram negative bacteria are 1) bioluminescence production by *Vibrio fisheri* 2) *Pseudomonas aerogionosa* synthesis and release of virulence factors 3) Conjugal transfer of genetic material by *Agrobacterium tumefaciens* and 4) antibiotic production by *Erwiniacarotovora* and *Pseudomonas aureofaciens* . Gram positive bacteria also regulate activities by quorum sensing often using an oligopeptide pheromone signals. Good examples are mating in *Enterococcus faecalis*, competence induction in *Streptococcus pneumonia*, and stimulation of sporulation by *Bacillus subtilis* , and production of many toxins and other virulence factors by *Staphylococcus aureus*. Another important function of quorum sensing is to promote the formation of mature biofilms by the pathogen, *Pseudomonas aeroginosa* , as it may play a role in cystic fibrosis. Quorum sensing is an example of what might be called multicellular behaviour in that many individual cells communicate and coordinate their activities to act as a unit. Other examples of such complex behaviour are pattern formation in colonies and fruit body formation in the Myxobacteria.

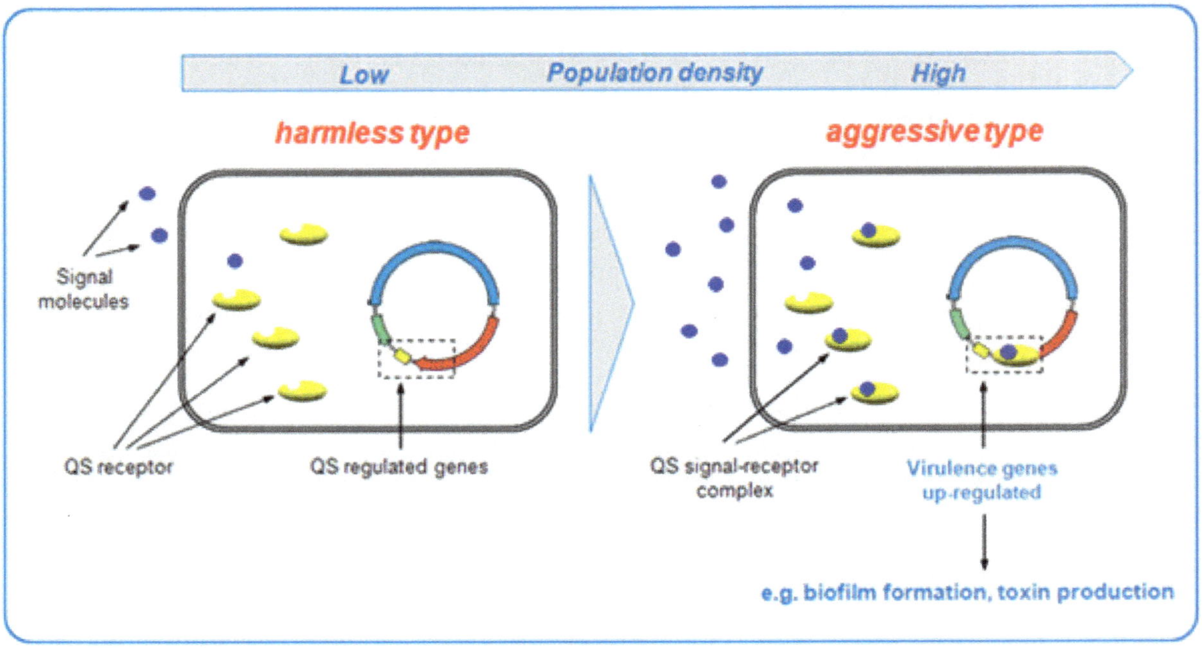

Quorum sensing

References

- Hubbell, S. (2003). "Modes of speciation and the lifespans of species under neutrality: a response to the comment of Robert E. Ricklefs.". Oikos. 100 (1): 193–199. doi:10.1034/j.1600-0706.2003.12450.x
- Sezonov, G.; Joseleau-Petit, D.; D'Ari, R. (28 September 2007). "Escherichia coli Physiology in Luria-Bertani Broth". Journal of Bacteriology. 189 (23): 8746–8749. doi:10.1128/JB.01368-07
- Britannica Educational Publishing (2011). Fungi, Algae, and Protists. The Rosen Publishing Group. ISBN 978-1-61530-463-9. Retrieved 2016-12-21
- Angert, Esther R. (March 2005). "Alternatives to binary fission in bacteria" (PDF). Nature Reviews Microbiology. 3 (3): 214–224. doi:10.1038/nrmicro1096. PMID 15738949. Retrieved 23 August 2016
- Bišová, K.; Zachleder, V. (17 January 2014). "Cell-cycle regulation in green algae dividing by multiple fission" (PDF). Journal of Experimental Botany. 65 (10): 2585–2602. doi:10.1093/jxb/ert466. PMID 24441762. Retrieved 23 August 2016

- P.Puranik, Asha Bhate (2007). Animal Forms And Functions: Invertebrata. Sarup & Sons. ISBN 978-81-7625-791-6. Retrieved 2016-12-21
- Rokney, Assaf; Shagan, Merav; Kessel, Martin; Smith, Yoav; Rosenshine, Ilan; Oppenheim, Amos B. (September 2009). "E. coli Transports Aggregated Proteins to the Poles by a Specific and Energy-Dependent Process". Journal of Molecular Biology. 392 (3): 589–601. doi:10.1016/j.jmb.2009.07.009. PMID 19596340
- Pecková, H.; Lom, J. (1990). "Growth, morphology and division of flagellates of the genusTrypanoplasma (Protozoa, Kinetoplastida) in vitro". Parasitology Research. 76 (7): 553–558. doi:10.1007/BF00932559
- Yoshinori Tanada; Harry K. Kaya (1993). Insect pathology. Gulf Professional Publishing. ISBN 978-0-12-683255-6. Retrieved 2016-12-21
- North, R. J. (1 June 1993). "Mycobacterial virulence. Virulent strains of Mycobacteria tuberculosis have faster in vivo doubling times and are better equipped to resist growth-inhibiting functions of macrophages in the presence and absence of specific immunity". Journal of Experimental Medicine. 177 (6): 1723–1733. doi:10.1084/jem.177.6.1723

Permissions

All chapters in this book are published with permission under the Creative Commons Attribution Share Alike License or equivalent. Every chapter published in this book has been scrutinized by our experts. Their significance has been extensively debated. The topics covered herein carry significant information for a comprehensive understanding. They may even be implemented as practical applications or may be referred to as a beginning point for further studies.

We would like to thank the editorial team for lending their expertise to make the book truly unique. They have played a crucial role in the development of this book. Without their invaluable contributions this book wouldn't have been possible. They have made vital efforts to compile up to date information on the varied aspects of this subject to make this book a valuable addition to the collection of many professionals and students.

This book was conceptualized with the vision of imparting up-to-date and integrated information in this field. To ensure the same, a matchless editorial board was set up. Every individual on the board went through rigorous rounds of assessment to prove their worth. After which they invested a large part of their time researching and compiling the most relevant data for our readers.

The editorial board has been involved in producing this book since its inception. They have spent rigorous hours researching and exploring the diverse topics which have resulted in the successful publishing of this book. They have passed on their knowledge of decades through this book. To expedite this challenging task, the publisher supported the team at every step. A small team of assistant editors was also appointed to further simplify the editing procedure and attain best results for the readers.

Apart from the editorial board, the designing team has also invested a significant amount of their time in understanding the subject and creating the most relevant covers. They scrutinized every image to scout for the most suitable representation of the subject and create an appropriate cover for the book.

The publishing team has been an ardent support to the editorial, designing and production team. Their endless efforts to recruit the best for this project, has resulted in the accomplishment of this book. They are a veteran in the field of academics and their pool of knowledge is as vast as their experience in printing. Their expertise and guidance has proved useful at every step. Their uncompromising quality standards have made this book an exceptional effort. Their encouragement from time to time has been an inspiration for everyone.

The publisher and the editorial board hope that this book will prove to be a valuable piece of knowledge for students, practitioners and scholars across the globe.

Index

A

Agar Plate, 12, 238, 248

Algae, 1-2, 7-10, 13, 19, 26, 44, 63-82, 90, 92, 94-95, 98-99, 113, 124, 126, 128, 131, 171, 230, 236-237, 242, 251-254, 258

Alginates, 73

Apicomplexans, 64, 66, 127, 242

Applied Microbiology, 14

Archaebacteria, 13, 26-27, 136, 141

Asexual Phase, 69

Asexual Reproduction, 52, 69-70, 105, 134, 239, 242

Atomic Force Microscope, 184-185, 201, 217

B

Bacterial Fission, 241

Bacterial Transformation, 38, 151, 155, 173

Bacteriochlorophyll, 47

Bacteriology, 13, 19, 43, 58, 219, 224, 227, 258

Bioremediation, 15-16, 62, 75, 121, 129

C

Carboxysomes, 46-47, 133, 163, 166, 169-170

Cell Wall, 6, 32-33, 45, 48, 58, 79, 82-84, 98-99, 103, 112, 133, 141-147, 149, 151-152, 155, 158-159, 221-222, 225, 227, 240-241, 246, 251

Cellular Structure, 46

Chemical Cycling, 40

Chemotherapeutics, 117

Chloroplasts, 7, 44, 46, 64-65, 75, 82, 85, 92, 98-99, 124, 137-138, 161, 168

Clonal Fragmentation, 242

Confocal Microscopy, 180-182, 198, 215

Coral Reefs, 69

Counterstaining, 222

Culture, 13, 19, 22-23, 32, 40, 59, 177, 220, 222, 224, 236-238, 241, 245-246, 250, 253

Cyanobacteria, 26, 47, 50-51, 63-65, 69, 74, 113, 131, 133, 150, 154, 159, 170-171, 224, 230-231, 236, 242

Cytoplasm, 32, 46-47, 49-50, 57, 85, 103, 132-133, 141, 145, 158-164, 169, 172-173, 178-179, 242, 251

Cytoplasmic Inclusions, 47, 161

Cytoskeletal Sieving, 167

D

Deconvolution, 182

Differential Staining, 222, 224

Dispersion Staining, 177

E

Endosymbiosis, 44, 79, 95, 137

Eukaryotes, 5, 7-8, 11, 26, 28-29, 31-32, 34-35, 37-38, 43-44, 46-47, 54, 58, 98-99, 103, 117, 126, 133, 135-138, 159, 161-162, 166-167, 173, 232-233

Extracellular Structures, 48-49

F

Feeding, 110, 126

Female Gametes, 69, 91-92

Fermenting, 16, 60, 238

Flagella, 9, 33-34, 49, 56-57, 82, 92, 103, 126, 141-142, 147, 153, 156-157, 214

Folk Medicine, 118

G

Gene Transfer, 2, 29, 31-32, 38, 54-55, 57-58, 95, 134, 151, 155, 173

Genetic Exchange, 38, 134

Genome Sequencing, 94

Gentle Agitation, 248

Gram Negative, 142, 144-145, 237, 258

Gram Staining, 147-148, 151, 153, 155, 220-222, 224, 226

Gram-positive Bacteria, 31-33, 48, 50, 58, 144-151, 153-154, 221-223, 257

I

Industrial Microbiology, 15, 23, 129-130, 250

Interference Reflection, 179

L

Laminaria, 76-77, 80

Lichens, 68-69, 108, 111, 113-114

Lipid Membranes, 46, 211

M

Macromolecules, 33, 161, 163-166

Membranes, 7, 26, 33-36, 46-47, 52, 57, 68, 75, 82, 98, 126, 133, 145, 149, 157-159, 162, 165-168, 211, 224, 228, 232, 249, 251-252

Metabolism, 7, 14, 24, 27, 35-37, 46, 50-51, 54, 57, 62, 69, 72, 89, 104, 129, 135, 162, 164, 167, 229, 241, 257

Metachromatic Granules, 47
Microorganisms, 1-7, 9-18, 20-24, 26-27, 36, 40, 42-44, 50, 96, 118, 120, 128-130, 132, 145, 173-174, 215, 219, 221, 224, 227-232, 235-236, 238-239, 245-246, 248-257
Microscopy, 13, 47, 106, 108, 117, 159, 162, 174-188, 191-192, 196-201, 205-206, 209-211, 215-220, 227
Mitochondria, 7, 44, 46, 84-85, 98, 110, 132, 137-138, 159, 162, 166, 239
Morphology, 2, 27, 32-33, 43-45, 58-59, 67, 76, 79, 88, 97-98, 102, 105, 111, 127, 134, 214, 219, 229, 239, 249, 257, 259
Motility, 49, 56-57, 92, 110, 123, 126, 156
Multiphoton Microscopy, 181
Mutualism, 9, 41, 60
Mycobacteria, 48, 58, 259
Mycology, 13, 96-97, 100-102, 112
Mycoparasites, 115

N
Nucleus, 6-7, 26, 32, 43, 46-47, 84-85, 95, 98, 132-133, 137, 159, 162, 166-168, 178-179, 242
Nutrition, 10, 74, 114, 129, 228, 237

O
Oblique Illumination, 177
Optical Microscopy, 175, 188, 198-199, 218

P
Parasexual Processes, 107
Parasitology, 13, 259
Pathogenesis, 14, 150
Pathogens, 9, 12, 15, 20, 22, 25, 27, 41, 48-49, 55, 58, 60-61, 97, 101, 104, 111, 115, 120, 129, 144, 155-156, 230, 237, 246, 253
Pellicle, 126
Peptidoglycan, 33, 48, 58, 138, 142-147, 149-151, 153-155, iv
221-224, 246
Phaeothamniophyceae, 79
Phase Contrast, 106, 176-178, 185, 189-190, 192, 196
Pigments, 47, 57, 75, 82, 88, 256
Plasmotomy, 242
Poisonous Species, 119
Polyhedral Organelles, 46
Polyhedral Protein Shells, 46
Prokaryotic Cytoskeleton, 32, 46, 133
Protein Complexes, 163, 166-167
Protein Structures, 49, 137
Protozoa, 1, 12-13, 18, 20, 26, 41, 44, 122-128, 254, 259
Pseudomonas Aeruginosa, 48, 57, 151-152, 154

R
Ribosomes, 47, 50, 98, 110, 137, 159, 161, 167-169, 228, 241, 245

S
Scanning Electron Microscope, 83, 93, 174, 201, 205, 207-209, 212, 214-215, 239
Sea Sponges, 69
Sexual Reproduction, 6, 64, 69-70, 84, 91, 103, 105-107
Spheroplast, 48, 145
Spore Dispersal, 107
Staining, 13, 48, 58-59, 142, 147-148, 151, 153, 155, 176-177, 192, 210-213, 219-224, 226
Staining Mechanism, 222
Sterile Medium, 250
Sub-diffraction Techniques, 182

T
Terrestrial Fungus, 80
Transmission Electron Microscope, 203, 205-207, 209, 212-213

U
Ultrasonic Force Microscopy, 184
Ultraviolet Microscopy, 184, 198
Unicellular Eukaryote, 79

V
Vibrio, 20, 45, 134, 141, 155-157, 168, 173, 252, 258
Virology, 6, 13, 19, 23
Volutin Granules, 47

X
Xanthophyceae, 79

Z
Zooplankton, 92